Cambridge Studies in Biological and Evolutionary Anthropology 63

Consanguinity in Context

An essential guide to this major contemporary issue, *Consanguinity in Context* is a uniquely comprehensive account of intra-familial marriage. Detailed information on past and present religious, social, and legal practices and prohibitions is presented as a backdrop to the preferences and beliefs of the 1100+ million people in consanguineous unions. Chapters on population genetics, and the role of consanguinity in reproductive behaviour and genetic variation, set the scene for critical analyses of the influence of consanguinity on health in the early years of life. The discussion on consanguinity and disorders of adulthood is the first review of its kind and is particularly relevant given the ageing of the global population. Incest is treated as a separate issue, with historical and present-day examples examined. The final three chapters deal in detail with practical issues, including genetic testing, education and counselling, national and international legislation and imperatives, and the future of consanguineous marriage worldwide.

ALAN H. BITTLES is Adjunct Professor and Research Leader in the Centre for Comparative Genomics, Murdoch University, and Adjunct Professor of Community Genetics in the School of Medical Sciences, Edith Cowan University, Australia.

Cambridge Studies in Biological and Evolutionary Anthropology

Series editors

HUMAN ECOLOGY
C. G. Nicholas Mascie-Taylor, University of Cambridge
Michael A. Little, State University of New York, Binghamton
GENETICS
Kenneth M. Weiss, Pennsylvania State University
HUMAN EVOLUTION
Robert A. Foley, University of Cambridge
Nina G. Jablonski, California Academy of Science
PRIMATOLOGY
Karen B. Strier, University of Wisconsin, Madison

Also available in the series

Consanguinity in Context

Alan H. Bittles

Murdoch University and Edith Cowan University, Australia

CAMBRIDGE
UNIVERSITY PRESS

University Printing House, Cambridge CB2 8BS, United Kingdom

One Liberty Plaza, 20th Floor, New York, NY 10006, USA

477 Williamstown Road, Port Melbourne, VIC 3207, Australia

314-321, 3rd Floor, Plot 3, Splendor Forum, Jasola District Centre, New Delhi - 110025, India

79 Anson Road, #06-04/06, Singapore 079906

Cambridge University Press is part of the University of Cambridge.

It furthers the University's mission by disseminating knowledge in the pursuit of education, learning and research at the highest international levels of excellence.

www.cambridge.org
Information on this title: www.cambridge.org/9781108822497

© Alan H. Bittles 2012

First published 2012
First paperback edition 2020

A catalogue record for this publication is available from the British Library

Library of Congress Cataloging in Publication data
Bittles, A. H. (Alan Holland), 1943–
Consanguinity in context / Alan H. Bittles.
    p.   cm. – (Cambridge studies in biological and evolutionary anthropology)
ISBN 978-0-521-78186-2 (hardback)
1. Human population genetics.   2. Consanguinity.   3. Consanguinity – Health aspects.   4. Kinship.   5. Incest – Psychological aspects.   I. Title.
GN289.B58   2012
306.83 – dc23   2012006052

ISBN 978-0-521-78186-2 Hardback
ISBN 978-1-108-82249-7 Paperback

Contents

Colour plate section between pp. 122 and 123.

v

1 *Consanguineous marriage, past and present*

Introduction

Major problems can arise when a term with a quite specific scientific definition becomes part of everyday speech. A prime example of this phenomenon is the word *mutation*, acknowledged within science as denoting a change in genetic structure and the driving force of evolution. But to members of the general public a mutation almost inevitably denotes a change that is at best disadvantageous and in many cases is life-threatening. Unfortunately, the terms *inbred* and *inbreeding* also fall into this category and, as a result, it has become virtually impossible to persuade members of the general public that inbreeding, and by extension marriage between biological relatives, can be anything other than harmful.

Yet in the animal kingdom there are many examples of deliberate inbreeding that have resulted in healthy and fertile stock, in particular the mouse strains which are routinely used in biomedical research. It has been claimed that all of the common laboratory strains of mice can be traced back to a single female (Ferris *et al.*, 1982), and after continuous brother–sister mating for a minimum of 20 generations, it was estimated that the animals would have inherited identical gene copies from each parent at approximately 98.6% of their loci (Beck *et al.*, 2000). Since some mouse strains have been maintained by sib-mating for more than 150 generations, in effect they now are genetically identical except for sex differences.

While there is no record of such sustained close inbreeding in human populations, even in Pharaonic Egypt, the anthropological literature contains ample evidence that unions between close biological kin have been commonplace and successful in many traditional human societies. Thus, in the cross-cultural ethnographic tabulations established by G.P. Murdock of the University of Pittsburgh, 353 of the 763 societies listed either permitted or favoured first- and/or second-cousin marriage (Murdock, 1967).

The continuing popularity of consanguineous unions in many present-day rural and urban populations is apparent from the detailed information presented in the Global Consanguinity DataBase (www.consang.net). Intra-familial unions between couples related as second cousins or closer are

1

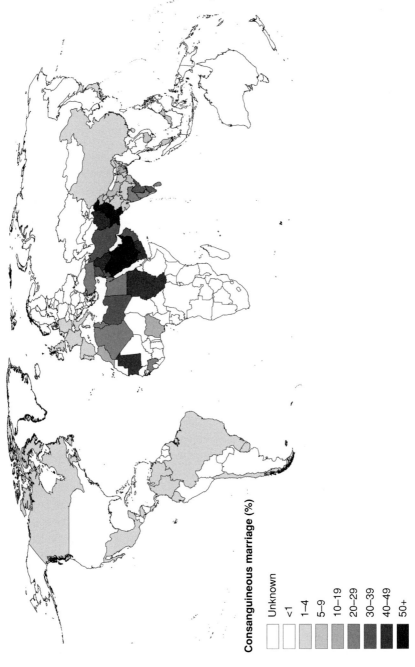

Consanguineous marriage (%)

Unknown
<1
1–4
5–9
10–19
20–29
30–39
40–49
50+

Figure 1.1 Global distribution of marriages between couples related as second cousins or closer ($F \geq 0.0156$). *Source:* www.consang.net

Table 1.1 *Current global prevalence of consanguineous relationships*

Consanguinity in population (%)	Percentage of global population	Population size (millions)
<1	15.5	1068
1–9	43.9	3026
10–19	0.5	35
20–29	6.5	448
30–39	2.1	145
40–49	3.2	221
50+	3.3	227
Unknown	25.1	1730

Sources: Global Consanguinity Website, www.consang. net; PRB (2011)

especially favoured in regions such as North and Sub-Saharan Africa, the Middle East, and Central and South Asia, and among the many emigrant communities from these regions now resident in Europe, the Americas and Oceania (Figure 1.1). In these populations 20% to more than 50% of marriages are contracted between couples who are related as second cousins or closer, with first-cousin marriage by far the most common form of consanguineous union. As will be discussed in Chapters 2–5, the rates and types of consanguineous union often vary according to historical, religious, legal and societal norms, but currently in excess of 1100 million people live in countries where consanguinity is highly favoured (Table 1.1).

These data should come as no real surprise, as even a cursory consideration of the size and structure of early human societies reveals that close kin mating must have been near-obligatory. It has been estimated that the Out-of-Africa migration of our human ancestors some 60 000–70 000 years ago involved potential breeding populations of as few as 700 individuals, to a maximum of some 10 000 persons (Harpending *et al.*, 1998; Zhivotovsky *et al.*, 2003; Liu *et al.*, 2006a; Tenesa *et al.*, 2007). Given their hunter-gatherer lifestyle, subdivision into separate small kindred groupings, and the suggestion that they exited Africa in two distinct waves (Rasmussen *et al.*, 2011), extensive inbreeding was well-nigh inevitable. Yet in the course of just 2400–2800 generations, their descendants who are scattered across the globe currently total some 5.9 billion, with an additional 1.1 billion relatives whose forebears had opted to remain in Africa (PRB, 2011). Statistics of this nature both indicate that all humans are genetically related to some degree and strongly suggest that close kin mating is not inevitably associated with an unfavourable health or reproductive outcome.

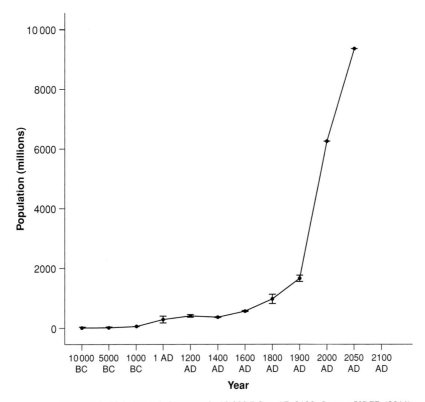

Figure 1.2 Global population growth, 10 000 BC to AD 2100. *Source:* USCB (2011)

Early urban development and social stratification

With urbanization and the establishment of increasingly sophisticated city states in Mesopotamia, Egypt and the Indus Valley around 3000 years ago, formal marriage was instituted and social stratification gradually became the norm. The net outcome of these changes was to restrict mate choice and to encourage endogamy and consanguinity within different social classes and strata. Thus, despite the slowly growing global population which, as indicated in Figure 1.2, is estimated to have numbered approximately 310 million by AD 1000 (USCB, 2011), most individuals married and reproduced within quite restricted local communities.

The Tribes of Israel provide a relevant and well-documented historical example of the influence of tribal subdivision dating back some 3000–4000 years, with the land and prescribed social and religious obligations subdivided between the 12 sons of Jacob and their descendants. Intra-familial consanguineous unions were favoured, a pattern set by Jacob himself as both of his wives, Leah and Rachel, were his first cousins (Genesis 28–29). It is, however, difficult to estimate the degree to which historical clan and/or tribal endogamy resulted in genomic homozygosity in such early, numerically small populations. Especially since, as in the example of Jacob, while six of his sons were born to Leah and two sons to Rachel, the remaining four founding males of the Tribes of Israel were the sons of Jacob's two concubines, Zilpah and Bilhah, who formerly had been maidservants to Leah and Rachel (Genesis 30).

Some indication of the population structure and dynamics of these earlier human groups can be gained through the study of present-day societies, such as the Kel Kummer Tuareg tribe of the southern Sahara, which was founded in the seventeenth century and by the 1970s numbered approximately 300 persons. Among the Kel Kummer, strict tribal endogamy has been maintained and marriage between a man and his mother's brother's daughter is regarded as obligatory (Degos *et al.*, 1974). By comparison, in other larger tribes, there is marked population stratification, with individuals and families born into traditional patrilineal clans and tribes. For example, the Abbad tribe in Jordan, which was established some 250 years ago, now comprises approximately 120 000 individual members divided into 76 male lineages of between 250 and 2000 individuals, with 47% of all marriages intra-clan and 90% contracted within the tribe (Nabulsi, 1995).

Human mating as a genetic continuum

Rather than inbreeding and outbreeding being regarded as separate and opposite reproductive strategies, it is more logical and credible to consider human mating as a genetic continuum that ranges across:

(i) Random mating: which is a rare event despite its incorporation in the Hardy–Weinberg principle that specifies an equilibrium relationship between gene frequencies and genotype frequencies within populations.

(ii) Positive assortative mating: with marriage occurring between couples who live in a specific geographical area and often in the same village or town; are of the same generation, and who share religious, educational and socioeconomic backgrounds.

(iii) Endogamous marriage: between partners preferentially and often obligatorily drawn from the same clan and tribe and therefore lineal descendants of a common male ancestor.

Table 1.2 *Human family and genetic relationships*

Biological relationship	Genetic relationships	Coefficient of relationship (r)	Coefficient of inbreeding (F)
Incest	First degree	0.5	0.25
Parent–child			
Sibling			
Half-sibling	Second degree	0.25	0.125
Uncle–niece			
Double first cousin			
First cousin	Third degree	0.125	0.0625
First cousin once removed	Fourth degree	0.0625	0.0313
Double second cousin			
Second cousin	Fifth degree	0.0313	0.0156
Second cousin once removed	Sixth degree	0.0156	0.0078
Double third cousin			
Third cousin	Seventh degree	0.0078	0.0039

(iv) Consanguineous marriage: in which the partners are known to share close biological ancestry, usually involving intra-familial marriage(s) within the preceding two to three generations.

It is apparent that positive assortative mating, endogamous marriage and consanguineous marriage are all examples of 'inbreeding', the differences among them being principally a question of degree. However, it could additionally be argued that while assortative mating is largely character-specific and so may be dependent on genes that determine attributes such as external appearance or temperament, consanguinity and to a lesser extent endogamy can influence the entire genome (Lewontin *et al.*, 1967).

Basic measurements of consanguinity in human populations

As will be described in detail in Chapter 6, in all forms of consanguineous unions the partners share genes inherited from one or more common ancestors and, for example, in first-cousin marriages, the spouses are predicted to have 1/8 of their genes in common, described as the coefficient of relationship (r). This means that on average, their progeny will be homozygous at 1/16 of gene loci, i.e. they will have inherited identical gene copies from each parent at this fraction of sites in their genome. As shown in Table 1.2, an individual's level of consanguinity is conventionally expressed as the coefficient of inbreeding (F), which for first-cousin offspring is 0.0625. By comparison, in second-cousin marriages, the equivalent figures are that they have 1/32 of their genes in common and therefore an F value for their progeny of 0.0156, while for first-degree (incestuous) relationships, the couple share $1/2$ of their genes and their progeny will have an F value of 0.25. If the same mutant gene is inherited from both parents, an individual will express the equivalent recessive disorder, prenatally, at birth, or later in life depending on the nature and site of the mutation, thus contributing to the phenomenon of inbreeding depression.

Western attitudes to consanguineous marriage

In contemporary Western society, the term *inbred* is widely used as a term of denigration, and marriages between biological relatives often are treated, at best, with suspicion and frequently with embarrassed astonishment. Once again, this is somewhat surprising because long lists of eminent and highly successful persons can be compiled who either contracted consanguineous marriages (from the Latin *con sanguineus*, of the same blood) or had long-standing relationships with a close biological relative. Among males who married a first cousin are the musicians Edvard Grieg and Sergei Rachmaninov, the scientists Charles Darwin and Albert Einstein, literary figures ranging from the doyen of Victorian horror stories Edgar Allen Poe to H.G. Wells and Mario Vargas Llosa, and the pre-eminent free-market economist Friedrich Hayek.

Despite the successful lives of these notable persons, and the evidence from the Out-of-Africa migrations indicating that consanguinity can be compatible with successful reproduction and population expansion, there remains a suspicion that inbreeding is necessarily deleterious. But, if so, why do so many contemporary societies continue to favour consanguineous marriage? Conversely, if consanguinity is not especially harmful, then why are marriages between cousins frequently a source of mirth and derision in Western societies, even though they are legal in virtually all countries? Whether any single volume could answer even these very basic questions is dubious. But as a starting-point, by examining and analysing the plentiful information that is available on the prevalence of consanguineous unions and their outcomes in terms of partner compatibility, reproductive success and the health of their children, it should at least be possible to identify the roots of the prejudices that seem to surround the entire subject of close kin marriage.

There is convincing evidence that consanguineous marriage was quite widely prevalent in Europe prior to the middle of the nineteenth century (Huth, 1875), and indeed the theme of cousin marriage initially was strongly favoured by many eminent Victorian novelists, including Charles Dickens, Anthony Trollope, Emily Brontë and William Thackeray, with Thackeray, John Ruskin and Lewis Carroll themselves the progeny of cousin marriages (Anderson,

Table 1.3 *Consanguineous marriage within geographical isolates in the Americas and Europe*

The Americas		
Canada	Québec	Phillipe & Gomila (1972)
	Newfoundland	Bear *et al.* (1988)
USA	Tangier Island, Virginia	Mathias *et al.* (2000)
North Atlantic	Flores, Azores	Smith *et al.* (1992)
Caribbean	St Thomas, US Virgin Islands	Leslie *et al.* (1978)
	St Barthélémy, Antilles	Serre *et al.* (1982)
South Atlantic	Tristan da Cunha	Bailit *et al.* (1966)
		Roberts (1968)
Europe		
France	Arthez-d'Asson, Pyrenees	Serre *et al.* (1985)
	Vallouise, Briançon	Boëtsch *et al.* (2002)
Hungary	Ivad	Nemeskéri & Thoma (1961)
Italy	Upper Bologna Apennines	Pettener (1985)
Scotland	Inner Hebrides	Sheets (1980)
	Orkney Islands	Roberts *et al.* (1979)
Spain	Formentera, Balearic Islands	Valls (1969)
	Western Pyrenees	Abelson (1978)
	Sigüenza-Guadalajara	Caldéron *et al.* (1998)
	Gredos Mountains, Avala	Fuster *et al.* (2001)
Switzerland	Alpine isolates	Morton *et al.* (1973)

1986). The fictional narratives, where a heroine falls in love with her 'dearest coz', are therefore at odds with the popular belief that consanguineous unions were restricted to population isolates and arose only because of a shortage of marriageable unrelated partners.

Marriage in geographical, social and religious isolates

As previously described under positive assortative marriage, it is nevertheless true that the choice of a marriage partner will necessarily be restricted in small isolated populations with few potential spouses and, in such communities, all of the members may be related to some degree. Geographical isolates of this nature exist in the Americas and in many European countries (Table 1.3), and similar patterns are seen in enclosed religious communities, such as the North American Anabaptist sects, the Amish (Khoury *et al.*, 1987; Dorsten *et al.*, 1999), Hutterites (Mange, 1964; Martin *et al.*, 1973) and Mennonites (Allen & Redekop, 1987; Moore, 1987); in a Brazilian Jewish community (Freire-Maia & Krieger, 1963); and in the Middle East among the Samaritans (Roberts & Bonné-Tamir, 1973), the Druze (Shlush *et al.*, 2008), and the Mandaean community in Iraq which reveres John the Baptist.

In addition, ethnic and social groups frequently choose to marry within their own community, exemplified by the Ramah Navaho and Hopi Native American tribes (Spuhler & Kluckhorn, 1953; Woolf & Dukepoo, 1969); or are subject to social stigmatization by co-resident majority populations, as in Gypsy/Roma populations (Williams & Harper, 1977; Thomas *et al.*, 1987; Kalaydjieva *et al.*, 1996; Martin & Gamella, 2005) and Irish Traveller communities (Flynn, 1986; Gordon *et al.*, 1991).

In some isolates, there is evidence of consanguinity avoidance, probably reflecting prevailing religious restrictions (Hussels, 1969; Leslie *et al.*, 1981; Magalhães & Arce-Gomez, 1987). Whereas in others, the pattern of concordance between consanguineous marriage and religious practice seems to be more finely tuned, e.g. with second-cousin but not first-cousin marriages contracted in Eriskay, a Scottish Roman Catholic island community (Robinson, 1983).

Anecdotal tales of close kin unions – licit and illicit

In his essay entitled *The Great Revolution in Pitcairn*, Mark Twain provided an example of the deliberately humorous approach to inbreeding in an isolated human settlement, in this case the community on Pitcairn Island in the southern Pacific which was founded by the mutineers of HMS Bounty in 1789 (Huth, 1875; Twain, 1899). At the time of writing, Pitcairn had a total population of 90 persons, comprising 16 men, 19 women, 25 boys and 30 girls. According to Twain's fictional traveller to Pitcairn, on talking with a male islander, he remarked, 'You speak of that young woman as your cousin; a while ago you called her your aunt'. To which the islander replied, 'Well, she is my aunt, and my cousin, too. And also my stepsister, my niece, my fourth cousin, my thirty-third cousin, my forty-second cousin, my great-aunt, my grandmother, my widowed sister-in-law – and next week she will be my wife'. All good knockabout fun, at least for the majority of the world's population not resident on Pitcairn, which in the 1990s gained a rather less acceptable reputation associated with the sexual exploitation of young females by adult male islanders.

Although every Western country seems to have at least one region where, allegedly, inbreeding is rife and the adverse physical and mental outcomes are obvious, few studies have been published to support this belief. During the late 1940s, social attitudes towards consanguineous marriage were, however, investigated in a remote mountain community in eastern Kentucky (Brown, 1951), a region where according to popular account many families were married 'through and through'. It transpired that consanguineous marriages occurred both among families who the author defined as 'high-class' and others who were 'low-class'.

Because of the shortage of suitable spouses from other families of equal social status, high-class families did sometimes resort to marriages between partners who were second cousins or second cousins once removed but, as described by the author, they were: '... long-resident families of good background, *moral athletes*, hard workers and good livers, less isolated and more modern than other families in the area and people who emphasized self-improvement and who participated more widely in neighborhood affairs'. By comparison, their much more highly inbred 'low-class' counterparts: '... tended to be newcomers with *shady pasts*, morally lax, economically insecure, not ambitious, old-fashioned and *backward*, and people who participated relatively little in many neighborhood activities'. (Original author's emphases in both quotations.)

One large 'low-class' family was particularly notorious for sexual misconduct, drunkenness and behaviour discreetly described as 'other deviations from accepted norms', and the patriarch allegedly had fathered a child with his step-daughter. Not surprisingly, they experienced difficulties in arranging marital alliances with other families in the locale; as a result, 18 of the patriarch's 70 grandchildren had married close kin, with 15 of the 18 married to their first cousins. The situation was succinctly summarized by a 'low-class' woman who had actually married into the family: 'I reckon they married each other because they couldn't get nobody else' (Brown, 1951).

Consanguineous marriage among European royalty and other dynasties

Colourful anecdotal material of this nature has tended to typify and reinforce attitudes against consanguineous marriage within Western society, but with double standards applied. Consanguineous marriage seems to be acceptable if families are well-to-do and generally regarded as pillars of society, especially if such marriages have been between more remote relatives. Hence, perhaps, the unquestioned acceptability of multi-generational intra-familial unions within many of the Royal Houses of Europe (Darlington, 1960). For example, with His Most Catholic Majesty King Philip II of Spain (1527–98) sequentially bound in matrimony with his double first cousin Maria of Portugal ($F = 0.125$), Mary Tudor a first cousin once removed ($F = 0.0313$), Elizabeth of Valois a non-relative, and finally with his niece Anne of Austria ($F = 0.125$).

The Spanish Habsburgs
Recently, the problems that can arise due to multi-generational consanguinity have been illustrated by a pedigree analysis of the Spanish Habsburg dynasty, conducted over 16 generations and comprising some 3000 individuals (Alvarez

et al., 2009), with the royal line finishing because of the early death and apparent impotence or infertility of Charles II (1661–1700). A cumulative coefficient of inbreeding (α) of 0.254 – i.e. equivalent to the level of homozygosity in the offspring of an incestuous relationship – was calculated for the unfortunate Charles II. Based on portraiture and contemporary medical records, the authors suggested that the king had inherited two rare recessive diseases: combined pituitary hormone deficiency and distal renal tubular acidosis. However, as discussed in Chapter 12, clinical diagnoses that are largely based on portraiture can be misleading, and it also is pertinent to note that the Habsburg line appears not to have started in a promising manner, with the marriage in 1516 of Philip the Fair to Joanna the Mad.

The Rothschild banking family

The highly influential Rothschild banking family provides a contrary example of a commercial dynasty in which consanguineous marriage was both strongly encouraged and prevalent, with more than a third of all marriages contracted between cousins during the late eighteenth and the early nineteenth centuries (Ferguson, 1999), but apparently with no resultant adverse health or intellectual effects. Successful pedigrees of this nature often tend to be overlooked in discussions on consanguineous marriage. Instead, in Western societies, consanguinity has been more commonly used to explain and define the supposed genetic shortcomings of families whose levels of achievement or social standards have failed to match public expectation (Bittles, 2003).

The Lubavicher Hasidim

A third example of preferential consanguineous marriage concerns a modern-day religious dynasty, the Lubavicher Hasidim, which was founded in eighteenth-century Belarus by the first rebbe Shneur Zalman (1745–1812). In Hasidic Judaism, the title 'rebbe' signifies a hereditary leader and the title is transmitted in a dynastic manner, as opposed to the more usual title of 'rabbi', which is based on the scholastic background and achievements of an individual (Hecht & Sandberg, 1987).

In keeping with Levitical regulations, among the Lubavicher there has been a strong tradition of cousin marriage from the third to the seventh generation of rebbes, all of whom have been members of the Schneerson family. Indeed, both the seventh rebbe and his wife were direct descendants of the founder of the sect, Shneur Zalman. The seventh rebbe Menachem Mendel Schneerson, who was the son-in-law of the sixth rebbe Joseph Isaac Schneerson, died in 1994 without an heir and the leadership of the movement decided that no successor would be chosen, with some members of the community believing that the seventh rebbe was the Messiah. Given the dynastic tradition of the Lubavicher sect, if an eighth rebbe is chosen, there would seem to be a strong possibility that the successful candidate would once again be selected from the Schneerson family (Hecht & Sandberg, 1987).

Commentary

From the brief examples provided, it can be seen that consanguineous marriage is frequently presented and discussed in a negative manner. The intention of the following chapters is therefore to redress that imbalance in a detailed but not necessarily turgid manner. Neither, it should be emphasized, with any intention of promoting consanguineous marriage but rather to provide an appropriately extensive framework of information within which the subject can be rationally and dispassionately evaluated.

2 Religious attitudes and rulings on consanguineous marriage

Introduction

There is wide diversity in the attitudes of the different major world religions and philosophies towards consanguinity. In general terms, consanguineous marriage is permitted within Judaism, in some branches of Christianity, Islam, Dravidian Hinduism, Buddhism, the Zoroastrian/Parsi religion, and the Confucian tradition (Table 2.1). However, the prevalence and specific types of marriage permitted vary according to the precepts and traditions of each religion and denomination and, in some cases, these characteristics appear to have altered significantly through time. In all of the major religions, there also are communities whose marriage practices are at variance with the beliefs and concepts of the majority, which in some cases suggests a carry-over from earlier times and systems of worship.

Consanguinity and religion

Judaism

There are many descriptions in the Bible of marriages between close biological relatives with, for example, Abraham and Sarah identified as half-brother and sister (Genesis 20:12); Jacob being simultaneously married to the sisters Leah and Rachel who were his first cousins (Genesis 29:9–30); and Amran and Jochebed, the parents of Aaron and Moses, who were related as nephew and aunt (Exodus 6:20). However, as indicated in Table 2.2, quite specific regulations governing marriage were subsequently codified in Leviticus 18:7–18, and while uncle–niece marriages ($F = 0.125$) are permitted, half-sib and aunt–nephew marriages (also $F = 0.125$) were prohibited under the Levitical guidelines, despite the example set by Amran and Jochebed.

Historical sources suggest that uncle–niece and first-cousin marriages were contracted in both Ashkenazi and Sephardi communities, but from the late nineteenth century, consanguineous unions progressively became rare among the Ashkenazim. Even in the Sephardi, the majority of whom have migrated

Table 2.1 *Religious attitudes to consanguineous marriage*

Judaism	Sephardi	Permissive
	Askenazi	Permissive
Christianity	Coptic Orthodox	Permissive
	Greek and Russian Orthodox	Proscribed
	Roman Catholic	Diocesan approval required
	Protestant	Permissive
Islam	Sunni	Permissive
	Shia	Permissive
Hinduism	Indo-European	Proscribed
	Dravidian	Permissive
Buddhism		Permissive
Sikhism		Proscribed
Confucian/Taoist		Partially permissive
Zoroastrian/Parsi		Permissive

Table 2.2 *Specific Biblical prohibitions on affinal and consanguineous marriages*

Prohibited in the previous generation
Mother, father's wife, father's brother's wife, father's sister, mother's sister
Prohibited in the same generation
Full sister, half sister, wife's sister
Prohibited in the following generation
Daughters, son's daughter, daughter's daughter, son's wife

Source: Leviticus 18:7–18

to Israel from the Middle East and North Africa, consanguineous unions have significantly declined in popularity since the levels recorded in the 1950s and 1960s when 15.0% to 20.7% consanguineous marriage was reported in couples from countries such as Iraq and Yemen (Tsafrir & Halbrecht, 1972), and with the levels of uncle–niece marriage low in most Sephardic communities (Goldschmidt *et al.*, 1960).

Christianity

Present-day attitudes towards consanguineous marriage also are notably mixed within the Christian churches and, as indicated in Table 2.1, first-cousin marriages are banned by the Greek and Russian Orthodox Churches but are permitted and are widely contracted in the older Coptic Orthodox Church, which dates its establishment back to St. Mark between AD 55 and AD 68. Likewise, among the Western creeds, first-cousin unions are subject to Diocesan approval in the Roman Catholic Church but are almost universally permissible within the various Protestant denominations.

Legislation on consanguineous marriage within the early Christian Church

Following the acceptance of Christianity as the official religion of the Eastern Roman Empire centred in Constantinople in AD 325, a ban on first-cousin marriage was promulgated by Theodosius the Great (AD 378–395) which, as will be indicated in Chapter 12, was contrary to Greek tradition. This law was, however, repealed by his son Arcadius (AD 395–408) and the validity of cousin marriage under secular Roman law was confirmed in the *Institutes* of Justinian (AD 527–565). But, by AD 692, first-cousin marriages were once again prohibited by the Greek Orthodox Church and this ruling remains in place (Knight, 2003). In the interim, the Coptic Orthodox Church had been accused of monophysitic heresy during the Fifth Ecumenical Council held in Chalcedon in AD 451. Perhaps because of the resulting theological division, while the other branches of the Orthodox Church banned consanguineous unions, the Copts were not bound by this decision and so have continued to permit first-cousin marriage.

Pre-Christian Roman attitudes towards consanguineous unions were more proscriptive, and even cousin marriages were regarded with disfavour (Shaw, 1992). Despite the capture of Rome by the Visigoths in AD 410 and the loss of Roman political power, the Christian Church remained an important institution and it became closely associated with the early Visigothic kings and the later Merovirigian and Carolingian dynasties (Glendon, 1989, p. 23). Further, the Church claimed and was effectively accorded exclusive jurisdiction over marriage, which it regarded as not only a natural institution and a contract between spouses but also a divine sacrament (Davidson & Ekelund, 1997).

Regulation of consanguinity in the Latin Church

An early judgement regarding consanguineous marriage within the Latin Church was recounted by the Venerable Bede in his *Ecclesiastical History of the English People* (*c.* AD 731). According to Bede, around AD 597, Augustine as the first Archbishop of Canterbury wrote to Pope Gregory I (AD 540–604) to seek guidance on the subject of close kin marriage. In his response, the Pope indicated that: 'Unions between consanguineous spouses did not result in children'. Further, quoting Leviticus 18:6: 'None of you shall approach to any that is near kin to him, to uncover their nakedness: I am the Lord', Gregory I advised that consanguineous unions were not permitted under sacred law.

Table 2.3 *Genetic and religious classification of consanguinity*

Biological relationship	Genetic relationship	Roman classification	Germanic classification
First cousin	Third degree	Fourth degree	Second degree
Second cousin	Fifth degree	Sixth degree	Third degree
Third cousin	Seventh degree	Eighth degree	Fourth degree

The latter claim was an interesting example of selective citation since, as previously noted (Table 2.2), the precise types of permitted and proscribed unions were formally listed in Leviticus 18:7–18 with both uncle–niece and first-cousin marriages permitted. Furthermore, apart from a strongly condemnatory epistle by the apostle Paul regarding a Church member who apparently was in an affinal relationship with his deceased father's wife: 'It is reported commonly that there is fornication among you . . . ' (1 Corinthians 5:1–5), intrafamilial marriage does not seem to have been a topic of particular significance to the authors of the New Testament.

For several centuries, there was confusion within the Latin Church as to exactly which types of consanguineous union were permitted or prohibited because of the different methods used to calculate degrees of biological relationship. Under the Roman system of calculation, the distance between relatives is counted by summing the number of links from each related individual to a common ancestor (Goody, 1985, pp. 134–46; Ottenheimer, 1996; Cavalli-Sforza *et al.*, 2004). By comparison, in the Germanic system, the number of links between just one of the relatives and the common ancestor is used, and if there is an unequal number of links between the two relatives and their common ancestor, the longer connection is used to calculate the 'genetic' distance. Therefore, according to the Roman method of consanguinity classification, first-cousin ($F = 0.0625$), second-cousin ($F = 0.0156$) and third-cousin ($F = 0.0039$) unions were described as 4th, 6th and 8th degree relationships, respectively, but under the Germanic system, they were 2nd, 3rd and 4th degree relationships (Table 2.3).

The problem posed by the co-existence of the Roman and the Germanic systems of classification was solved in a canon issued by Pope Alexander II in 1076, with an obligatory shift to the Germanic system (Goody, 1985, pp. 136–42). Prior to this date, marriage had been permitted between persons related beyond the 8th degree (Roman), i.e. with dispensation required for marriages between couples related as third cousins or closer. However, with adoption of the Germanic system, in which third cousins were related at the 4th degree, the restrictions on consanguineous unions became significantly more stringent and even couples related as sixth cousins had to obtain dispensation to marry. Given the small population sizes of the time and the high levels of village endogamy, in practice this prohibition would either have resulted in mass celibacy or more probably it was conveniently disregarded as unenforceable.

Although the proscription on consanguineous unions was restated by the Fourth Lateran Council in 1215, it was reduced to 4th degree relationships, i.e. restoring the right of couples who were in a relationship beyond third cousins to marry without prior Church dispensation (Adam, 1865a; Goody, 1985, p. 144; Worby, 2010). This level of regulation was confirmed by the post-Reformation Council of Trent (1545–63), with dispensation also required for second marriages to the first, second or third cousins of a deceased spouse (Merzario, 1990). Discretion was, however, allowed in the enforcement of the proscriptions, with the ban reduced to 2nd degree relationships (couples related as first cousins or closer, $F \geq 0.0625$) for South American Amerindians in 1537, later for the indigenous population of The Philippines, and for Black populations in 1897 (Goody, 1985, p. 144; New Advent Catholic Encyclopedia, 2011). After 1917, the consanguinity prohibition was reduced in all populations, initially to second cousins or closer and in 1983 to first cousins or closer (Cavalli-Sforza *et al.*, 2004). Somewhat surprisingly, multiple pathways of consanguinity – e.g. where a couple are related as both first cousins ($F = 0.0625$) and second cousins ($F = 0.0156$) – were excluded from these revisions (Cavalli-Sforza *et al.*, 2004), despite the fact that they quite frequently occur in small endogamous communities.

Even prior to 1917, dispensations for cousin marriages were freely available in various Roman Catholic dioceses, with a notably relaxed stance towards consanguineous marriage adopted in some Ibero-American countries. Perhaps reflecting the specific Church allowances made for Amerindians, in Costa Rica during the latter half of the nineteenth century, consanguinity dispensations were granted for 16.9% of marriages in the Spanish community (i.e. between third cousins or closer), equivalent to a mean coefficient of inbreeding (α) of 0.0043 (Madrigal & Ware, 1997). In part, this may have been necessitated by a localized lack of numbers of marriageable partners. However, there were highly significant differences between the numbers of dispensations obtained by different priests (Madrigal & Ware, 1997), a finding also observed in late-nineteenth-century post-Famine Ireland (Bittles & Smith, 1994), and suggestive of mixed clerical attitudes towards consanguineous marriage within the Church.

During the nineteenth and early twentieth centuries, consanguineous unions seemed to be more prevalent in the Roman Catholic countries of southern Europe than in the northern mainly Protestant countries (McCullough & O'Rourke, 1986), where no civil or religious barriers existed to their practice. More recent investigations in countries such as Sweden have, however,

Table 2.4 *Grounds for dispensation for consanguineous marriage in the Roman Catholic Church*

1.	Limited size of the locality and population, and/or with a restricted number of potential partners within the Roman Catholic faith
2.	Advanced age of a female applicant, i.e. between 24 and 50 years of age and unmarried
3.	Small size or lack of a female applicant's dowry
4.	Poverty of an applicant widow with children
5.	Where a consanguineous marriage will end serious ligation, hatred or animosity between families
6.	If the applicant female is an orphan
7.	Deformity, physical imperfections or sickness of the applicant female
8.	Validation of a Church wedding celebrated in good faith that had violated the Church Dispensation requirement
9.	Imminent wedding whose cancellation would result in severe moral and economic damage to the applicants
10.	Marriage between older persons (both over 50 years of age)
11.	Where a wedding has already been announced whose cancellation would generate derogatory suspicions
12.	To favour the well-being of children, where one or both applicants are widowed and their marriage would support, educate and assist orphan minors
13.	Mutual familiarity of the applicants
14.	Suspicions cast on the applicant female resulting from her engagement that could severely affect her future chances of marriage
15.	Suspicious cohabitation in the same house that cannot be easily interrupted
16.	Determination of two engaged persons to marry
17.	Danger of an incestuous concubinal relationship
18.	Risk of a civil wedding, to be used when a civil dispensation has been obtained or requested or when a civil wedding has already been celebrated
19.	Pregnancy, with a need to make the child legitimate
20.	Removal of a public scandal or of a well-known concubinal relationship
21.	Loss of virginity by the bride with a person other than the proposed groom
22.	The applicant woman having been born illegitimate
23.	Following elopement; to apply only if the woman has been returned to a safe place and freely consents to the marriage

Source: Cavalli-Sforza, Moroni and Zei (2004)

suggested that the apparent lack of consanguineous unions in northern Europe may simply have been due to a lack of reporting, since first-cousin unions were perfectly legal and therefore regarded as unexceptional (Bittles & Egerbladh, 2005; Egerbladh & Bittles, 2008).

There is no doubt that in strongly Roman Catholic countries in southern Europe, first-, second- and third-cousin marriages were quite common, e.g. as a means of maintaining the family patrimonies in Italy during the seventeenth to the nineteenth centuries (Merzario, 1990). As detailed in Table 2.4, dispensation for consanguineous marriage could be granted on a wide variety of grounds, with the limited size of the village of residence and/or the lack of a dowry quite widely accepted in remote rural communities (Cavalli-Sforza *et al.*, 2004). It also has been argued that the marked increase in consanguineous marriages recorded in Italy during the nineteenth century mainly resulted from the introduction of new civil codes which required that family property should be divided among all heirs (Merzario, 1990).

Specific local circumstances may have been important, and in the Biella area of north Italy, where 30% of first marriages were between biological relatives, the local weavers had taken employment in urban centres and kin marriage was used to keep the family patrimonies intact (Merzario, 1990). By comparison, in Calabria and Sicily in the south of the country, geographical isolation resulted in very high levels of village endogamy. But consanguineous marriages were generally rare until the late nineteenth century when first-cousin unions became more popular despite improved communication between neighbouring communities (Danubio *et al.*, 1999; Cavalli-Sforza *et al.*, 2004).

Local needs, customs and circumstances also seem to have been important in Spain, where uncle–niece unions were largely restricted to the northwestern regions of Galicia and the Basque country (Calderón *et al.*, 1993; Varela *et al.*, 1997; Fuster, 2003). Although even in central Spain, consanguineous unions were common (Calderón, 1989) and, as in Italy, much higher rates of consanguineous unions were reported in remote regions, with 24.2% consanguineous marriage ($\alpha = 0.0026$) in the mountain isolate of La Cabrera in Léon province during the latter decades of the nineteenth century (Blanco Villegas *et al.*, 2004).

A further problem that arose from the Church regulations on kin marriage was the failure to distinguish between biological relationships and affinal relationships, i.e. between persons related via a common biological ancestor rather than through marriage (Plates 2.1 and 2.2). In the eyes of the Church, both types of relationship were equivalent in terms of 'consanguinity'. While this approach was perhaps understandable from a social perspective, it makes little or no sense if the rationale that underpinned the proscription of marriages between different degrees of relatives was allegedly based on adverse biological outcomes, as originally claimed by Pope Gregory I. In the absence of partners from other areas, the failure to differentiate between the two types of relationship also led to considerable problems in identifying possible spouses in remote rural settlements, and in England resulted in some affinal relationships being condemned as incestuous (Morris, 1991).

The attitudes of Post-Reformation Protestant denominations to consanguineous marriage

As the Christian Church grew in stature and importance, the provision of financial support by parishioners for the poor and needy was commonly regarded as following the teaching of Christ and as a form of 'salvation through donation' (Durey, 2008). By comparison, it has been suggested that the active discouragement of intra-familial marriages by the imposition of fees for dispensation to marry within the prohibited degrees of consanguinity were primarily devised as sources of income for the Latin Church, to help maintain financially non-viable monasteries and convents (Goody, 1985, pp. 134, 181–2). The fact that Gregory I was the first monk to be elected as Pope and throughout his reign he was a strong supporter of monasticism lends some weight to this suggestion.

As part of his criticism of Church practices, Martin Luther had condemned the requirement for consanguinity dispensation payments as a regulation devised by man and not in keeping with Divine Intention, since as revealed in Leviticus 18:7–18 there should be no impediment to marriage between first cousins (Table 2.2). Luther's identification of the requirement for consanguinity dispensation as an abuse by the Church indicates that marriage between persons related as third cousins or closer had remained popular in at least some parts of Europe, with defined schedules of fees for dispensation that varied according to the degree of consanguinity, and with the nobility paying commensurately higher sums (Davidson & Ekelund, 1997). It also explains the subsequent widespread adoption of the Levitical regulations on consanguinity by the emergent Protestant denominations, with marriages up to and including first-cousin unions permissible under Reformed Church law.

Where differences exist between the Protestant denominations, they mainly represent doctrinal differences between Martin Luther, who favoured the outright adoption of the Levitical guidelines, and John Calvin and his followers, who regarded the guidelines as illustrative rather than prescriptive (Goody, 1985, p. 176). An exception to this generalization is provided by the State Lutheran Church of Sweden, which until 1680 refused to recognize first-cousin unions, and from 1680 to 1844 approval was required from the King of Sweden before a first-cousin union could proceed (Bittles & Egerbladh, 2005; Egerbladh & Bittles, 2008).

The Anglican schism
The situation in England was both individual and more complex in that the stance on consanguinity adopted by the emergent Church of England was largely determined by the marital problems of Henry VIII. Henry's first marriage in 1509 was to his sister-in-law Catherine of Aragon, the widow of his deceased elder brother Arthur. Because this marriage failed to provide a male heir, after eighteen years of marriage, Henry petitioned the Pope for an annulment on the grounds that since their marriage was affinal, it had contravened Leviticus 18:6 (Goody, 1985, pp. 168–73).

Although Henry had been awarded the Papal title of 'Defender of the Faith' in 1521 for his counter-Reformation support of the Church, his petition was refused despite support from English and Continental legal authorities. Annulment of the marriage to Catherine was nevertheless approved by the

English Parliament, allowing Henry to marry Anne Boleyn in 1533. This union provoked Henry's excommunication by Pope Clement VII in 1533, to which the King responded by having Parliament pass the Act of Supremacy in 1534 which recognized Henry as 'the only Supreme Head of the Church of England'.

Unfortunately, Anne Boleyn and two further wives also failed to provide Henry with a male heir. To fulfil his aim of a son and heir, Henry wished to marry a fifth time, but had to resolve the problem that Catherine Howard was a first cousin of the late Anne Boleyn and hence his affinal relative. To enable his marriage to Catherine Howard and in his role as Supreme Head of the Church of England, in a statute of 1540, Henry legalized all forms of first-cousin unions, consanguineal as well as affinal. By so doing, he effectively adopted the Levitical guidelines on consanguineous marriage which, somewhat ironically, were supported by Luther and most of the Reformed Churches. Following the Act of Supremacy in 1534, dispensations on the grounds of consanguinity received from Rome had been declared invalid. It appears that this ruling had little real effect, however, since in the Faculty Office Registers held in Lambeth Palace for the period 1534–40, no first-cousin unions were recorded and all other forms of consanguineous marriages were rare (Smith *et al.*, 1993).

The 1540 statute recognizing first-cousin unions was revoked by the Roman Catholic Mary Tudor, the daughter of Henry VIII and Catherine of Aragon, on her accession to the throne in 1553. Mary's marriage to Phillip II of Spain, her first cousin once removed ($F = 0.0313$), was childless. Following Mary's death in 1558, her half-sister Elizabeth I, the daughter of Henry and Anne Boleyn, commissioned Matthew Parker as Archbishop of Canterbury to tabulate formal marital rules for the Church of England.

The intention was that the rules should replace the often beautiful and highly decorative, but somewhat obscure, figurative representations of degrees of consanguinity and affinity used by the pre-Reformation Catholic Church (Plates 2.1 and 2.2), and that they would be readily comprehensible to both the clergy and literate laity. The resultant *Tables of Kindred and Affinity of the Church of England* first appeared in 1560, were revised in 1563 and included in the Book of Common Prayer in 1681, and with minor modifications the guidelines remained in force for the global Anglican/Episcopalian community until 1940 (Goody, 1985, pp. 174–80).

Islam

From its origins in the Arabian Peninsula, Islam has increasingly become a global and multi-ethnic religion, with believers now drawn from many regions of the world. However, at its core are the revelations of the Quran which were initially revealed to the Prophet Muhammad in Mecca AD 610. Many of the social concepts that are central to Islam therefore reflect the structure of Arab societies, which traditionally have focused on the family and, by extension, the patrilineal kin relationships that are typical of clans and tribes (Bittles & Hamamy, 2010).

It is believed that in the pre-Islamic era, *bint 'amm* marriage (i.e. between paternal cousins) was infrequent within Arab society, with inter-tribal marriage used as a means of preventing or quelling tribal feuds (Stein, 1939). After conversion of the Arab tribes to Islam, the subsequent high prevalence of consanguineous marriage may have been encouraged by the reform of the laws of inheritance introduced in the Quran, whereby daughters were entitled to inherit half of the amount received by sons, and wives received a determinate share from their husbands (Bittles & Hamamy, 2010). A dower (*mahr*) also is specified as part of the marriage arrangement, with the goods transferred to the bride at marriage (Khuri, 1970; Tucker, 1988). Under these circumstances, it could be ensured that a woman's share of her family wealth would be retained within the clan by marriage to her paternal cousin.

With one exception, the prohibition of uncle–niece marriage (Quran 4:23), the permitted degrees of consanguinity in marriage within Islam closely match the Levitical guidelines, with double first-cousin marriage ($F = 0.125$) accounting for up to 5% of unions in some Islamic communities. While the prohibition on uncle–niece marriage is rigorously observed in Middle Eastern Arab communities (Teebi & Marafie, 1988), in different regions of India, small numbers of uncle–niece marriages have been reported in Muslim communities (Sanghvi, 1966; Basu & Roy, 1972; Malhotra *et al.*, 1977; Roychoudhury, 1980; Bittles *et al.*, 1991; Bittles & Hussain, 2000; Bhasin & Nag, 2002; Iyer, 2002), possibly reflecting pre-Islamic marital customs among converts from the Hindu or Buddhist faiths.

Contrary to widespread Western opinion, there is no specific guidance in the Quran to encourage consanguinity, even though the prevalence of close kin marriage can exceed 50% in some Muslim countries of the Middle East and Pakistan (Jaber *et al.*, 1992; Hussain, 1999). In fact, the overall attitude to consanguineous marriage within Islam is somewhat ambiguous since a *hadith* (oral pronouncement) of the Prophet Muhammad stated: 'Do not marry cousins as the offspring may be disabled at birth' (Hussain, 1999; Akrami & Osati, 2006). Further, the second Caliph, Omer Ibn Al-Khatab reputedly advised members of the Bani Assayib tribe to: 'Marry from far away tribes, otherwise you will be weak and unhealthy' (Albar, 1999). On the other hand, two of the six wives of the Prophet were biological relatives and the Prophet Muhammad married his daughter Fatima to Ali, his paternal first cousin and ward (Armstrong, 1991). Thus, for Muslims, the practice of cousin marriage could potentially be interpreted as following the example provided by the *sunnah* (deeds) of the Prophet.

In Palestine during the eighteenth and nineteenth centuries, the prevalence of consanguineous marriage varied according to social class, with the highest rates of paternal-cousin marriage contracted within the ruling group or other affluent families (Tucker, 1988). Strong regional differences in the preference or proscription of consanguinity also are found in different Muslim populations and in the types of first-cousin marriage contracted (Table 6.2). The differences observed in non-Arab populations may indicate pre-Islamic traditions with, for example, marked avoidance of consanguineous unions in the European Muslim populations of Albania, Bosnia and Kosovo, and the apparent rarity of consanguineous marriage in many of the islands of Indonesia where Islam is the predominant religion.

In the Indian subcontinent, differential rates of consanguineous marriage have been observed according to the branch of Islam (e.g. whether Sunni or Shia), the particular School of Islamic Jurisprudence favoured by an individual, by ethnic affiliation, and by membership of specific occupational or social groupings (i.e. *biraderi* or *zat/qoums*) (Shami *et al.*, 1994; Bhasin & Nag, 2002; Hussain, 2005). Higher rates of cousin marriage are generally reported in Shia than Sunni communities in India (Basu, 1975, 1978), but the reverse is true in Iran (Abbasi-Shavazi *et al.*, 2008). Despite their Arab origins, the Mappilla community, which has been resident in the southwest Indian state of Kerala since the eighth century AD, generally avoids marriages between biological relatives (Thapar, 1986; Mathur, 1997–1998; Bittles & Hussain, 2000), as also is the case in the Maldives, the Muslim island republic in the Indian Ocean.

In the Middle East, both Arab Muslim communities and the Druze, who separated from orthodox Islam in the eleventh century (Ayoub, 1959; Shlush *et al.*, 2008), are characterized by a preference for parallel first-cousin marriage, specifically between a man and his father's brother's daughter (FBD) (Ayoub, 1959; Murphy & Kasdan, 1959; Korotayev, 2000). FBD marriage also is favoured in neighbouring non-Arab Muslim populations such as the Kurds (Barth, 1954) and in an Israeli Jewish community that originated in the Tripolitania region of Libya (Goldberg, 1967). Although opinion varies as to the specific reason for this preference, there seems to be agreement that marriage between the children of two brothers is especially propitious, in part since they will benefit from the near-equal social status of the fathers (Barth, 1954).

Hinduism

Hinduism is the majority religion in India, with more than 900 million adherents, and in most overseas Indian communities. Marriage regulations within Hinduism are complicated and embody three sets of rules, one of which promotes endogamy whereas the other two require exogamy. At the uppermost level of differentiation, all Hindu communities are hierarchically structured into four *varnas*: Brahmin, Kshatriya, Vysya and Sudra, and into many thousands of castes and subcastes each of which has its own set of rights, duties and privileges (Rao, 1984). Becuase individuals are born into their caste (*jati*), caste membership is regarded as predetermined and therefore inviolable.

Approximately 16% of the population, while mainly Hindu, are effectively positioned outside the caste system. Formerly termed 'Untouchables' because by personal contact with a person whose traditional role in society was deemed to be polluting to caste Hindus (e.g. leather work or latrine cleaning) (Padhy *et al.*, 2006), the term *dalit* is now widely used to describe these socially disadvantaged communities and they are designated by the Government of India, as Scheduled castes. In the 2001 Census of India 166 million respondents were categorized as Scheduled castes and a further 85 million individuals were listed as members of Scheduled tribes, believed to have been the first permanent inhabitants of the subcontinent.

With few exceptions, caste and Scheduled caste endogamy is obligatory within Hinduism. Castes are further subdivided into *gotras* (subcastes) which are male lineages. The patrilineal *gotras* can be considered exogamous since marriage between the children of two brothers is prohibited (Sanghvi, 1966). In addition, according to *sapinda* regulation, persons related to each other within certain boundaries of relatedness are forbidden to marry (Kapadia, 1958). The net result is that virtually all Hindu marriages continue to be contracted within caste boundaries, and *gotra* regulations also are observed on a near-universal basis.

At the *sapinda* level, major differences are observed between the peoples of North and South India (Bittles, 2002). Although consanguineous marriage appears originally to have been quite widely practised by Hindus throughout India, the situation had radically changed by the seventh century AD (Kapadia, 1958). According to the *Manusmriti* (the Codes of Manu), which are thought to have been written around 200 BC, among the Indo-European peoples of North India, marriage of a man either with the daughter of his father's sister, or his mother's sister, or his mother's brother was unacceptable. With some specific customary exceptions (Ghurye, 1936), pedigrees are examined over an average of seven generations for males and five generations for females to ensure avoidance of such consanguineous unions (Kapadia, 1958).

By comparison, uncle–niece marriage and first-cousin unions between a man and his maternal uncle's daughter (mother's brother's daughter) have a long tradition in Dravidian South India (Kapadia, 1958) and were mentioned by Baudhayana in the *Dharmasutra*, which was written some 2000 years ago (Sanghvi, 1966; Sastri, 1976). Given their customary status, Baudhayana regarded

Table 2.5 *Consanguineous marriage in India by state, 1992–3*

Region	State	Consanguineous marriage (%)	Mean coefficient of inbreeding (α)
North	Delhi	4.3	0.0023
	Haryana	1.0	0.0004
	Himachal Pradesh	0.8	0.0003
	Jammu and Kashmir*	8.0	0.0049
	Punjab	0.9	0.0006
	Rajasthan	1.3	0.0006
Central	Madhya Pradesh	4.1	0.0025
	Uttar Pradesh	7.5	0.0044
East	Bihar	5.0	0.0032
	Orissa	5.7	0.0035
	West Bengal	5.0	0.0030
Northeast	Arunachal Pradesh	3.9	0.0029
	Assam	1.7	0.0010
	Manipur	2.1	0.0013
	Meghalaya	2.7	0.0018
	Mizoram	0.5	0.0002
	Nagaland	1.5	0.0009
	Tripura	1.9	0.0010
West	Goa	10.6	0.0066
	Gujarat	4.9	0.0029
	Maharashtra	21.0	0.0131
South	Andhra Pradesh	30.8	0.0212
	Karnataka	29.7	0.0180
	Kerala	7.5	0.0042
	Tamil Nadu	38.2	0.0255
All India		11.9	0.0075

*Data collected in Jammu only due to civil unrest.
Sources: International Institute for Population Sciences (1995); Bittles (2002)

consanguineous unions as acceptable for the people of southern India, i.e. those living south of the sacred Narmada River which rises in Madhya Pradesh and flows west through the state of Gujarat into the Indian Ocean (Kapadia, 1958). There also is some evidence of grandfather–granddaughter marriage in the historical past (Aiyappan, 1934), but these seem to have been very unusual events.

As indicated in Table 2.5, the marked subdivision in marriage differences between the majority Hindu populations of North and South India has continued. The popularity of first-cousin marriage, especially between a man and his mother's brother's daughter, and uncle–niece marriages, with a man marrying his elder sister's daughter, has continued in South India, especially in the states of Andhra Pradesh, Karnataka and Tamil Nadu (Bittles *et al.*, 1993; IIPS, 1995),

Table 2.6 *Consanguineous marriage in India by religion, 1992–3*

Religion	Consanguineous marriage (%)	Mean coefficient of inbreeding (α)
Hindu	10.6	0.0068
Muslim	23.3	0.0141
Christian	10.3	0.0068
Sikh	1.5	0.0009
Jain	4.3	0.0024
Buddhist	17.1	0.0107
Others	8.7	0.0053

Sources: International Institute for Population Sciences (1995); Bittles (2002)

and less commonly in Kerala and southern Maharashtra. However, some recent decline in consanguineous marriage may have occurred among specific groups within the wider South Indian community (Audinarayana & Krishnamoorthy, 2000) and, given the reduction in family sizes and specific preferences in age differences between males and females at marriage, it seems probable that this trend will accelerate in urban populations.

Sikhism

Since the Sikh religion originated in North India in the sixteenth century AD, with converts largely drawn from the majority Hindu population, regulations forbidding close kin marriage similar to those in the Indo-European branch of Hinduism are followed in most Sikh communities (Table 2.6). Consanguineous marriage is, however, permitted in some branches of Sikhism, with 21.0% first- or second-cousin unions ($\alpha = 0.0128$) in a Sikh community in Pakistan (Wahab & Ahmad, 2005), where first-cousin marriage is widely preferential in the majority Muslim population (Hussain & Bittles, 1998).

Buddhism

Buddhism originated in North India in the third century BC and the Prince Buddha is reputed to have married his first cousin. Perhaps for this reason there is no overall proscription on consanguineous marriage within the Buddhist tradition. As indicated in Chapter 5, first-cousin marriage remains common in India (IIPS, 1995), the mainly Buddhist South Asian countries of Nepal

(Fricke *et al.*, 1993 Fricke & Teachman, 1993) and Sri Lanka (Reid, 1976), but it is proscribed among the Buddhist Bodh community in the Ladakh region of Jammu and Kashmir (Bhasin & Nag, 2002), possibly reflecting the strong local Tibetan Buddhist influence, which proscribes cousin marriage.

The situation is less clear in East and Southeast Asia because of a lack of published information on consanguinity, even though occasional reports have suggested that first-cousin marriage is widely practised in at least some communities. Consanguineous marriage has become less prevalent among contemporary Buddhist communities in China and Japan. In the case of China, this probably reflects the introduction of prohibitive civil legislation in 1951 and 1981 (see Chapter 3), whereas in Japan, the change has coincided with the rapid urbanization and industrialization in the country since the 1950s and the consequent expansion of marriage pools (Imaizumi *et al.*, 1975; Imaizumi, 1986).

Confucian/Taoist tradition

Consanguineous marriage is permitted in the Chinese Taoist/Confucian tradition, with close kin unions described among the majority Han population as early as the second century BC (Dull, 1978), although its popularity does not appear to have approached the levels reported in other regions of Asia. As in other societies, the types of consanguineous unions contracted were specifically defined. Thus, the traditionally favoured *biao* or *paio* cross-cousin marriages were between couples not related through male lineages; i.e. they include mother's brother's daughter, mother's sister's daughter and father's sister's daughter unions (Meijer, 1971; Cooper, 1993; Cooper & Zhang, 1993). Parallel-cousin marriages between couples with the same surname were regarded as being equivalent to incest and, as such, they were strictly prohibited.

Zoroastrian/Parsi

The religious significance of incestuous marriage within the Zoroastrian religion is discussed in Chapter 12, although there is no surviving information to indicate how prevalent such marriages may have been. Zoroaster (Zarathushtra), the founding prophet, is believed to have lived in the region of present-day Iran around 1000 BC (Dhalla, 1938), and small Zoroastrian communities still exist in Iran (Farhud *et al.*, 1991) and in neighbouring parts of Pakistan. There also are small but economically significant Parsi communities in India and other parts of the world, founded by Zoroastrians who initially migrated to western India after the Muslim Arab invasion of Iran in AD 651.

Religious endogamy is obligatory for Parsis, and in contemporary Indian Parsi communities there is a continuing tradition of consanguineous marriage between couples related as first cousins, first cousins once removed or second cousins (Sanghvi *et al.*, 1956; Undevia & Balakrishnan, 1978). To some extent, this practice may reflect the small sizes of many of these communities, their favourable economic stature, and the desire to retain their religious and social traditions.

Commentary

The global prevalence and preferred patterns of consanguineous marriage have changed in Western societies over the course of the past 150 years, but religious ordinances still exert a powerful influence on marriage preferences in many other populations, especially in more traditional rural areas. Besides major differences between religions in attitudes towards consanguinity, substantial inter-denominational differences also exist, typified by doctrinal divisions within Christianity, Hinduism and Buddhism. As will be discussed in Chapter 5, social change and educational advancement among females predictably will exert increasing influence on marriage choices during the course of the present century. What is less clear is the degree to which these choices may be countered by the current increasing levels of religious fundamentalism in many countries.

3 Civil legislation on consanguineous and affinal marriage

Introduction

In virtually all countries, civil legislation on consanguinity is currently limited to the prohibition of incestuous relationships rather than restricting consanguineous marriage per se. As indicated in Chapter 2, attitudes towards consanguineous marriages often have been strongly influenced by the prevailing religious beliefs and mores of individual societies. Even in England, where first-cousin marriage had been recognized by the Church of England in 1540, until 1908 ecclesiastical courts could rule on the legality of marriages between closely related individuals and order their annulment if deemed appropriate.

The law in Great Britain on consanguinity and affinity

A late-nineteenth-century compilation of books and papers on cousin marriage published in Europe from the sixteenth century to the start of the nineteenth century suggested limited interest in the outcomes of consanguineous marriage (Huth, 1875). In Great Britain there was, however, considerable ambiguity surrounding the legality and desirability of affinal marriages (i.e. unions between individuals related through marriage) as opposed to consanguineous relationships (Plates 2.1 and 2.2). This problem came to a head with the presentation to the House of Lords by Lord Lyndhurst, the Lord Chancellor, of a bill to restrict the time-period during which marriages within the prohibited degrees of affinity could be annulled to two years. Parliament decided that the entire concept of voidable marriages should be discarded and instead passed a revised form of the bill which recognized the validity of all marriages within the prohibited degrees of affinity, as defined in the *Tables of Kindred and Affinity of the Church of England*, that had been contracted before 31 August 1835 (Anderson, 1982). This ruling did not, however, apply to marriages within the prohibited degrees of consanguinity, thus opening a division under civil law between consanguineous and affinal unions (Ottenheimer, 1990).

Table 3.1 *England and Wales Marriage Law 1949*

Prohibited degree of relationship	
Male	Female
Mother	Father
Daughter	Son
Father's mother	Father's father
Mother's mother	Mother's father
Son's daughter	Son's son
Daughter's daughter	Daughter's son
Sister	Brother
Wife's mother	Husband's father
Wife's daughter	Husband's son
Father's wife	Mother's husband
Son's wife	Daughter's husband
Father's father's wife	Father's mother's husband
Mother's father's wife	Mother's mother's husband
Wife's father's mother	Husband's father's father
Wife's mother's mother	Husband's mother's father
Wife's son's daughter	Husband's son's son
Wife's daughter's daughter	Husband's daughter's son
Sons' son's wife	Son's daughter's husband
Daughter's son's wife	Daughter's daughter's husband
Father's sister	Father's brother
Mother's sister	Mother's brother
Brother's daughter	Brother's son
Sister's daughter	Sister's son

At the same time, the revised bill incorporated a clause which made void marriages of both affinity and consanguinity contracted after the nominated date of 31 August 1835. This additional clause met with strong opposition from widowers who wished to marry their sisters-in-law and resulted in the introduction into Parliament in 1842 of the Marriage with a Deceased Wife's Sister Bill, with the aim of exempting a wife's sister from the list of prohibited degrees of affinity (Morris, 1991). The seemingly innocuous bill met with significant opposition inside and outside Parliament and, despite its repeated presentation to Parliament, formal approval of the legislation did not pass until 29 August 1907, with a subsequent extension to include marriage with a deceased brother's widow passed in 1921, and the inclusion of nieces or nephews by marriage in 1931 (Anderson, 1982). Rather than continue to rely on these somewhat ad hoc arrangements, the legal status of both consanguineous and affinal marriages were subsequently ratified by Parliament in the Marriage Act of England and Wales 1949 (Tables 3.1 and 3.2).

Table 3.2 *England and Wales Marriage Law 1949*

Statutory exceptions from prohibited degree of relationship	
Male	Female
Deceased wife's sister	Deceased sister's husband
Deceased brother's wife	Deceased husband's brother
Deceased wife's brother's daughter	Father's deceased sister's husband
Deceased wife's sister's daughter	Mother's deceased sister's husband
Father's deceased brother's wife	Deceased husband's brother's son
Mother's deceased brother's wife	Deceased husband's sister's son
Deceased wife's father's sister	Brother's deceased daughter's husband
Deceased wife's mother's sister	Sister's deceased daughter's husband
Brother's deceased son's wife	Deceased husband's father's brother
Sister's deceased son's wife	Deceased husband's mother's brother

Countries with civil legislation restricting consanguineous marriage

Three major sets of countries have passed specific laws to prohibit or restrict consanguineous marriage: the USA at state but not national level; the Peoples Republic of China, and Taiwan; the Republic of Korea and the Democratic People's Republic of Korea. Legislation governing consanguineous marriage also is current in the Philippines as part of wider laws governing marriage and, according to the terms of the Hindu Marriage Act of 1955, uncle–niece marriages were banned in India (Kapadia, 1958). However, given the popularity of this form of marriage in South India, the legislation was amended in the Hindu Code Bill of 1984 with uncle–niece unions recognized as legal in communities in which they were customary.

United States of America

A majority of the early European migrants to the American colonies were of Nonconformist British stock whose attitudes towards consanguinity generally reflected the Levitical marriage guidelines (Table 2.2) and the regulations imposed by the Church of England in 1540, with first-cousin marriage freely permissible. During the formative post-colonial period in the USA, legislation to control or prohibit marriages between relatives was mostly concerned with affinal unions, in particular between a man and his dead wife's sister, thus echoing the long-standing controversy that had been taking place in Great Britain (Anderson, 1982).

According to Ottenheimer (1996), 14 of the 16 states that joined the Union in the eighteenth century adopted regulations against affinal marriages, compared with 6 of the 14 states that joined during the first half of the nineteenth century and just 4 of the 20 states that joined thereafter. Criminal sanctions continue to be applied to step-parent/stepchild relationships in 21 states. But in Colorado, a Native American male can marry his stepdaughter, while in Alabama, Texas and Utah the step-parent/stepchild prohibition applies only during the existence of the marriage creating the relationship (otherwise, presumably, it would be bigamous), and in Illinois and Washington the prohibition ceases once the stepchild reaches 18 years of age (Bratt, 1984).

As detailed in Chapter 4, concerns over the desirability or otherwise of consanguineous marriage arose at a later date and appear to have been initiated in response to medical and scientific opinion, as opposed to the social concerns that had been the driving force behind the legislation on affinal marriage. The Calvinist Puritans had expressed opposition to cousin marriage as early as the seventeenth century, and opposition continued into the mid-nineteenth century among some Presbyterian congregations in the northeastern states of the USA (Ottenheimer, 1990, 1996). This is somewhat surprising as many of their members would have been eighteenth- and early-nineteenth century Scots-Irish migrants, coming from a background where first-cousin marriage was acceptable and practised (Bittles & Smith, 1994) and given the scripture-based reforms introduced by the Presbyterian Church of Scotland in the sixteenth century (Smout, 1980).

Indeed, some Scots-Irish migrants steadfastly retained their preference for first-cousin marriage as they moved across the USA from their original settlement in Pennsylvania in 1720, and through Virginia, Kentucky and Tennessee to Missouri and Illinois in 1830 (Reid, 1988). From 1725 to 1849, 29 of the 142 marriages contracted by a group of Scots-Irish families were between first cousins (of all four types: mother's brother's daughter, father's sister's daughter, mother's sister's daughter and father's brother's daughter). By comparison, in the following 50 years, just 1 of 36 marriages in the community was between first cousins, a shift that seems to have been reflective of the changing attitude within the country as a whole towards consanguineous unions and certainly within the Presbyterian community (Reid, 1988).

As shown in Figure 3.1, first-cousin marriage is currently a criminal offence in nine states – Arizona, Mississippi, Nevada, North Dakota, Oklahoma, South Dakota, Texas, Utah and Wisconsin – and it is subject to civil sanction in a further 22 states, with double first-cousin marriage ($F = 0.125$) but not first-cousin marriage ($F = 0.0625$) banned in North Carolina (Farrow & Juberg, 1969; Bratt, 1984; Ottenheimer, 1996; Paul & Spencer, 2008). The introduction of legislation prohibiting first-cousin marriage and other degrees of consanguineous

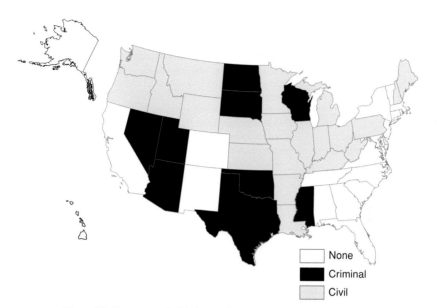

Figure 3.1 Current state legislation on first-cousin marriage in the USA. *Source:* NCSL (2011)

relationship commenced in Kansas in 1858, with Nevada, North Dakota, South Dakota, Washington, New Hampshire, Ohio and Wyoming following suit in the 1860s; by the beginning of the twentieth century, a further five states, Arkansas, Colorado, Illinois, Montana and New Hampshire, had passed laws banning first-cousin unions (Ottenheimer, 1996; Paul & Spencer, 2008).

In Colorado, the ban on first-cousin marriage was subsequently rescinded (Ottenheimer, 1996), whereas in some other states, the prohibition of first-cousin marriage is of recent origin, including Maine in 1985 and Texas in 2005, with an attempt to prohibit first-cousin unions also unsuccessfully introduced into the Maryland legislature in 2000/2001. Allegedly, the Texas legislation was introduced in an attempt not specifically to prohibit first cousins from marrying but rather primarily to prevent underage marriage and polygamy in a community which had moved across the state border from Utah. The legal situation in Georgia is anomalous since it is based on Leviticus 18, with uncle–niece unions therefore permitted; and because Leviticus 18 does not explicitly prohibit marriage between a man and either his daughter or his grandmother, such unions theoretically also could have been legal in Georgia (Farrow & Juberg, 1969).

In a majority of cases, there is no evidence to indicate that the decisions taken by state legislatures to prohibit first-cousin marriage were founded on credible biological evidence. However, as listed in Table 3.3, there are some

Table 3.3 *Exceptions to prohibitions on first-cousin marriage in individual US states*

Illinois	If both partners are >50 years or one is unable to reproduce
Wisconsin	If the woman is ≥55 years or one is unable to reproduce
Indiana	If both partners are ≥65 years
Utah	If both partners are ≥65 years or if both are ≥55 years and one is unable to reproduce
Arizona	If both partners are ≥65 years or one is unable to reproduce
Maine	On presentation of a physician's certificate of genetic counselling when their intention to marry is announced

Source: NCSL (2011)

exceptions which are based on chronological age and/or the inability of one or both partners to reproduce (Bratt, 1984; Ottenheimer, 1996; NCSL, 2011). Exceptions were also granted to specific religious or ethnic groups to marry a biological relative, for example, with members of the Rhode Island Jewish community free to contract an uncle–niece marriage (Bratt, 1984).

The situation with respect to consanguineous marriage in the USA remains confused and confusing, with little evidence of consistency or logic in the legislation. Perhaps the most surprising facet is the persistence of prohibitions on first-cousin marriage, which is contrary to a unanimous recommendation by the National Conference of Commissioners on Uniform State Laws in 1970 that these marriages should be freely permitted in the USA, a recommendation that subsequently was approved by the American Bar Association in 1974 (Glendon, 1989). The Commissioners had additionally recommended that uncle–niece marriages should be permitted in indigenous cultures where they were customary, with Native American, Alaskan and Hawaiian communities specifically identified. But once again, little action appears to have been initiated on these recommendations.

Peoples Republic of China and Taiwan

Through the course of Han Chinese history, patrilateral parallel first-cousin marriages (e.g. between a man and his FBD) were deemed to be *tang* (within the lineage) and as such were regarded as unacceptable because they were considered as equivalent to sibling marriage (Hsu, 1945; Zhaoxiong, 2001). By comparison, matrilateral and patrilateral cross first-cousin marriages – i.e. between a man and his mother's brother's daughter (MBD) and a man and his father's sister's daughter (FSD) – were *baio* or *zhong-baio* (remote); therefore, both were acceptable within the Confucian/Taoist tradition. MBD marriage was the more favoured form of consanguineous union because it was believed to result in a better relationship between a mother-in-law and her daughter-in-law (Fei, 1939; Hsu, 1945; Gallin, 1963; Geddes, 1963; Cooper & Zhang, 1993; Zhaoxiong, 2001). However, attitudes towards consanguineous marriage changed under different dynasties, and while patrilateral cross-cousin unions were taboo in some communities, they could be permissible in others (Croll, 2010).

Consanguineous marriage was also a long-established tradition in many of the 55 officially recognized minority populations (*minzu*) within the Peoples Republic of China (www.consang.net). This was especially the case among the 10 Muslim *minzu*, the Hui, Uygur, Dongxiang, Salar, Ba'on, Tatar, Kazakh, Kirgiz, Tajik and Uzbek, who are predominantly resident in the provinces of Gansu and Xinjiang and are estimated to have a combined population of approximately 100 million (Du *et al.*, 1981; Wu, 1987; Wang *et al.*, 2003; Black *et al.*, 2006).

According to the autonomous marriage regulations adopted by the Chinese Communist Party in 1931 while they were still engaged in a civil war in the 'red zone', and incorporated into Article 5 of the Marriage Law of 1934, marriage between patrilineal descendants was prohibited within five generations, which equated to third cousins (Meijer, 1978; Müller-Freienfels, 1978). Following the establishment of the Peoples Republic of China (PR China) in 1948, a new Marriage Law was introduced in 1950. Besides bans on marriage between direct lineal relations, and between sibs and half-sibs, Article 5 stated: 'The question of prohibiting marriage between collateral relatives by blood (up to the fifth degree of relationship) is to be determined by custom'.

This statement led to considerable confusion as to whether traditional *biao* marriage between cross cousins (i.e. between a man and either his FSD or his MBD), were permitted or prohibited (Meijer, 1978). To avoid possible ambiguity, Article 6 of the revised Marriage Law of 1981 explicitly prohibited first-cousin unions: 'Marriage is not permitted . . . where the man and woman are lineal relatives by blood or collateral relatives by blood (up to the third degree of relationship)'.

The underlying health intent of the legislation was made equally clear: '. . . marriages between collateral relatives by blood often produce children with congenital defects'. However, this openly eugenic measure also stipulated that: 'Because of certain traditional customs, especially in certain remote mountainous areas this provision should be implemented gradually. Crude and summary methods should not be used' (Government of PR China, 1982). In effect, the Government of PR China regards the consanguinity regulations within the 1981 Marriage Law as legislation which is on the statute book, but it does not necessarily expect that its statutes will be pursued with zeal.

At least in the major urban populations, the ban on first-cousin marriage incorporated in the Marriage Law of 1981 was unnecessary, since in practical terms it had largely been rendered redundant by the prior implementation of the One Child Certificate Programme introduced in 1979. The One Child Certificate Programme encouraged couples to restrict their reproduction to a single child by offering various incentives, including a monetary bonus and preferential housing and educational opportunities (Arnold & Zhaoxiang, 1986), in combination with economic and social disincentives for unapproved births (Doherty *et al.*, 2001). The resultant small family sizes, exacerbated by highly significant male-biased sex imbalances, effectively negated the likelihood of first-cousin marriages in many communities, especially in the urban Han population. Among the Man (Manchu) ethnic minority in northeast China, there also is evidence that improved communications in rural areas resulted in reduced village endogamy and a lower prevalence of first-cousin unions (Wang *et al.*, 2002).

It is appropriate to consider the anti-consanguinity provisions of the Marriage Law 1981 as part of a suite of laws enacted by the Government of PR China to preclude the birth of children with inherited defects. This intention was made even more explicit in the Law on Maternal and Infant Health Care enacted in 1994, which in Article 10 included the requirement that if the partners were diagnosed as having '. . . a certain disease of a serious nature which is considered to be inappropriate for child-bearing from a medical point of view' then they '. . . may be married only if both sides agree to take long-term contraceptive measures or to take ligation operation for sterility' (Ministry of Health, PR China, 1994; Bittles & Chew, 1998).

It has been stressed that the primary aim of the Law on Maternal and Infant Health Care was to ensure the birth of healthy infants (*yousheng*) rather than being eugenic in intent (Su & Macer, 2003). However, in the 2001 revision to the Marriage Act of 1981, Article 6 was renumbered as Article 7 and additional explanatory text was added to the prohibition of first-cousin marriage: '. . . if the man or the woman is suffering from any disease which is regarded by medical science as rendering a person unfit for marriage' (Government of PR China, 2001).

The status and regulation of consanguineous marriage in Taiwan has proceeded along very similar lines to the legislation introduced in PR China, with the initial legislation promulgated by the Nationalist Republic of China in 1930. In 2002, the Republic of China Civil Code, Part IV Family, was amended and, according to Article 983, marriages between couples related as first cousins or closer were prohibited, with detailed guidance as to the types of banned consanguineous unions provided in Articles 968–970 of the Code (Republic of China, 2002).

The Republic of Korea and the Democratic People's Republic of Korea

According to Article 809 of the Civil Act of the Republic of Korea that was adopted in 1957: 'A marriage may not be allowed between blood relatives, if both surname and origin are common to the partners'. A major problem that arose from this legislation was that by 1985, there were some 3.0 million persons with the surname Kim in the city of Kimhae, 2.4 million persons with the surname Lee in Chunju, and 2.7 million persons with the surname Park in Milyang, for all of whom marriage with a partner bearing the same surname was illegal (Constitutional Court of Korea, 2001). As a result, by the 1990s, an estimated 200 000 couples had chosen to ignore the ruling and were living in de facto relationships (Constitutional Court of Korea, 2001).

The legality of this situation was examined in the *Same-Surname-Same-Origin Marriage Ban Case*, accepted by the Constitutional Court of Korea in 1997; in 1998, the Court ruled in favour of an amendment to Article 809 to allow persons of the same surname and general lineage to marry (Constitutional Court of Korea, 2001). Legislation still exists to prohibit marriage between individuals related as first, second or third cousins, even though there is no evidence that consanguineous marriage was traditionally favoured in Korea. Indeed, a study of Koreans living in PR China reported a total avoidance of consanguineous unions (Du *et al.*, 1981). Somewhat surprisingly, the anti-consanguinity legislation has been applied to non-Korean couples married in another country who wish to visit Korea, as in the case of a Pakistani citizen whose wife was refused an entry visa on the grounds of their premarital relationship as first cousins (*Korea Times*, 2009).

Despite their often fraught political relationships, the two parts of the Korean peninsula behave as one with regard to consanguineous marriage and, according to Article 10 of the Family Law of the Democratic People's Republic of Korea, adopted by Decision No. 5 of the Standing Committee of the Supreme People's Assembly in 1990: 'Marriage between blood relatives up to and including third cousins, or between relatives by marriage up to and including first cousins, is prohibited'. Perhaps the most interesting aspect of Article 10 is the distinction drawn between the prohibited types of biological and affinal marriage, although with no specific rationale indicated.

The Philippines

According to Article 38 of the Family Code of the Philippines (1987), Title 1, Chapter 3: Void and voidable marriages: '1. Marriages between collateral relatives, whether legitimate or illegitimate, up to the fourth civil degree shall be regarded as void for reasons of public policy', i.e. marriages between first cousins are not legal. However, Article 38 includes a number of other quite diverse categories of non-biological relationship which also are banned, for example, including: '2. Between step-parents and stepchildren'; '7. Between an adopted child and a legitimate child of the adopter'; and, rather more sinisterly: '9. Between parties where one, with the intention to marry the other, killed that other person's spouse, or his or her own spouse'. It therefore would appear that in this section of the Philippines Family Code, concerns other than genetic health were in the minds of the draftees of the legislation.

Consanguineous marriage and international law

Arguably, the maintenance of civil legislation in the USA, PR China and Taiwan, and South and North Korea banning first-cousin unions could be held to be in contravention of international human rights conventions on the right to marry (Bittles, 2003). Article 16 of the United Nations Declaration of Human Rights (1948) proclaimed that: 'Men and women of full age, without any limitation due to race, nationality or religion, have the right to marry and to found a family'. Similarly, according to Article 12 of the European Convention on Human Rights: 'Men and women of marriageable age have the right to marry and found a family, according to the national laws governing the exercise of that right'. Article 14 further states: 'The enjoyment of the rights and freedoms set forth in this Convention shall be secured without discrimination on any grounds such as sex, race, colour, language, religion, political or other opinion, national or social origin, association with a national minority, property, birth or other status' (Council of Europe, 1953). More specifically in genetic terms, Article 6 of the Universal Declaration on the Genome and Human Rights (UNESCO, 1997) states: 'No one shall be subjected to discrimination based on genetic characteristics that is intended to infringe or has the effect of infringing human rights, fundamental freedoms and human dignity'.

Given our present knowledge of the quite limited adverse effects of consanguinity on health at the population level, which will be discussed in detail in Chapters 7–11, it is difficult to discern how laws prohibiting first-cousin marriage can continue to be required or justified. Nevertheless, the reservations surrounding consanguineous marriage in Western societies were apparent in a Policy Statement issued by the American Society for Human Genetics (ASHG) in October 1998 on *Eugenics and the Misuse of Genetic Information to Restrict Reproductive Freedom*. In the introduction, the Board of Directors of the ASHG reaffirmed its commitment to the fundamental principle of reproductive freedom and unequivocally declared its opposition to coercion based on genetic information. But in discussing the support of governments for programmes, mandatory or otherwise: '...to improve the odds that children will be healthy...', the Policy Statement indicated that 'Laws forbidding first cousin marriages and other consanguineous unions...' were acceptable and did not involve the misuse of genetic information.

The possible mandatory prohibition of consanguineous marriage apparently approved by the Drafting Committee and accepted by the ASHG Board of Directors contradicts the 1970 conclusion of the US National Conference of Commissioners that state laws banning first-cousin marriage lacked scientific foundation and should be repealed. It also is at odds with the 2002 Recommendations by the US National Society of Genetic Counselors that, other than supplemental neonatal screening by tandem mass spectrometry at one week of age and hearing screening at three months, the progeny of first- or second-cousin couples did not require additional preconception, prenatal or postnatal testing (Bennett, Motulsky *et al.*, 2002).

Commentary

While first-cousin marriage may be subject to specific religious sanctions in Western societies, no laws restricting first-cousin marriage are in force at the national level. The legislation banning first-cousin marriage in China, Taiwan, both Koreas, and the Philippines appears to be somewhat anachronistic, the more so since equivalent laws have not been deemed necessary to restrict the numbers of children born with genetic defects elsewhere in the world. This observation equally applies to state-based legislation in the USA. Given the changing demographic profile of the USA, with many millions of immigrants from countries where consanguineous marriage remains popular, and the weight of evidence that points to the limited adverse effects of first-cousin marriage on offspring health, the current state laws on consanguineous unions could appropriately be re-addressed.

4 Consanguinity: the scientific and medical debates of the nineteenth and early twentieth centuries

Introduction

To understand and appreciate the major attitudinal change towards consanguineous marriage that occurred in the latter half of the nineteenth century, and continues to influence thinking in the twenty-first century, it is important to identify the persons and factors largely responsible for this major shift in public perception. In 1855, the Revd Charles Brooks of Medford, Massachusetts, delivered a paper to the 9th Meeting of the American Association for the Advancement of Science (AAAS) entitled 'Laws of reproduction, considered with particular reference to inter marriage of first cousins'. In his presentation, Brooks called for a survey into the outcomes of consanguineous marriages to determine whether: '. . . the marriage of first-cousins is forbidden of God' (Brooks, 1856), a topic that in the twenty-first century would generally be considered more theological than scientific in nature and, as indicated in Chapter 2, a query that already seemed to have been divinely established by the matrimonial rulings provided in Leviticus 18:7–18.

The information that Brooks presented to the meeting largely consisted of a series of anecdotal tales on physical and mental problems experienced by the progeny of first-cousin unions. Thus: 'Mr E.S. and wife, of N., Mass., are first-cousins, both of sound mind and robust health. They had seven daughters and one son. Three daughters are deranged, and the rest are nervous'. Which could be considered an understandable reaction on the part of the four non-deranged daughters. Similarly: 'Mr E. of M., Mass., married his first-cousin. They have five daughters and three sons. One daughter is an idiot, and two are feeble-minded. One son has run away with the town's money'. The experience of other clergymen also was cited, as in: 'Revd Mr Wisnor, of Boston, asked a friend: "Do you know Mr C., who attends my church; he looks queerly, but he thinks more queerly still?" His parents were cousins'. Which apparently was regarded as case proven with respect to the expected outcome of a first-cousin marriage.

On the basis of these rather eccentric case studies, Brooks concluded that a stand had to be taken on the question of first-cousin marriage, stating: 'The improvement and prosperity of thousands of families depend on its solution:

and, in a degree the safety and elevation of society' (Brooks, 1856). With hindsight, this statement appears greatly exaggerated, since US marriage data on Mormons and their non-Mormon relatives indicate that in the period in question, 1840–1859, only 0.59% of marriages were between first cousins, having decreased from a peak of 1.17% during 1720–1739 (Woolf *et al.*, 1956). Nevertheless, Brooks' call to arms seems to have galvanized medical, public and political opinion in the USA on the topic of consanguinity, with a number of states rapidly introducing prohibitions on marriage between cousins (Chapter 3).

Early medical studies on consanguinity in the USA

Dr Samuel Bemiss and the health outcomes of consanguinity

A swift response to Brooks' appeal came from Dr Samuel Bemiss of Louisville, Kentucky, who had collected information from medical colleagues in 25 states of the Union, from Maine to Mississippi, on the outcomes of 873 consanguineous unions ranging from incest to marriages between third cousins (Bemiss, 1858). As controls, Bemiss collated information on 125 marriages between couples who were neither themselves related nor the descendants of blood relatives. The author displayed appropriate caution in noting that these data: '... however correct they may be in selection and statement, are not ample enough in number to justify me in offering them as the average results of marriage where no influence of consanguinity prevails'.

Bemiss was also aware of the possibility of ascertainment bias and cautioned: 'It is natural for contributors to overlook many of the more fortunate results of family intermarriage, and furnish those followed by defective offspring or sterility. The mere existence of either of these conditions would prompt inquiry, while the favorable cases might pass unnoticed'.

According to his summary of the collated results, consanguineous unions were indeed harmful (Bemiss, 1858). There are, however, a number of problems in the analysis of the data set. For example, in categorizing the information on incest, the author clumped together outcome data on four brother–sister and parent–child matings ($F = 0.25$), five half-sib matings ($F = 0.125$), and a single grandfather–granddaughter relationship ($F = 0.125$). The data also included 61 cases listed simply as: 'Marriage between cousins themselves the offspring of kindred parents', 27 cases of 'Double cousins', and 30 cases of 'Marriages irregularly reported, all first cousins', without qualifying or specifying the exact types of relationship(s).

Table 4.1 *Consanguinity and early death*

Relationship	Cases	Coefficient of inbreeding	Mean no. of offspring	Early death
Incest	4	0.25	1.3	0%
Brother/half-sister	5	0.125	4.8	0
Uncle/niece, aunt/nephew	12	0.125	4.8	40.4
First cousin	600	0.0625	4.8	23.1
Second cousin	120	0.0156	4.6	17.3
Third cousin	13	0.0039	5.5	12.7

Source: Bemiss (1858)

For the 754 remaining couples, data on fertility and on offspring who 'died young' have been summarized in Table 4.1. Mean family size was much higher in the non-consanguineous group ($n = 6.7$) than in any of the consanguineous sub-groups but, otherwise, with the exception of the incestuous couples ($n = 4$), consanguinity did not appear to influence fertility (range, $n = 4.6$–5.5). Apart from the small number of sib and half-sib ($n = 5$) matings in which no deaths were reported, there was a striking positive relationship between the level of inbreeding and mortality. However, this relationship is less convincing when the cause of death is examined. Taking uncle–niece/aunt–nephew relationships as an example, a cause of death was presented for the offspring of just 6 of the 12 couples. In two of these families, the cause of death was 'unknown'; in a further two families, marasmus (i.e. malnutrition due to severe prolonged caloric deficit) was given as cause of death, including one family in which all nine offspring had died; and for the remaining two families, deaths were either 'accidental' or caused by typhoid fever.

In fact, a cause of death was available for just 475 of 883 deaths in the combined consanguineous families (53.8%) and 69 of 134 deaths in the controls (51.5%); therefore, given the large numbers of missing data, any analysis of mortality could be significantly biased. The discrepancy in deaths from tuberculosis (TB) is striking with 22.7% TB deaths in the combined consanguineous families versus 8.7% in the non-consanguineous controls, and with nine of ten persons dying of tuberculosis in a single 'Double cousin' family. During the mid-nineteenth century, TB was the major killing disease in all industrializing countries and principally affected the poorest socioeconomic sectors of society. On the basis of these incomplete data, it appears that a significant proportion of the deaths in the consanguineous families sampled by Bemiss and his collaborators could more probably be ascribed to their poor living conditions rather than resulting from specific genetic disorders.

Table 4.2 *Consanguinity and cause of early death*

Relationship	Total deaths	Diagnosis available	TB as diagnosed cause of death
Combined consanguineous offspring	883	475	22.7%
Non-consanguineous offspring	134	69	8.7

Source: Bemiss (1858)

Understandably, the report by Bemiss was widely read and discussed, and it formed a useful basis for the anti-consanguinity legislation introduced in a number of states. A more cautious stance on the findings was, however, adopted by a Committee of the Medical Society of the State of New York established to investigate the impact of consanguinity on health. In the considered opinion of the Committee: 'If two cousins are healthy and see fit to marry, there is as much reason to believe that their children would be healthy, as if they were not connected by cousinship or consanguinity at all' (Newman, 1869).

Lewis Henry Morgan and the classification of biological relationships

By comparison, strong support for Bemiss' work was provided by Lewis Henry Morgan, an influential lawyer and anthropologist from the state of New York. Morgan had produced a remarkable tome on consanguinity, mainly based on his field studies into the marriage patterns of Native American tribes (Morgan, 1871). He also sought to place these findings into a global context by supplementing the data he himself had gathered with information on the specific types of consanguineous marriages favoured in other parts of the world. He undertook this task through correspondence with academics, businessmen, missionaries and others, including a letter forwarded from London to Dr David Livingstone who was then pursuing his religious vocation in southern Africa (White, 1957).

Morgan discovered that the classification system used by the North American Seneca-Iroquois tribes in designating biological relatives was virtually identical to that of South Indian Tamils and Telegus, inhabitants of the present-day states of Tamil Nadu and Andhra Pradesh. On this basis, he argued that the similarity of the two kinship systems was evidence of strong genetic and historical relationships and hence indicated an Asiatic origin of Native Americans (White, 1957).

Despite misgivings from his Presbyterian clergyman in Rochester, New York, in 1851 Morgan had married Mary Elizabeth Steel, who was his mother's brother's daughter and thus a first cousin (Trautmann, 1987), and he appears to have been initially content with his choice of partner. But after the birth of an intellectually disabled son and the deaths of both of their daughters from scarlet fever (Ottenheimer, 1990), Morgan seems to have concluded that each of these sad events was a direct outcome of his consanguineous marriage and thereafter he became an increasingly vigorous opponent of close kin marriage. Elected a member of the USA National Academy of Sciences and successively President of the Section of Anthropology of the American Association for the Advancement of Science (AAAS) and then in 1879 President of the AAAS itself, his views would have carried considerable weight in persuading state legislatures in the USA to pass laws limiting or banning first-cousin marriages.

Alexander Graham Bell, consanguinity and the education of the deaf

A third highly influential figure in the consanguinity debate within the USA was Alexander Graham Bell, now best known as inventor of the telephone but renowned during his lifetime as an educator of deaf children. In November 1883, Bell presented a paper to the USA National Academy of Sciences meeting in New Haven, provocatively entitled 'Upon the formation of a deaf variety of the human race'. Using data from two institutions, comprising 2106 pupils of the American Asylum for the Education of the Deaf and Dumb in Hartford, Connecticut, for the period 1817–1877, and 1620 pupils of the Illinois Institute for the Deaf and Dumb in Jacksonville, Illinois, for 1846–1882, Bell clearly demonstrated the over-representation of pupils with certain surnames, names that otherwise were quite uncommon in the general population (Bell, 1883).

In further analyses, he showed that 32.9% of the pupils in the Connecticut institution had relatives who were deaf-mute, with several pupils coming from families with up to 15 deaf-mute members. There also was compelling evidence that a majority of the spouses of deaf-mute persons were themselves deaf-mute, and that the proportion of such marriages had increased significantly during the course of the early to mid nineteenth century – encouraged by the social opportunities afforded by their co-education in institutions catering only for those with major hearing defects and by their ability to communicate via sign language – a trend that has continued to the present day (Arnos *et al.*, 2008).

Concerned by the increasing numbers of individuals with congenital deaf-mutism, Bell suggested that it might be appropriate to introduce legislation to ban consanguineous marriages in families with deaf-mute members (Bell, 1883). However, in noting the prevalent impression that deafness, blindness, idiocy and insanity were often due to consanguinity, to avoid ascertainment bias it was proposed that in the Eleventh Census of the USA, scheduled for 1890, the parental relationships of persons with these disorders should be noted (Bell, 1889), with particular emphasis on first-cousin unions (Pyia, 1889).

George Arner and consanguineous marriage in the USA

In 1908, George Arner of Columbia University considered the published data on consanguinity in the USA and estimated that on a national basis, no more than 1% of late-nineteenth-century marriages had been between first cousins, although in some isolated mountain or island communities the prevalence may have been more than 5% of unions. Arner also concluded that the numbers of consanguineous marriages were declining across the country due to improved communications. While supportive of the work by Bell on deafness and deaf-mutism, and with data that showed a significant positive association between consanguinity and blindness due to retinitis pigmentosa, Arner criticized as fallacious the study design and the conclusions drawn by Bemiss (1858). Further, with respect to the often-cited link between consanguinity and the high prevalence of deaf-mutism in some villages in Martha's Vineyard, Massachusetts (Jenkins, 1891), Arner noted that two of the earliest European settlers into the area in the seventeenth century had been deaf-mute (Arner, 1908), a very early recognition of the importance of founder effect in genetic disorders.

According to Arner, 'By far the greater part of the literature of consanguineous marriage is of a controversial rather than of a scientific nature, and a search for statistical evidence for either side of the discussion reveals surprisingly little that is worthy of the name. Yet men of high scientific standing have repeatedly made most dogmatic assertions in regard to the results of such unions, and have apparently assumed that no proof was necessary'. Twelve years later, a similar scathing criticism of the consanguinity debate in the USA was made by Raymond Pearl, founder of the journal *Human Biology* and an eminent scientist who in his lifetime published more than 700 papers and 17 books, including the first study to demonstrate the adverse effect of cigarette smoking on longevity (Goldman, 2002). In a remarkably unrestrained book review published in the journal *Science*, Pearl noted, 'It is safe to say that no phase of biology has been enveloped in such a fog of superstition, old wives' tales, and other sorts of misapprehension as has inbreeding' (Pearl, 1920).

Early consanguinity studies in Great Britain and Ireland

During the same period, the controversy surrounding consanguineous marriage also had become a topic of wide interest in Great Britain and Ireland. The initial impetus seems to have been the extended and acrimonious debate surrounding the permissibility of a marriage between a widower and his dead wife's sister discussed in Chapter 3 (Anderson, 1982). But, as in the USA, through time, the debate had switched from the rights and wrongs of affinal marriage to whether or not marriages between first cousins should be permitted.

Sir William Wilde, cousin marriage and deaf-mutism

The first major investigation into the possible adverse effects of consanguinity was conducted by Sir William Wilde, a renowned ophthalmologist and father of the author and playwright Oscar Wilde. As Deputy Commissioner of the Post-Famine 1851 Census of Ireland, Wilde studied the possible relationship between consanguinity and deaf-mutism and collected information on the numbers and family circumstances of the deaf and dumb (Wilde, 1854).

Some initial difficulty had been encountered with the returns of individual enumerators, with one enumerator recording all children less than one year of age as deaf-mute on the premise that they could not speak nor, apparently, did they respond to speech. These problems notwithstanding, 4747 deaf-mutes were enumerated in the total population of 6.55 million, i.e. an overall prevalence of 1 per 1380. In 170 cases (3.6%), the parents were related as first, second or third cousins, although Wilde considered this figure to be an underestimate. He also recorded 525 families in which there were between two and nine children who had been born deaf-mute and very presciently commented on the lack of correlation between deaf-mutism in parents and their progeny.

Following Wilde's pioneering work, physicians and anthropologists in Great Britain and continental Europe conducted studies into the prevalence of deaf-mutism and other familial defects (Boudin, 1862; Mitchell, 1862, 1864; Child, 1863; Dally, 1864; Voisin, 1865; Darwin, 1875a). In many cases, these studies were conducted in geographically remote populations where, it was reasoned, high rates of consanguineous marriage would be the norm. A lively debate ensued in the medical and lay press between those who were strongly opposed to first-cousin marriage (Crossman, 1861; Shuttleworth, 1886) and others who saw no reason for its control (Gardner, 1861; Child, 1863; Huth, 1875). In retrospect it seems curious, if circumspect, that no one appears to have publicly mentioned the marriage of the reigning monarch, Queen Victoria, to her first cousin Prince Albert. However, sensitivity to the Queen's prolonged distress following the death of her husband from typhoid fever in December 1861 may have quelled speculation on the advisability, or otherwise, of the union.

Dr Arthur Mitchell and cousin marriage in Scots fishing villages

Dr (later Sir) Arthur Mitchell, who held the position of Deputy Commissioner in Lunacy for Scotland, played an important although somewhat idiosyncratic role in these discussions. He published extended, and closely matched, reports on the health of the progeny of consanguineous marriage in the *Memoirs of the Anthropological Society of London* (Mitchell, 1862) and the *Edinburgh Medical Journal* (Mitchell, 1864–5). In particular, Mitchell focused on the purported adverse effects of consanguinity on mental capacity in fishing communities between Fife and Caithness in north-east Scotland where 11 of the 119 married couples were related as first cousins and 16 as second cousins (i.e. 22.7% consanguinity, equivalent to a mean coefficient of inbreeding $(\alpha) = 0.0079$).

In the absence of specific measures of intelligence, Mitchell demonstrated that male hat sizes in the region were on average $6^{7/8}$ to 7, equivalent to head circumferences of $21^{5/8}$ to 22 inches, whereas in neighbouring areas, the average hat size was $7^{1/8}$ ($22^{1/4}$ inches circumference). Although the author regarded this difference as clear evidence that the fishermen were 'below par in intellect', in accompanying articles, he was much more circumspect in assessing the claimed adverse effects of cousin marriage in rural Scotland: 'If we carefully study the literature of the subject we shall find that it abounds in unsupported assertion, and that important conclusions are very often made to rest on a basis which is undefined or clearly too narrow' (Mitchell, 1864–5, p. 781). Further: 'Startling illustrations of calamitous sequences to cousin-marriages have been detailed, and pointed at with a finger of warning, *the relation of cause and effect being assumed*' (Mitchell, 1864–5, p. 783, original author's emphases).

Mitchell sought to instil some degree of balance into reports on the adverse effects of consanguineous marriage. He collated evidence to demonstrate that 'unsoundness of mind' and deaf-mutism were more common in the progeny of cousin marriages but at the same time noted that: 'There are many causes of idiocy which are undoubtedly of greater power than kinship or parentage. (W)hooping cough, scarlatina and measles, for instance, produce a large amount of the idiocy in Scotland, as they do probably of other countries' (Mitchell, 1862, pp. 419–420).

Although local sentiment in general was against consanguineous marriage, on asked for her view on the effects of close kin unions on children's health, an elderly local lady commented in the vernacular: '... I'll tell ye what, Doctor, bairns that's hungert i' their youth aye gang wrang. Tha's far waur nor sib marriages' (Mitchell, 1864–5, p. 913). An opinion which matched Mitchell's own overall conclusions that: 'Consanguinity in the parents *may* very decidedly *tend* to injure the offspring, yet it by no means follows that every defect in the children born to blood-related parents is an expression of this tendency, for the general causes of defect will exist among them as among other children, and will give results at least equally disastrous' (Mitchell, 1862, p. 783, original author's italics).

Charles Darwin, cousin marriage and family concerns

The intervention of Charles Darwin, who had married his first cousin Emma Wedgwood in 1839, was of great importance in the debate on consanguinity. For most of his adult life, Darwin suffered periodic bouts of debilitating ill-health (Katz-Sidlow, 1998), which he worried might be transmitted to his children. Like Morgan in the USA, Darwin seems to have decided that consanguinity was disadvantageous and, according to Kuper (2009), his concerns had been provoked by the book *Intermarriage: or the Mode in Which and the Causes Why, Beauty, Health, and Intellect Result from Certain Unions, and Deformity Disease and Insanity from Others*, written by Alexander Walker and published in 1838, the year before Darwin's marriage. The manner in which Darwin's concerns were expressed was decidedly peculiar, being presented as the concluding sentence in a book on the avoidance of self-fertilization in orchids: '...that marriage between near relations is... in some way injurious, – that some unknown great good is derived from the union of individuals which have been kept distinct for many generations' (Darwin, 1862).

The timing of these concerns was equally unusual since the book was published six years after the birth of his last child, the tenth his wife had borne in 17 years of marriage and delivered when she was 48 years of age. Two of their children, including the last infant who from a photograph appears to have the characteristic facial features of Down syndrome, died in infancy and a third, his favourite daughter Annie, had died aged ten years probably of tuberculosis (Fenner *et al.*, 2009). Conversely, three of his sons George, Francis and Horace became eminent scientists in their own right and were elected Fellows of the Royal Society of London, while Leonard, after a successful army career, became a Member of Parliament and President both of the Royal Geological Society and the Eugenics Society.

At Darwin's suggestion, his friend and neighbour Sir John Lubbock, an anthropologist and Member of Parliament for Maidstone, proposed that the prevalence of first-cousin marriages should be determined in the 1871 Census of Great Britain and Ireland. The precise reasons given by Lubbock to the Parliamentary Select Committee for this proposal were that: 'It was believed that consanguineous marriages were injurious throughout the whole vegetable and animal kingdoms, and it was desirable to ascertain whether that was not the case with the whole human race' (HMSO, 1870).

Even though, as detailed in Chapter 2, first-cousin unions had been legal in England since the sixteenth century, the proposal was rejected by a Parliamentary Select Committee on a 92-to-45 vote. The main reasons given by those opposing Lubbock's proposal were that the subject was of particular sensitivity and any formal enquiry into close kin marriage would be unacceptably intrusive (HMSO, 1870). However, it is clear from the intemperate nature of the remarks made by certain Members that the topic had been highly controversial, with a Mr Lock opining that: 'This was a piece of the grossest cruelty ever thought of'. While Mr Gathorne Hardy stated that he: '. . . did not see the desirability of holding up families where such marriages had taken place, and the children being anatomised for the benefit of science'. Furthermore: 'Such children would be held up as discreditable' (HMSO, 1870).

Darwin's annoyance at the Parliamentary rejection was equally apparent. Writing in the *The Descent of Man* (1871), he stated: 'When the principles of breeding and of inheritance are better understood, we shall not hear ignorant members of our legislature rejecting with scorn a plan for ascertaining by an easy method whether or not consanguineous marriages are injurious to man'.

George Darwin and isonymy

To circumvent the lack of data, Charles Darwin's son George, a Fellow of Trinity College, Cambridge, devised a method subsequently termed *isonymy* for calculating the percentage of first-cousin unions in a population from the numbers of couples who shared the same premarital surname. As sources, George Darwin used information on marriages announced in the socially impeccable *Pall Mall Gazette* from 1869–1872 and marriages recorded in the equally prestigious *Burke's Landed Gentry* and the *English and Irish Peerage*. He also mailed some 800 circulars to members of the 'upper middle and upper classes' and consulted the Registrar General's Annual Report on marriages for 1853 (Darwin, 1875a). First-cousin unions were most common in the peerage (4.5%), with 3.8% and 3.5% first-cousin marriage among the landed gentry and middle classes, respectively. The data from the Registrar General's Annual Report showed lower levels of first-cousin marriage, perhaps reflecting social class differences, and a rural–urban cline with 2.3% first-cousin unions in Rural

Table 4.3 *First-cousin marriage in Great Britain by residence*

Residence	No. studied	Same-name marriage (%)	First-cousin marriage (%)
London Metropolitan	33 155	0.55	1.5
Urban districts	22 346	0.71	2.0
Rural districts	13 391	0.79	2.3

Source: Darwin (1875a)

Table 4.4 *First-cousin marriage and insanity in Great Britain and Ireland*

Country	No. studied	Offspring of First cousins (%)
England and Wales	4 308	3.5
Scotland	514	5.3
Ireland	651	0.8

Source: Darwin (1875a)

Districts, 2.0% in Urban Districts and just 1.5% in the London Metropolitan area (Table 4.3).

As with previous authors, including Arthur Mitchell (1864–5, pp. 781–794), George Darwin also analysed data from asylums to see if a positive association between consanguinity and insanity or idiocy could be determined. But, as indicated in Table 4.4, there was no major evidence for an increased prevalence of consanguinity among asylum inmates by comparison with the levels of first-cousin marriage in the general public, and more especially with members of the peerage (Darwin, 1875a). In a private letter to his son commenting on a draft of these results, Charles Darwin advised: 'For Heaven's sake put a sentence in some conspicuous place that your results seem to indicate that consanguineous marriage, as far as insanity is concerned, cannot be injurious in any very high degree' (Darwin, 1874).

A further investigation conducted by George Darwin on members of the rowing eights of Colleges of Oxford and Cambridge Universities (excluding coxswains), described by Darwin as 'a picked body of athletic men', indicated limited adverse effects of consanguinity on physical performance. Among the boating men, 2.4–2.8% had parents who were first cousins versus 3.0–3.5% first-cousin parentage among their social peers, findings which were summarized by Darwin (1875b): '. . . these numbers appear, to some extent, to justify the belief that offspring of first cousins are deficient physically, whilst at the same time they negate the views of alarmist writers on the subject'.

The findings of George Darwin's composite investigations into the effects of consanguinity on mental health and physical fitness were readily accepted by Francis Galton, guiding spirit of the Eugenics movement and his father's half-cousin, who in a letter to his younger relative dated 10 November 1875 commented: 'I . . . have read your cousin-paper with very great interest. You have certainly exploded most effectually a popular scare'.

Presumably with tongue firmly in cheek, he also suggested that publication of a pamphlet on the topic presented a very promising commercial opportunity. According to Galton's calculations: 'There are, say, 200,000 annual marriages in the kingdom, of which 2,000 and more are between first cousins. You only have to print in proportion, and in various appropriate scales of cheapness or luxury:

> "WORDS of Scientific COMFORT
> and ENCOURAGEMENT
> To COUSINS who are LOVERS"

then each lover and each of the two sets of parents would be sure to buy a copy; i.e. an annual sale of 8,000 copies!! (Cousins who fall in love and don't marry would also buy copies, as well as those who think that they *might* fall in love)' (Pearson, 1924, original author's emphases).

When the second edition of Charles Darwin's book on self-fertilization in orchids was published, his original statement on the injurious effects of marriage between near relatives had been omitted (Darwin, 1877). Possibly in acknowledgement of his privileged personal circumstances, Darwin had previously rationalized his son's findings as indicative of the fact that: ' . . . the widely different habits of life of men and women in civilized nations, especially amongst the upper classes, would tend to counter-balance any evil from marriages between healthy and somewhat closely related persons' (Darwin, 1876). Despite Darwin's acceptance that marriage to his first cousin had not been harmful to their children, the possible ill-effects of consanguinity on their subsequent reproductive capacity (Golubovsky, 2008) and the health of the wider Darwin–Wedgwood family across four generations has been the subject of recent speculation (Berra *et al.*, 2010).

Cousin marriage in decline

The prevalence of first-cousin unions rapidly reduced in Great Britain from the 4.7% calculated in a survey of 1600 respondents recruited from the readership of the *British Medical Journal* and covering the mid to late nineteenth century (Pearson, 1908). This was a somewhat higher level than the 1.1% first-cousin marriage rate reported for approximately the same time period by members of the Society of Genealogists (Smith, 2001). By the beginning of the twentieth century, 0.9% of children in Great Ormond Street Hospital, London, had first-cousin parents (Pearson, 1908), and a survey of genealogists' families in the 1920s indicated 0.3% first-cousin marriage (Smith, 2001).

To conclusively establish the rates of consanguineous parentage among hospital inpatients, a comprehensive hospital-based survey was organized in Great Britain from 1935 to 1939 with all types of consanguineous marriage from uncle–niece to second cousin and beyond noted. The survey indicated an overall consanguinity prevalence of 1.0% or less ($\alpha = 0.0005$) in London Teaching Hospitals and it was lower at 0.5% ($\alpha = 0.0002$) in the parents of the patients in Children's Hospitals, suggesting that the urban decline in consanguineous marriage was continuing. Even in two rural general practices, only 3.1% of marriages ($\alpha = 0.0019$) were consanguineous. This downward trend continued after World War II, and in the most recent hospital-based survey in the English Midlands, just 0.2% of UK-born patients were married to a first cousin (Bundey *et al.*, 1990).

Eugenics and the consanguinity debate

Consanguinity and the application of biometric analysis

Early in the twentieth century, investigations into the effects of first-cousin marriage had been initiated in the newly established Francis Galton Laboratory for National Eugenics housed in University College London, with Karl Pearson as its inaugural Director. As a mathematician, Pearson believed that correlation coefficients could be used to determine the levels of genetically determined similarity, defined as 'resemblance', between persons related as sibs, grandparents–grandchildren, uncles–nieces, and first cousins. Indeed, in a series of debates held in the Royal Society of Medicine in London in November 1908, Pearson had argued forcefully against the views of William Bateson (Bateson, 1909) and others that Mendelian analysis could adequately describe the inheritance of conditions such as albinism: 'My own standpoint is that there is no definite proof of Mendelism applying to any living form at present; the proof has got to be given yet' (Pearson, 1909).

In Pearson's opinion, biometric analysis would more credibly provide the requisite answers to pedigree analysis in albinism and other human disorders, and the larger the data set the more likely that an appropriate solution would be

Table 4.5 *Family eugenic histories and characteristics*

Characteristic	Categories
Present age, or age at death	To be specified
Ailments in life	To be specified
Cause of death, if dead	To be specified
General health	Very robust
	Robust
	Normally healthy
	Delicate
	Very delicate
Ability	Mentally defective
	Slow dull
	Slow
	Slow intelligent
	Fairly intelligent
	Distinctly capable
	Very able
Temper	Sullen temper
	Quick temper
	Even temper
	Weak temper
Temperament	Reserved, Expressive or Betwixt
	Sympathetic, Callous or Betwixt
	Excitable, Calm or Betwixt
Success in life	Marked success
	Prosperous career
	Average career
	Difficult career
	Failure

Sources: Elderton & Pearson (1907), Elderton (1911)

obtained (Pearson, 1909). As indicated in Table 4.5, in the Galton Laboratory, a series of 'physical and psychical characteristics of families' were collected to measure resemblance between relatives, all of whom were assessed as adults (Elderton & Pearson, 1907; Elderton, 1911).

The data collected were predominantly subjective in nature, although the authors provided notes to assist in the assessment of those tested. It therefore was claimed that differentiation could be made between persons who were defined in terms of ability as being: 'Slow: very slow in thought generally, but with time understanding is reached'; versus 'Slow intelligent: slow generally, although possibly more rapid in certain fields: quite sure of knowledge when once acquired'. Or, 'Distinctly capable: A mind quick in perception and in reasoning rightly about the perceived'; versus 'Very able: Quite exceptionally able intellectually, as evidenced either by the person's career or by consensus

Table 4.6 *'Resemblance' between first cousins*

Cousin type	Health	Intelligence	Success in Life	Temper	Temperament	Mean values
Male	0.31	0.34	0.19	0.18	0.26	0.26
Female	0.33	0.34	0.26	0.19	0.23	0.26
Male–female	0.30	0.34	0.26	0.25	0.27	0.28
Mean values	0.31	0.34	0.24	0.21	0.25	0.265

Source: Elderton & Pearson (1907), Elderton (1911)

of opinion of acquaintances'. Assuming, of course, that in the latter example the aforesaid acquaintances were themselves of sufficient intellectual calibre to provide reasoned judgement on the matter.

In measuring resemblance between first cousins, 10 separate categories of cousin were identified, including separate male/female entries, and contingency tables drawn up for each subcategory of variable by specific type of cousin relationship. Approximately 5800 cousin pairs were available for comparison in terms of their Health, Intelligence and Temper, but with significantly fewer results entered for the Success in Life and Temperament categories.

As shown in Table 4.6, based on the results of 70 contingency tables (i.e. with ten types of first cousins each tested for seven characters), the mean between-cousin resemblance was calculated to be 0.265 ± 0.008, with a standard deviation of 0.093 ± 0.006. The average resemblance of first cousins was therefore taken to be between 0.25 and 0.30 by comparison with a mean resemblance between sibs of 0.51, which was explained on the grounds that first cousins have two common grandparents whereas sibs have four. In the authors' opinion, the results for Temperament were the least reliable and resulted in much wider confidence limits, which they ascribed to the recorders' lack of experience in scoring this particular variable.

Sir Francis Galton, consanguinity and paradoxical facts

Initially, the biometric data derived from these 'Family Records' were presented only for sibs, grandparents–grandchildren, and for first-cousin offspring because the by-now Sir Francis Galton quite reasonably regarded as improbable the claim that uncle–nieces and aunt–nephews had slightly lower mean levels of genetic distance, 0.24 for uncle–nieces/aunt–nephews versus 0.265 for first cousins (Elderton, 1911). Citing the rationale previously adopted by Elderton and Pearson, Galton supposed that logically uncle–nieces and aunt–nephews would have three rather than two grandparents in common and so should exhibit a greater level of resemblance than first cousins. In addition, the resemblance data conflicted with Galton's findings in *Hereditary Genius*, which had shown that the uncles and nephews of eminent persons were significantly more likely to have shared their special talents than had their cousins (Galton, 1869).

Galton's blocking objection to these 'paradoxical facts' was effectively removed by his death in January 1911; later that year, the mean values for resemblances between first cousins (0.265) and uncles and their nieces (0.24) were duly published. More contentiously, on the basis of these findings, it was argued that since uncle–niece marriages were illegal in Great Britain, first-cousin marriages should likewise be restricted (Elderton, 1911). Conversely, according to an eminent ophthalmologist whose work on retinitis pigmentosa had been cited by William Bateson in the 1908 Mendelism versus Biometrics debates in the Royal Society of Medicine: 'I think . . . we may conclude that marriages between cousins are as safe from the eugenic point of view as any other marriages, provided the parents and stock are sound' (Nettleship, 1914, p. 137) – a conclusion that was very similar to the findings of the Committee of the Medical Society of the State of New York some 50 years earlier in response to the survey reported by Samuel Bemiss (Newman, 1869).

A somewhat different stance was adopted by Mr Macleod Yearsley FRCS, Senior Surgeon of the Royal Ear Hospital and Medical Inspector to the London County Deaf Schools. In a study published in *The Eugenics Review* drawing attention to the over-representation of consanguineous progeny among deaf-mute children in the London Deaf Schools, Yearsley was of the opinion that: 'Every congenital deaf-mute should be sterilized', at the same time acknowledging, ' . . . this is a bold statement, but I do not fear to say it here' (Yearsley, 1911).

Eugenics and purging of the gene pool

Highly variant opinions were expressed by other eugenicists following publication of the theoretical prediction that in the case of rare, detrimental recessive genes, inbreeding would facilitate their expression and hence their elimination from the population gene pool (Jacob, 1911). Echoing the revised opinion of his father Charles, and of Nettleship (1914), in 1926 Leonard Darwin, President of the Eugenics Society of Great Britain, wrote: 'If the stock is good, the chances of any harm arising from cousin marriages are very small'. However, choosing to err on the safe side: ' . . . it is only neutral or bad qualities which come to the surface when recessive genes are brought together, and prudence in regard to the next generation does, therefore, dictate the avoidance of cousin marriages where the common pedigree shows definite signs of hereditary defects.'

The latter comment seems to have been derived from another aspect of the work of the Galton Laboratory in which it was estimated that in a brother–sister mating, the outcomes would be '25% hale, 50% latent evil and 25% patent evil', whereas in a first-cousin mating, the result would be '56.25% hale, 37.5% latent evil and 6.25% patent evil'. No consideration was given to the possibility that these percentages, whether hale or evil, might vary between communities in which consanguineous marriage was customary and the effects therefore cumulative, versus those where cousin marriage had occurred only in a single generation (Chapter 6).

Displaying his political skills, Leonard Darwin also commented on a more contentious aspect of contemporary eugenic thinking: 'We could never actually advocate cousin marriages on the ground that such unions would increase the number of patently defective children and would thus facilitate the elimination of certain hidden racial defects'. Others displayed no such inhibitions and in its most extreme form, their dogma was summarized as: 'Inbreeding canalizes health and desirable qualities, just as it canalizes and isolates ill-health and undesirable qualities'. Whereas: 'Cross-breeding conceals and spreads ill-health and undesirable qualities, and thus contaminates desirable stock' (Ludovici, 1933–4).

The recommended solution to the problem was therefore to discourage marriages between unrelated individuals and to permit first-degree (incestuous) relationships as a means of purging deleterious recessive genes from the gene pool. The author did, however, concede that: '... deaths from disease and insanity would be heavy', and: '... a scheme of canalization of disease and health would not be practicable without artificial selection accompanied by legalized infanticide ...'. With these considerations in mind: '... it would probably be wise to delay for a generation or two an immediate recourse to the closest consanguinity in sound stocks, because of the fear of over-rapidly isolating strains with a too limited set of desirable qualities' (Ludovici, 1933–4). By arguing the case for successive generations of incestuous relationships, and with marriage to non-relatives discouraged, it seems probable that Ludovici succeeded only in persuading members of the medical profession, the clergy and the general public of the lack of commonsense or even sanity behind his proposals.

Although approached from quite a different starting-point, the comments of Lord Horder of Ashford (1933) on consanguineous marriage are nevertheless worth noting. The eminent physician made it clear that he viewed consanguinity as a part of the eugenic challenge to be faced by general medical practitioners in twentieth-century Great Britain, although his perspective in the section of his paper on altruism and eugenics is at considerable variance with the views on human altruistic behaviour developed by W.D. Hamilton some 30 years later and discussed in Chapter 14.

Commentary

As with many aspects of life, the gift of hindsight is an invaluable attribute when summarizing the scientific opinions and beliefs of earlier generations. This is especially so in addressing the more extreme views on the adverse effects of consanguinity espoused by commentators such as Brooks (1856) and Pearson (1909). Conversely, an impressive aspect of the scientific and medical debates that took place in Great Britain and the USA was the preparedness of a major public figure like Charles Darwin to radically alter his opinion on a matter of major personal as well as scientific significance. Together with the acuity of investigators such as William Wilde, Arthur Mitchell, Alexander Graham Bell and George Arner in their observation and recording of the familial patterns of inheritance in deaf-mutism and other inherited disorders that predated the emergence of medical genetics as a recognized discipline.

5 Demographic and socioeconomic aspects of consanguineous marriage

Introduction

As was discussed in Chapter 1, the predominant Western stereotype of consanguineous marriage being restricted to small groups living in remote regions of the planet, or to religious and social isolates, is far from the truth. While religious belief, and to a lesser extent civil legislation, does determine who one may marry, unions between close biological kin occur in many parts of the world and they continue to be willingly contracted by many millions of couples (Table 1.3).

The global prevalence of consanguineous marriage

Close kin marriage is preferential in many major populations, with the influence of religion apparent in the major regional differences in consanguinity prevalence across the globe (Figure 1.1). Although anthropological and medical genetics reports consistently indicate a preference for consanguineous marriage in many sub-Saharan African populations, for example, in the Sotho and Tswana peoples of southern Africa (Kromberg & Jenkins, 1982; Christianson *et al.*, 1994; Lund, 1996), and in Guinea (Chantrelle & Dupire, 1964), Nigeria (Scott-Emuakpor, 1974), Burkina Faso (Hampshire & Smith, 2001), and the Gambia (Bennett, Lienhardt *et al.*, 2002) in West Africa, there is little quantitative information on consanguinity from these regions. Likewise, while close kin marriage has been recorded in many communities within populous Asian countries including Bangladesh and Indonesia (Murdock, 1967), to date the data available are few in number and unrepresentative of the national populations.

Despite this major shortcoming, current data indicate that more than 1100 million people live in countries where 20–> 50% of marriages are between couples related as second cousins or closer ($F \geq 0.0156$), with some 10.4% of the 7.0 billion global population married to a biological relative ($F \geq 0.0156$) (Bittles & Black, 2010a). Although the global prevalence of consanguineous marriage seems to be declining, the data are quite inconsistent and in some countries the present-day rates of consanguinity exceed those of the preceding generation, possibly reflecting greater overall survival to adulthood which in turn increases the numbers of marriageable biological relatives (Bittles, 2008).

The patterns and changes in consanguineous marriages observed during the past 50 years can be conveniently subdivided in regional terms.

Western societies

There has been a significant general decline in consanguineous marriage throughout Western Europe, North and South America and Oceania from the mid nineteenth century onwards (Chapter 4), with first-cousin marriage rates of 0.6% or less in the current populations of these regions (Bundey *et al.*, 1990; Freire-Maia, 1968; Liascovich *et al.*, 2001; Port & Bittles, 2001). Precise data for the USA are difficult to obtain given the various legal prohibitions against consanguinity enacted in many individual states (see Figure 3.1), and where data have been collected, the information often refers only to Roman Catholic communities in which Diocesan dispensation for first-cousin unions has been a long-standing pre-requisite to a Church marriage ceremony (see Chapter 2).

Large-scale immigration from Asia and Africa has been a significant demographic feature of all Western countries during the course of the past 50 years. Current data suggest that in Western Europe there are at least 10 million resident migrants from regions where consanguineous marriage is preferential. Although a decline in first-cousin marriage has been observed in the small Norwegian Pakistani community (Grjibovski *et al.*, 2009), no similar trend seems to have occurred in the approximately one-million-strong UK Pakistani population (Shaw, 2000) or in the substantial Turkish or Moroccan communities in Belgium (Reniers, 1998), and a rapid reduction in the preference for consanguineous unions by first- and second-generation immigrant families in Europe currently appears improbable.

In 2010 an estimated 64 000 Muslim migrants entered the UK, representing 28.1% of all migrants, and the equivalent estimates for France were 66 000 Muslim migrants who formed 68.5% of the immigrant total for that country (Pew Research Center, 2011). By 2030, Muslims are predicted to form 8.3% of the UK population, 9.3% in Austria, 9.9% in Sweden, 10.2% in Belgium and 10.3% in France (Pew Research Center, 2011). Thus, unless there is a marked shift in the patterns of Muslim marriage preferences (Chapter 2), these data strongly suggest that consanguineous marriage will continue to be a topic of major interest in Western Europe well into the future.

Japan

Since World War II, there has been a marked decline in consanguineous marriage in Japan from an estimated 5.9% (Hiroshima) and 8.0% (Nagasaki) in the late 1940s and early 1950s (Schull, 1958) to 5.7% in the 1970s (Imaizumi *et al.*, 1975) and 3.9% in the 1980s (Imaizumi, 1986a). Religion has been an important differentiating factor, with the highest rates of consanguinity in the Buddhist community and the lowest among Roman Catholics and those declaring no religion (Imaizumi, 1986a). Consanguineous marriage was more common among couples with both a lower level of education and lower occupational status (Imaizumi, 1986a), and the decline in the prevalence of consanguineous marriage has been more marked in urban than rural areas (Hosoda *et al.*, 1983), and for first-cousin as opposed to second-cousin marriages (Imaizumi & Shinozaki, 1984).

While increased personal mobility in education and employment has led to enhanced opportunities to meet potential marriage partners from different regions and backgrounds (Imaizumi, 1986b), familial aggregation of consanguineous marriages was still apparent in the 1980s (e.g. in Fukue-Shi, Nagasaki Prefecture) (Imaizumi, 1988). Given the past importance of parental advice on consanguineous marriage (Imaizumi, 1987), a higher prevalence of marriage between biological kin may persist in some more rural parts of the country. Examples include the island of Hirado, also in Nagasaki Prefecture, where in the mid 1960s, 14.7% consanguineous unions were reported (Schull *et al.*, 1970), and in the small island community of Hosojima, part of Hiroshima Prefecture, where in the 1950s 29 of the 45 recorded marriages (64.4%) had been consanguineous (Ishikuni *et al.*, 1960). An additional general factor that merits consideration is the decline in marriage rates in Japan and other East and South-east Asian countries, with highly educated females increasingly reluctant to marry (*The Economist*, 2011).

South Asia

South India
Until the 1970s and 1980s, no overall decline was observed in the prevalence of consanguineous marriage in the four Dravidian states of South India, Andhra Pradesh, Karnataka, Kerala and Tamil Nadu (Rao & Inbaraj, 1977a; Rao, 1983; Bittles *et al.*, 2003), which have a combined population of some 230 million, and in neighbouring Maharastra with a population of 100+ million (ORGCC, 2006). However, during the 1990s, evidence of declining consanguinity was reported (Audinarayana & Krishnamoorthy, 2000; Krishnamoorthy &

Audinarayana, 2001), ascribed both to rising ages at marriage and to increasing levels of female education (Audinarayana & Krishnamoorthy, 2000).

As predicted some 30 years ago (Radha Rama Devi *et al.*, 1982), the decline in consanguinity appears to be greatest in uncle–niece marriages, which in a majority of cases are contracted between the eldest or second eldest daughter of the oldest female in a family and one of her mother's younger brothers (Mohan Reddy & Malhotra, 1991). Thus, for uncle–niece marriages in particular, smaller family sizes and the consequent difficulty of arranging marriages between spouses within socially acceptable age differences, which in South India average between five and eight years, create considerable demographic problems.

In earlier generations, marked differences were reported in the overall prevalence of consanguineous unions, with religion, caste, ethnicity, geographical origin, urban versus rural residence, education and socioeconomic status all identified as important variables (Rao *et al.*, 1971; Rao, 1983). The changes in the prevalence and types of consanguineous marriage so far reported, therefore, may not be uniform across all of the highly diverse communities in South India.

Pakistan

By comparison in Pakistan, a nationally representative study conducted in the 1990s indicated that 61.2% of marriages were between first or second cousins (mean coefficient of inbreeding, $\alpha = 0.0332$), with little or no evidence of any change in prevalence during the preceding 40 years (Ahmed *et al.*, 1992). Other forms of kin-exchange marriage also are common, e.g. *watta satta* marriage, which is the agreed exchange of marriage partners between two families and usually involves the marriage of a brother–sister pair from each household. In rural regions of the provinces of Sindh and southern Punjab, approximately 33% of marriages are *watta satta* and an estimated 48% of these rural *watta satta* unions are consanguineous, most commonly between first cousins (Jacoby & Mansuri, 2007, 2010). Less commonly, two sisters are contracted to marry two brothers or other male relatives from another household, in which case the sisters also become sisters-in-law on marriage.

While to Western eyes *watta satta* unions may at first sight appear unusual, in the nineteenth century, marriage exchanges of this type were regarded as desirable in Great Britain, and prior to Charles Darwin's marriage to his first cousin Emma Wedgwood in 1839, his elder sister Caroline had already married Emma's brother Josiah Wedgwood III in 1837 (Bittles, 2009a). Sibling exchange marriages, some between first cousins, also were quite commonly contracted among land-owning families in northern Sweden during the mid to late nineteenth century (Egerbladh & Bittles, 2011).

Table 5.1 *Changing prevalence of consanguineous marriage in the Middle East and North Africa*

Country	Increased	Decreased	Reference
Lebanon		+	Khlat (1985)
Morocco		+	Lamdouar Bouazzaoui (1994)
United Arab Emirates	+		Al-Gazali *et al.* (1997)
Saudi Arabia		+	al-Abdulkareem & Ballal (1998)
Kuwait		+	Radavanovic *et al.* (1999)
Israel, Arab		+	Jaber *et al.* (2000)
Palestinian Territories	+		Pedersen (2000)
Israel, Arab		+	Zlotogora *et al.* (2002)
Jordan		+	Hamamy *et al.* (2005)
Mauritania		+	Hammami *et al.* (2005)
Yemen	+		Jurdi & Saxena (2003)
Qatar	+		Bener & Allali (2006)
Israel, Arab		+	Sharkia *et al.* (2007)
Palestinian Territories		+	Assaf & Khawaja (2009)

The Middle East

The contemporary prevalence of consanguineous marriage in the Middle East and North Africa appears to vary quite considerably, with secular declines reported in Lebanon, Saudi Arabia, Kuwait, Jordan, Israeli Arab communities, and the Palestinian Territories. Conversely, in the United Arab Emirates (UAE), Yemen and Qatar, the overall levels of consanguineous marriage and the prevalence of first-cousin unions appear to have increased (Table 5.1), as is the case in neighbouring Iran (Akrami *et al.*, 2009).

To some extent, these latter increases in the prevalence of consanguineous marriages may reflect the larger family sizes of recent generations and hence the greater availability of potential cousin spouses. However, as will be discussed in Chapter 6, caution is warranted in interpreting and comparing the prevalence of consanguineous marriage across time because of differences in the composition of the populations sampled; the variant study protocols employed; and the major social, economic and educational changes which have occurred in most Arab countries during the second half of the twentieth century (Bittles, 2008). From a genetic perspective, it also is important to note that even if intra-familial marriage declines in prevalence, marriage probably will continue to be preferentially contracted within clan and tribal boundaries, thus facilitating the expression of rare recessive genes that may be unique to individual communities.

In other diverse populations, the observed pattern of consanguineous marriage appears to be largely determined by local issues with, for example, a decline in cousin unions in mountainous regions of Morocco (Lamdouar Bouazzaoui, 1994), probably reflecting improved means of communication with regional urban centres. By comparison, higher rates of consanguineous marriage were observed in the Fulani tribe of Burkina Faso following severe droughts which had drastically reduced their cattle numbers and wealth (Hampshire & Smith, 2001).

Societal and regional preferences in types of consanguineous marriage

Variations in the specific types of marriage contracted are seen in different human societies, e.g. with first-cousin unions between a man and his father's brother's daughter (in Arabic, *bint 'ammi*) preferred in Arab Muslim communities (Khlat, 1997). By comparison, in such disparate populations as the Dravidian Hindu states of South India (Rao & Inbaraj, 1977a), the Han Chinese (Wu, 1987) and the Tuareg of North Africa (Degos *et al.*, 1974), marriage between a man and his father's brother's daughter is regarded as being equivalent to a brother–sister union. Therefore, a first-cousin union between a man and his mother's brother's daughter is strongly preferred (Bittles, 1994), although additional regional and local customary influences can apply. Equally, as indicated in Chapter 2, although uncle–niece unions can be contracted within Judaism, they are not permissible to Muslims, although in Islam, double first-cousin marriages which represent the same genetic distance ($F = 0.125$) are allowed (Bittles, 1994).

Demographic, social and economic correlates of consanguinity

Consanguinity, education and socioeconomic status

Besides various social and economic explanations, in general terms women married to a close biological relative or whose family has a tradition of consanguineous unions are more favourably disposed to the practice (Khlat *et al.*, 1986; Jaber *et al.*, 1996; Hussain, 1999). The fact that the espoused partners would have met at family gatherings before betrothal is held to be especially helpful, and it also is useful in promoting and achieving harmony between a bride and her future in-laws to whom she is related (Khlat *et al.*, 1986).

In virtually all countries and societies, consanguineous marriage is most prevalent in rural communities (Table 5.2), which seek to maintain their

Table 5.2 *Demographic, social and biological correlates of consanguineous marriage*

Rural residence
More traditional lifestyle
Lower socioeconomic status
Family tradition of consanguineous marriage
Low level of maternal literacy
Younger maternal and paternal ages at marriage
Lower spousal age differences
Younger maternal age at first birth
Shorter birth intervals
Extended maternal reproductive span
Larger completed family sizes

cultural traditions by following more traditional lifestyles, e.g. in India (Bittles *et al.*, 1991), Pakistan (Bittles, 1994; Hussain, 1999), Turkey (Tunçbílek & Koc, 1994) and the Middle East (Al-Salem & Rawashdeh, 1993; Al-Gazali *et al.*, 1997; Radavanovic *et al.*, 1999; Zaoui & Biémont, 2002; El-Mouzan *et al.*, 2007; Joseph, 2007; Weinreb, 2008; Hamamy & Bittles, 2009). The highest prevalence of consanguineous marriage is therefore usually reported among families in rural areas (Al-Mazrou *et al.*, 1995), and with the lowest standard of living (Saedi-Wong *et al.*, 1989; Tunçbílek & Koc, 1994; Assaf & Khawaja, 2009), and wives in consanguineous marriages mostly have a lower level of education (Al-Thakeb, 1985; Khlat, 1988; Al-Mazrou *et al.*, 1995; Sureender *et al.*, 1998; Jurdi & Saxena, 2003; Alper *et al.*, 2004; Donbak, 2004; Akbayram *et al.*, 2009; Joshi *et al.*, 2009).

These generalizations are by no means uniform, and historical records from eighteenth- and nineteenth-century Palestine indicated a higher prevalence of cousin marriage in upper-class Arab society (Tucker, 1988). Similarly, members of the present-day ruling classes, and families with large land-holdings and/or significant personal wealth, frequently contract consanguineous unions as a means of maintaining their privileged status (Al Thakeb, 1985; Bittles 1994; Bittles, 2008). Mixed outcomes also have been reported in terms of education, with a positive association between consanguinity and education among Afghans resident in Pakistan (Wahab *et al.*, 2006), no significant difference in the prevalence of consanguineous marriage across the spectrum of female educational standards in the Palestinian Territories (Assaf & Khawaja, 2009), whereas in the UAE (Bener *et al.*, 1996) and Yemen (Jurdi & Saxena, 2003), males with advanced educational backgrounds expressed a preference for intra-familial marriage.

Table 5.3 *Social and economic advantages of consanguineous marriage*

Simplified premarital negotiations
Assurance of marrying within the family and the strengthening of family ties
Assurance of knowing one's spouse before marriage
Avoidance of unexpected, and unwelcome, health issues
Reduced chances of marital maltreatment or desertion
Social protection due to greater compatibility of the bride with her husband's family, especially her mother-in-law
Reduced dowry or bridewealth payments, with the maintenance of family goods
Maintenance of the integrity of family land-holdings

Socioeconomic factors play an important, even critical, role in the preference for consanguineous unions, with particular emphasis on the strengthening of family relationships and the maintenance of family property, including land-holdings (Table 5.3). The maintenance of family wealth can take different forms, for example, with reduced or no dowry or bridewealth payment incurred in most close kin marriages in South India (Govinda Reddy, 1988) and Kuwait (Al-Nassar *et al.*, 1989). A similar observation was made with respect to the payment of bridewealth among resident and refugee Afghans in Pakistan (Wahab *et al.*, 2006). By comparison, in Turkey, bridewealth payments were higher in consanguineous marriages, although the enhanced levels of payment offered were somewhat illusory in real terms since the monies paid stayed within the family (Tunçbílek & Ulusoy, 1989).

It has been convincingly argued that although dowry payments are less frequent in first-cousin marriages, intra-familial unions are not contracted simply to minimize the outflow of family wealth through dowry or bridewealth payments. Rather, based on the results of a rural study in Bangladesh, it was concluded that women in consanguineous unions, and thus their husbands and children, were more likely to benefit from the post-marital transfer of gifts and property from their parents (Joshi *et al.*, 2009). Therefore, because of the reduced requirement for dowry or bridewealth and the optimization of subsequent levels of female inheritance, consanguineous marriage represents an especially attractive strategy for poorer families (Do *et al.*, 2011).

Prenuptial arrangements also are greatly simplified, and in communities where the bride moves to her in-laws' household after marriage, she has the considerable assurance of knowing and being related to her mother-in-law, thus reducing concerns regarding social incompatibility (Bittles, 1994; Qidwai *et al.*, 2003; Raz & Atar, 2004). There is the additional security of marrying a partner whose entire family background is known, so that any medical problems which might exist do not unexpectedly emerge after the wedding ceremony has been completed (Bittles, 1994; Khlat, 1997; Hussain, 1999). When a family is known to have a poor health record, as in an isolated Israeli Arab village in which two extended families had multiple intellectually disabled offspring, it may however prove impossible to arrange marriages outside the family circle (Basel-Vanagaite *et al.*, 2007b), thus increasing the likelihood of further intra-familial unions despite the greater implied risk of disease recurrence.

Consanguinity, age at marriage and reproductive span

In keeping with a more traditional lifestyle, and the potential problems of undesirable parent–child and sibling relationships that can arise in families living in one-room accommodation, consanguineous marriage is generally associated with younger maternal age at marriage (Basu, 1993; Khlat, 1988; Sureender *et al.*, 1998; Hussain & Bittles, 1999; Iyer, 2002; Donbak, 2004; Assaf & Khawaja, 2009; Joshi *et al.*, 2009), and younger parental age at first birth (Bittles *et al.*, 1991) (Table 5.4). In addition, consanguineous couples may continue child-bearing to later ages (Tunçbilek & Koc, 1994; Sueyoshi & Ohtsuka, 2003), and a positive association between consanguinity and age at menopause has been reported in the UAE (Bener *et al.*, 1998). The resultant enhanced female reproductive spans in consanguineous unions thus facilitate larger family sizes (Chapter 7).

Consanguinity, marital stability and marital violence

A small number of detailed studies have been conducted into the social outcomes of consanguineous marriage, although they mainly have involved female subjects only. An early report from Sudan indicated greater marital stability in consanguineous unions irrespective of the type of cousin relationship, with divorce in 3.6% of first-cousin marriages compared with 14.6% divorce in other types of consanguineous and non-consanguineous marriage (Hussien, 1971). This pattern also has been reported in non-Arab populations, possibly because of the highly disruptive effect of marriage failure on the stability of the extended family (Bittles, 2005; Joshi *et al.*, 2009). In general terms, divorce also can impact adversely on the health of children, particularly in infancy (Alam *et al.*, 2001).

One of the reasons given for *watta satta* marriages in Pakistan was that they helped to coordinate the actions of two sets of in-laws, each of whom wished to protect the wellbeing of their daughter by restraining any possible

Table 5.4 *Consanguinity and female age at marriage/first birth*

Country	Younger	No effect	Reference
Lebanon			
Beirut	+		Khlat (1985)
South India			
Karnataka	+		Bittles *et al.* (1991)
Pakistan			
Punjab	+		Bittles *et al.* (1993)
South India			
Tamil Nadu	+		Sivaram *et al.* (1995)
Tamil Nadu			Sureender *et al.* (1998)
Egypt			
Alexandria	+		Mohamed (1995)
UAE			
Al Ain	+		Al-Gazali *et al.* (1997)
Turkey			
National	+		Tunçbílek & Koc (1994)
Saudi Arabia			
Riyadh	+		al-Abdulkareem & Ballal (1998)
Kuwait			
National	+		Radavanovic *et al.* (1999)
Israel			
Arab	+		Jaber *et al.* (2000)
Israel			
Arab	+		Zlotogora *et al.* (2002)
Yemen			
Sanaa	+		Jurdi & Saxena (2003)
Jordan			
National	+		Hamamy *et al.* (2005)
Qatar			
	+		Bener & Allali (2006)
Israel			
Arab	+		Sharkia *et al.* (2007)
Tunisia			
	+		Kerkeni *et al.* (2007)
Lebanon			
Bekaa		+	Kanaan *et al.* (2008)
Palestinian Territories			
	+		Assaf & Khawaja (2009)

aggressive actions by their son-in-law. But, realistically, each family could only influence the behaviour of their own sons (Jacoby & Mansuri, 2008), and so it was expected that by arranging a *watta satta* union, a brother would act as his sister's protector (Das, 1973). Empirical evidence suggests that this tactic works because estrangement, domestic abuse and adverse effects on wives' mental health are significantly lower in *watta satta* unions than in 'conventional' marriages (Jacoby & Mansuri, 2007, 2008), although the extent to which this advantage might be due solely in *watta satta* relationships to unions that involve two sets of first cousins is unclear.

A higher degree of marital dissatisfaction may arise when there is a large age differential between consanguineous partners (El Islam, 1976). A positive although statistically non-significant association between the degree of consanguinity and marital discord was reported in Saudi Arabia (Chaleby, 1988), and somewhat surprisingly greater conflict with the extended family also occurred in consanguineous marriages in Turkey (Fisloglu, 2001). More recent investigations have centred on the highly sensitive issues of domestic violence and coerced sexual intercourse within marriage, with no significant advantage or disadvantage reported for consanguineous marriage among low-income women in Syria (Maziak & Asfar, 2003), Palestinian refugees in Lebanon (Khawaja & Tewtel-Salem, 2004; Khawaja & Hammoury, 2008), or in rural Bangladesh (Joshi *et al.*, 2009). However, both consanguinity and higher partner education levels were found to be protective against violence during pregnancy in Jordan (Clark *et al.*, 2009).

Consanguinity and dowry in India

Over the course of the past 50 years, dowry payments (i.e. the transfer of resources from a bride and her family to grooms and their families) have significantly increased in India (Rao, 1993). Marital violence in the form of 'dowry deaths' is both controversial and apparently widespread (Sharma *et al.*, 2005), with victims suffering severe burns, usually described as 'bride burning' (Kumar & Tripathi, 2004; Jutla & Heimbach, 2004; Shahi & Mohanty, 2006), or poisoning which may be self-administered (Kumar, 2004). Ongoing demands from husbands and their families for increased dowry payments have been identified as the major provoking factor behind the violence which frequently is fatal.

As one of the advantages offered by a consanguineous union is a negligible or reduced dowry (Table 5.3), not surprisingly, the phenomenon of dowry deaths has been largely reported in the northern states of India where consanguineous marriage is strenuously avoided (Chapter 2). In north Indian Hindu communities, the position of females within marriage may be further compromised by a number of factors, including the tradition of hypergamy, i.e. marrying into a family of superior social and/or economic status, by limiting daughters marrying within their own village, and restrictions on the remarriage of widows (Chakraborty & Kim, 2010).

Consanguinity and gender equality

Besides the generally positive effect of consanguinity on marital relationships in India, there is strong presumptive evidence that intra-familial marriage also exerts a significant beneficial influence on gender equality. Concern has long been expressed regarding the highly biased tertiary sex ratios in the northern states of India. Indeed, until the introduction of the 1870 Infanticide Act, female infanticide was routinely practised in North and West India, to the extent that at District level in Gujarat during 1836–1837, tertiary sex ratios among Jhareja Rajput communities ranged from 300 to 1967 males per 100 females (Pakrasi & Sasmal, 1971), whereas the expected sex ratio at birth is typically 105–106 males per 100 females (Bittles *et al.*, 1993). Although ameliorated by legal proscription, a male preponderance has been recorded in all censuses of India since their inception in 1872 (Bittles *et al.*, 1993), with the historical excess of males variously explained in terms of female infanticide, excess female mortality during and following childbirth, systematic under-enumeration of females, and sex-selective migration (Visaria, 1967; El-Badry, 1969).

While female infanticide now appears to be rare (Jha *et al.*, 2006), the emphasis on increasingly large dowry payments has encouraged the establishment of clinics for the determination of fetal sex, initially by amniocentesis or chorionic villus sampling but with sex determination now widely available by non-invasive ultrasound in privately operated mobile facilities for a fee of approximately 500 rupees ($\sim$ US$10) (Sharma & Taub, 2008). The ready availability of ultrasound services has meant that female feticide has become an established feature of life in many parts of India, especially in certain northern and western states (Jeffery *et al.*, 1984; Sachar *et al.*, 1990; Arnold *et al.*, 2002; Sahni *et al.*, 2008).

The 1992–1993 National Family Health Survey of India (NFHS, 1995) already demonstrated a very clear distinction between tertiary sex ratios in the northern states of India where consanguineous marriage is strictly prohibited in the majority Hindu community, and the southern states in which uncle–niece and first-cousin marriage has customarily been favoured. As shown in Figures 5.1 and 5.2, there was a highly significant inverse correlation between the percentage of consanguineous marriages and tertiary sex ratios at state level, apparently indicative of much lower levels of discrimination against females in states where the prevalence of consanguineous marriage was highest.

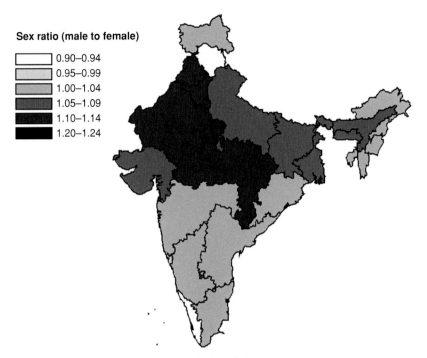

Figure 5.1 Tertiary sex ratios (M/F) in India by state

According to the 2001 Census of India, there had been little change in this situation with a national tertiary sex ratio of 107.2 males per 100 females, and in Fatehgarh District in Punjab the sex ratio among children under seven years of age was 132.6 males per 100 females (Sharma & Haub, 2008). This despite the implementation in 1996 of the Pre-Natal Diagnostic Techniques (Regulation and Prevention of Misuse) Act (PNDT) which banned sex-selective feticide throughout India. In an attempt to close loopholes in the Act and tighten control of sex-selective feticide, the PNDT was amended in 2002, along with the Medical Termination of Pregnancy Act (MTP) in 1971.

Although some equalization of the sex ratios at birth was subsequently claimed (Sharma & Haub, 2008), no significant effect on the odds of having a male infant were observed in the immediate pre- and post-PNDT periods (Sahni *et al.*, 2008; Subramanian & Selvaraj, 2009). Preliminary results of the 2011 Census of India have indicated a national tertiary sex ratio of 109.4 males per 100 females, and projections for 2026 produced by the Office of the Registrar General of India (ORGCC, 2006) indicate an expected national

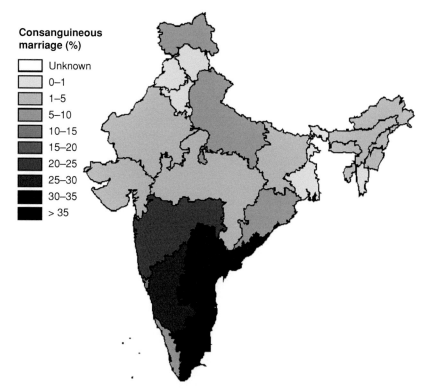

Figure 5.2 Consanguineous marriages ($F \geq 0.0156$; %) in India by state. *Source:* NFHS (1995)

tertiary sex ratio of 107.2, with even more extreme ratios during the next 15 years in some of the more affluent states, including Delhi, Haryana, Punjab, Gujarat and Maharashtra. In contrast, the southern states of Andhra Pradesh, Karnataka, Kerala and Tamil Nadu are predicted to continue to be significantly more gender-egalitarian with respect to their tertiary sex ratios than the states of North, Central and West India (Table 5.5).

Differential female autonomy in North and South India

The data support the general contention of greater female autonomy in the preferentially consanguineous southern states of India (Dyson & Moore, 1983; Iyer, 2002; Dommaraju & Agadjanian, 2009; Chakraborty & Kim, 2010), and a

Table 5.5 *Tertiary sex ratios* in India by state, 2001 and 2026*

Region	State	Sex Ratio 2001	Sex Ratio 2026
North	Delhi	121.8	126.7
	Haryana	116.1	119.2
	Himachal Pradesh	103.3	106.2
	Jammu and Kashmir*	112.1	107.2
	Jharkhand	106.3	104.9
	Punjab	114.2	119.0
	Rajasthan	108.6	109.3
	Uttaranchal	104.0	104.9
Central	Madhya Pradesh	108.8	109.4
	Uttar Pradesh	111.4	110.9
East	Chhattisgarh	101.1	100.7
	Bihar	108.8	105.4
	Orissa	102.9	101.0
	West Bengal	107.1	104.3
North-east	Assam	107.0	104.2
	North-East States[†]	106.4	104.7
West	Gujarat	108.7	113.4
	Maharashtra	108.5	110.6
South	Andhra Pradesh	102.2	100.1
	Karnataka	103.6	102.2
	Kerala	94.5	96.5
	Tamil Nadu	101.3	100.0
	All India	107.2	107.5

*Expressed as males per 100 females
[†]Comprises the states of Arunachal Pradesh, Manipur, Meghalaya, Mirozam and Tripura

much less biased level of female infant and childhood mortality (Arokiasamy, 2004). A further probably important factor that enhanced female status in the past is that in the southern states, bridewealth was traditionally paid on marriage, with a switch to dowry occurring only from the 1950s and 1960s and initially restricted to the wealthier merchant classes (Caldwell *et al.*, 1983; Govinda Reddy, 1988).

In addition to feticide, the higher death rates of female children in northern India have been ascribed to discrimination against girls in the provision of food, and more particularly in healthcare provision (Das Gupta, 1987; Basu, 1989; Das Gupta, 1990). Excess female childhood mortality was highest in families which already had one or more surviving daughters (Das Gupta, 1987; Arnold *et al.*, 1998; Retherford & Roy, 2003) and somewhat unexpectedly it was more pronounced among younger, educated mothers (Das Gupta, 1987; Retherford & Roy, 2003). The 1992–1993 National Family Health Survey also confirmed a regional sex bias in health care, with distinct son preference in the provision of BCG, DPT, polio and measles vaccinations in the northern states, whereas in the south of the country, vaccination uptake levels were virtually identical for boys and girls (Arokiasamy, 2004).

The overall picture with respect to female status in South India is not, however, universally positive and even during the 1980s and 1990s, female infanticide was reported in several districts in the state of Tamil Nadu, usually involving higher-birth-order female infants and especially among multiparous women with no living sons (George *et al.*, 1992; George, 1997; Nielsen *et al.*, 1997). In recognition of the issue, the then-Chief Minister of Tamil Nadu, Ms Jayalalitha, introduced the *Jayalalitha Protection Scheme for the Girl Child*, which provided monetary incentives for poor families with no sons but one or two girls, albeit on condition that one parent agreed to be sterilized (George, 1997).

Commentary

Globalization, improved access to higher education and economic advancement may lead to a gradual decline in consanguineous marriage by expanding the potential marriage pool for many individuals. Yet significant social advantages offered by consanguineous marriage have been identified in Middle Eastern and South Asian populations, with an economic advantage also apparent in terms of social capital (Joshi *et al.*, 2009). The economic flow-on effects are, however, complex and in India, rather than enhanced female autonomy and higher female social status being ascribed solely to consanguinity, econometric analysis has suggested that gender bias may be concentrated within particular caste and religious groups (Barooah *et al.*, 2009). This judgement also may apply in South Asian emigrant communities, for example, with a skewed secondary sex ratio favouring male births in third- and higher-order pregnancies among Indian mothers in Norway, but no such trend in the larger co-resident Pakistani community (Singh *et al.*, 2010). The ongoing advantage to female status offered by consanguineous marriage within the Pakistani migrant community may, however, be a more straightforward overall explanation for this observation.

6   A population genetics perspective on consanguinity

Introduction

Throughout recorded human history, marriages between males and females have been the predominant institution within which procreation occurred and genes were transmitted. Therefore, to understand genetic structure and composition at individual, community and population levels, it is necessary first to examine how and why marriage partners are chosen in different societies. To the present day, the populations of virtually all developing countries are divided into communities and sub-communities, in most cases with very limited inter-community marriage, even though some gene flow between communities may occur via marital alliances and miscegenation.

Historical accounts have indicated the existence of similar divisions within Western societies, e.g. with strict endogamy maintained within the Scottish clan system into the nineteenth century and close cousin marriage contracted in Highland districts during the later nineteenth and early twentieth centuries (Macpherson, 1968). Arranged kin marriage was virtually mandatory in the Breton region of France until the early twentieth century, the only choice being which of his future in-law's daughters a man might opt to marry (Segalen & Richard, 1986). More recently, genome-based investigations have revealed obvious genomic stratification in the present-day populations of major industrialized countries including Germany (Tian *et al.*, 2008), Japan (Yamaguchi-Kabata *et al.*, 2008) and the UK (McEvoy *et al.*, 2009), and even in the population of Iceland which generally has been considered to be genetically homogeneous (Helgason *et al.*, 2005).

During the mid to late twentieth century in the USA, Canada and Australia, newly arrived immigrants preferentially married within their own ethnic and/or religious communities in the first and second post-migration generations (Bittles, 2009b). While offering strong social and in many cases economic advantages, this tradition has important genetic implications, since it is probable that couples from the same national, ethnic or religious sub-community will have inherited a significant proportion of their genes in common, and therefore that their progeny are more likely to be homozygous for a detrimental recessive disorder than would be the case in a panmictic population (Bittles, 2002).

Sampling and analytical protocols in consanguinity studies

Criteria for data collection

The diverse methodologies used in data collection have proved to be a major impediment in the study of consanguinity and, more particularly, to inter-population comparisons. As illustrated in Table 6.1, data on the prevalence of consanguineous marriages have variously been derived from compulsory civil marriage registration, household, school and workplace surveys; blood-donor panels; obstetric inpatients; and Roman Catholic consanguinity dispensations. Similarly, the types of marriage surveyed and noted have ranged from first cousin only ($F = 0.0625$), to double first cousin and first cousin only, first cousin and second cousin only, and uncle–niece/aunt–nephew/double first cousin, first cousin, first cousin once removed, and second cousin ($F = 0.125$–0.0156).

As previously noted in Chapter 5, the widely variant nature of the data-collection procedures adopted for consanguinity studies can cause major comparability problems when the prevalence and specific types of consanguineous marriage are contrasted at different time-periods and under differing personal living circumstances. Unless essentially the same demographically and socially defined study populations are accessed, and identical categories of consanguineous marriage are investigated, there can be no guarantee that an observed difference in the prevalence of consanguineous marriage across time is not simply an artefact of the sampling procedure. Similarly, because the prevalence of consanguineous marriage usually is positively correlated with rural residence and is inversely proportional to a couple's socioeconomic status and level of educational achievement, failure to control for these basic demographic variables will substantially disrupt the validity of any time-based comparisons.

The genetic significance of specific types of consanguineous union

Table 6.1 shows that in studies on the health outcomes of consanguinity, the principal emphasis has been on the category and degree of marital relationship, with the twin expectations that: (i) the excess risk that an autosomal recessive disorder will be expressed in the progeny of consanguineous unions is inversely proportional to the frequency of the disease allele(s) in the gene pool (Table 6.2); and (ii) the closer the genetic relationship between parents, the greater the probability that they will conceive an affected child.

Whereas these relationships hold true for autosomal recessive disorders, the situation differs with respect to recessive disorders encoded on the X-chromosome where, for example, according to the type of first-cousin marriage

Table 6.1 *Prevalence (%), mean coefficient of inbreeding (α), and types of consanguineous marriage in different regions, countries and populations*

Region	Study setting	Participants	Number	%	Relationships	α	Authors
Western Europe							
UK	Birmingham	Obstetric inpatients	2431	0.2	1C,2C	0.0001	Bundey et al. (1990)
Norway	All-Norway	Civil registration	893 941	0.7	1C,2C	0.0002	Magnus et al. (1985)
Spain	Alava	RC dispensation	11700	2.5	UN,1C,1$^{1/2}$C,2C	0.0006	Calderón et al. (1998)
North America							
Canada	Québec	RC dispensation	21 874	1.3	UN,1C,1$^{1/2}$C,2C	0.0003	De Braekeleer & Ross (1991)
USA	National	RC dispensation	133 228	0.2	1C,1$^{1/2}$C,2C	0.0001	Freire-Maia (1968)
South America							
Argentina	National	Civil registration	212 320	0.4	1C	0.0002	Castilla et al. (1991)
Brazil	Rio Janeiro	RC dispensation	4070	1.5	UN/AN,1C,1$^{1/2}$C,2C	0.0008	da Fonseca & Freire-Maia (1970)
Paraiba	RC dispensation	9521	12.8	UN/AN,1C,1$^{1/2}$C,2C	0.0058	da Fonseca & Freire-Maia (1970)
Chile	National	Hospital births	6959	0.7	1C,1C,1$^{1/2}$C,2C	0.0004	Liascovich et al. (2001)
East Asia							
China	Zejiang (Han)	Household survey	17 381	2.5	1C,1$^{1/2}$C,2C	0.0014	Zhan et al. (1992)
	Xinjiang (Uygur)	Household survey	2553	8.2	D1C,1C,2C	0.0047	Ai et al. (1985)
	Laioning (Manchu)	Household survey	418	3.1	1C,1$^{1/2}$C,2C	0.0012	Wang et al. (2002)
Japan	Fukuoka	Household survey/ civil registration	45 230	7.6	D1C,1C,1$^{1/2}$C,2C	0.0029	Yamaguchi et al. (1970)
All-Japan		Household registration	9225	3.9	1C,1$^{1/2}$C,2C	0.0013	Imaizumi (1986a)

Northern Africa							
Morocco	National	Civil registration	4773	19.9	1C,2C	0.0089	Lamdouar Bouazzaoui (1994)
Egypt	National	Household/school/ workplace	26 554	29.0	D1C,1C,1½C,2C	0.0101	Hafez et al. (1983)
Sudan	Khartoum	Blood donors	4833	52.0	1C,2C	0.0302	Saha & El Sheikh (1988)
West Asia							
Lebanon	National	Household survey	1556	35.5	1C,2C	0.0204	Barbour & Salameh (2009)
Palestinian Territories	National	Household survey	4971	27.7	D1C,1C	0.0186	Assaf & Khawaja (2009)
Iraq	National	Household survey	23 937	33.0	D1C,1C	0.0219	COSIT (2005)
Turkey	National	Household survey	7435	20.1	1C,2C	0.0110	Koç (2008)
South Asia							
India–north	Lucknow (Hindu)	Obstetric inpatients	7955	0.1	UN,1C	0.0001	Agarwal et al. (1991)
India–south	Karnataka (Hindu)	Obstetric inpatients	86 448	33.5	UN,1C,2C	0.0333	Bittles et al. (1991)
Pakistan	National	Household survey	6611	61.2	1C,2C	0.0332	Ahmed et al. (1992)
Iran	National (Shia)	Household survey	169 310	30.0	D1C,1C,2C	0.0170	Saadat et al. (2004)
Oceania							
Australia	Western Australia	Civil marriages	62 376	0.2	1C,1½C,2C	0.0001	Port & Bittles (2001)

Abbreviations: UN, uncle–niece ($F = 0.125$); AN, aunt–nephew ($F = 0.125$); D1C, double first cousin ($F = 0.125$); 1C, first cousin ($F = 0.0625$); 1½C, first cousin once removed ($F = 0.0313$); D2C, double second cousin ($F = 0.0313$); 2C, second cousin ($F = 0.0156$)

Table 6.2 *Consanguinity and autosomal recessive gene expression*

	Affected offspring		
Gene frequency, q	Unrelated parents, q^2	First cousins, $q^2 + (1/16)pq$	Multiplier
0.1	0.01	0.0156	1.6
0.05	0.0025	0.00547	2.2
0.01	0.0001	0.000719	7.2
0.05	0.000025	0.000336	13.4
0.001	0.000001	0.000063	63.0

favoured, the coefficient of inbreeding can vary from $Fx = 0$ to 0.1875. As illustrated in Table 6.3, the preferred patterns of first-cousin marriage, and hence the expression of X-linked recessive alleles, vary markedly between populations. Despite the significant number of clinically important genes located on the X-chromosome, this topic has been under-investigated, even though reports have been published of a high prevalence of glucose 6-phosphate (G6PD) deficiency among females in communities where various types of first-cousin marriages were preferentially contracted (El-Hazmi & Warsy, 1989; El-Hazmi *et al.*, 1991; Ramadevi *et al.*, 1994).

Potential confounding issues in consanguinity studies

Besides its pejorative connotation, unqualified use of the term *inbreeding* within a human genetic context can be misleading because it potentially encompasses a number of quite different concepts, which include relationships between biological relatives; genetic drift; departure from panmixia in mating behaviour (comprising positive or negative assortative mating and endogamy); and subdivision of a population into several isolated groups (the Wahlund principle) (Jacquard, 1975). For present purposes, in considering potential confounding factors in consanguinity studies, the major emphasis will be on the influence of drift and founder effect, community endogamy, and the role of population subdivision.

Genetic drift and founder effect

Genetic drift is most simply defined as the influence of chance on gene frequencies in successive generations, and the probability of genetic drift is greatest in communities with small effective population sizes (i.e. with restricted numbers of individuals in the breeding pool). The latter situation can arise in several ways: e.g. through founder effect, when a sub-group of a population

Table 6.3 *Preferred patterns of first-cousin marriage in different societies*

Region and country	Type I (%)	Type II (%)	Type III (%)	Type IV (%)	All first-cousin unions (%)	Authors
Japan						
Population isolates	25.9	21.7	22.7	29.4	16.7	Kishimoto (1962)
Shizuoka	17.6	34.6	15.6	32.2	4.7	Komai & Tanaka (1972)
North Africa						
Sudan	66.3	7.2	13.3	13.3	49.5	Saha *et al.* (1990)
Middle East						
Lebanon	37.4	22.5	10.4	29.7	14.1	Khlat (1985)
Jordan	62.8	9.9	7.7	11.3	32.0	Khoury & Massad (1992)
United Arab Emirates	64.9	8.4	12.2	14.5	26.2	Al-Gazali *et al.* (1997)
Jordan	67.3	12.0	9.6	11.1	34.2	Sueyoshi & Ohtsuka (2003)
Yemen	48.9	18.1	14.2	18.8	29.4	Gunaid *et al.* (2004)
Qatar	66.0	9.4	10.7	13.7	26.7	Bener *et al.* (2007)
Palestinian Territories	47.9	17.6	20.1	14.4	14.4	Assaf & Khawaja (2009)
Israel	48.1	14.8	14.8	22.2	19.6	Zlotogora & Shalev (2010)
Pakistan						
Lahore, rural	20.8	24.6	18.5	36.1	34.2	Yaqoob *et al.* (1993)
Lahore, peri-urban	18.8	23.5	23.5	34.2	34.1	Yaqoob *et al.* (1993)
Lahore, urban	22.2	27.8	13.0	37.0	28.6	Yaqoob *et al.* (1993)
Swat, rural	40.9	18.6	18.6	21.9	21.1	Wahab & Ahmad (1996)
Swat, urban	31.8	21.9	16.7	29.6	22.9	Wahab & Ahmad (1996)

(cont.)

Table 6.3 (*cont.*)

Region and country	Type I (%)	Type II (%)	Type III (%)	Type IV (%)	All first-cousin unions (%)	Authors
South India						
Tamil Nadu, Hindu	1.0	0.9	35.1	63.0	25.0	Rao & Inbaraj (1977a)
Muslim	11.1	8.9	31.1	48.9	21.2	Rao & Inbaraj (1977a)
Christian	0	0	41.7	58.3	19.0	Rao & Inbaraj (1977a)
Andhra Pradesh, tribal	0	0	39.4	60.3	35.8	Pingle (1983)
South America						
Argentina	19.6	38.3	18.7	23.8	0.4	Liascovich *et al.* (2001)
Parallel-cousin marriage:						
Type I	father's brother's daughter,	$F = 0.0625$, $Fx = 0$				
Type II	mother's sister's daughter	$F = 0.0625$, $Fx = 0.1875$				
Cross-cousin marriage:						
Type III	father's sister's daughter,	$F = 0.0625$, $Fx = 0$				
Type IV	mother's brother's daughter	$F = 0.0625$, $Fx = 0.125$				

establishes a new breeding colony; via a demographic bottleneck following major disease- or disaster-related mortality; and in subdivided populations that comprise multiple, strictly endogamous sub-communities. The net effect is similar to positive assortative mating (Chapter 1), and the principal outcome is a higher, random probability of homozygosity across the genome resulting in the increased likelihood of recessive gene expression.

Finland provides a prime example of how founder effect and drift have resulted in a unique national genetic disease profile despite a general avoidance of consanguineous marriage. Described as the Finnish Genetic Heritage, some 36 mostly autosomal recessive diseases are present in the Finnish population whereas they are rare or absent in other populations (Norio, 2003a, 2003b). Conversely, recessive diseases such as cystic fibrosis and phenylketonuria that are common in most other European countries are rare in Finland. The emergence of the Finnish Genetic Heritage was initially facilitated by the dispersed patterns of human settlement, coupled with internal migration, and large-scale population losses caused by disease and famine leading to demographic bottlenecks. The genetic outcome has been that whereas some of the Finnish genetic diseases are distributed across all regions of the country, others are clustered within geographical sub-regions and linked to specific waves of colonization and settlement (Bittles, 2009b).

An example of the overlap among founder effect, drift and consanguinity is provided by the sub-Saharan country of Cameroon, which comprises 136 ethnic groups with the predominant Bamileke tribe accounting for approximately 20% of the total population. The consanguineous Bamileke are further subdivided into 131 endogamous kingdoms that in the late 1990s varied in number from 500 to 66 000 inhabitants (Puri *et al.*, 1997). Type 2 oculocutaneous albinism is common in the Cameroon population and, at the beginning of the twentieth century the eighth and ninth rulers of the Balengou kingdom both were albino; following tradition, they also were polygamist with 80–100 wives and 200–300 children. It is therefore not surprising that by the end of the century, 70% of the cases of albinism in Cameroon were reported in the Bamileke tribe (Aquaron, 1980, 1990).

Disentangling the influence of consanguinity and endogamy on health outcomes

With the Bamileke in mind, a further factor that needs to be addressed is the relationship between endogamy (i.e. marriage within a community) as opposed to consanguinity (i.e. marriage within the family) as determinants of health and disease. As previously discussed in Chapters 1, 2 and 5, marriage within clan, tribal, caste or *biraderi* boundaries remains the rule or expectation in most traditional human societies, and endogamous marriage also is strongly favoured in many of the migrant communities from Asia and Africa now resident in Western countries. Besides their effect on population genetic structure, these subdivisions continue to have powerful and well-recognized political connotations. As illustrated by 'The Troubles' in Northern Ireland from the 1970s to the 1990s, the ethnic/national/religious differences which re-emerged in the Balkans during the latter decades of the twentieth century and the influence of clan and tribal allegiances in the 'Arab Spring' of 2011.

The genetic consequences of endogamous marriage on disease prevalence are frequently overlooked or at least significantly under-estimated. Various factors mediate the influence of endogamy on community gene pools, including the number of founders and their genetic relationship, the age of the community and its size, the flexibility or rigidity of community marriage boundaries and regulations, interaction with other communities and community amalgamations through time, and religious conversions. Although each of these variables could at least potentially influence the structure of all endogamous populations, they are essentially community-specific in their action and hence in their impact on gene-pool structure and composition.

The genetic effects of population subdivisions

Endogamous subdivisions result in significantly greater intra-community genetic homogeneity and therefore an increased proportion of homozygotes in the population as a whole (i.e. the Wahlund effect), leading to the increased expression of recessive genes which could mistakenly be interpreted as arising from consanguineous marriage (Overall & Nicholls, 2001; Overall *et al.*, 2003; Bittles, 2005; Overall, 2009). In addition, with population stratification, a recessive founder or de novo mutation can rapidly increase to high frequency in a sub-community and result in the birth of an affected child, whether or not the parents recognize themselves as genetically related (Zlotogora *et al.*, 2006). Unqualified comparisons between the birth outcomes of couples categorized solely in terms of their consanguineous or non-consanguineous marital relationships may therefore prove to be invalid unless they all are members of the same sub-community (Bittles, 2008).

Measuring consanguinity at individual and population levels

Coefficients of relationship and inbreeding

As summarized in Table 1.2, two basic measures are widely applied in studies of human consanguinity to describe and quantify genetic relatedness, both originally derived for pedigree studies on British shorthorn cattle by Wright (1922). The first is the coefficient of relationship (r), which is the proportion of genes identical by descent (IBD) shared by two individuals and calculated according to the formula:

$$r = (1/2)^n$$

where n is the number of steps apart on a pedigree for these two individuals via their common ancestor. Thus, for two persons related as first cousins:

$$r = (1/2)^4 + (1/2)^4 = 1/8$$

The second measure is the coefficient of inbreeding (F), which describes the proportion of gene loci at which an individual is homozygous by descent. Incestuous relationships (i.e. between father–daughter, mother–son or brother–sister) are the closest form of human mating, with the partners sharing half of their genes ($r = 0.5$), and so any offspring would be homozygous at $1/4$ of gene loci ($F = 0.25$) (Table 1.2). The closest legally permissible consanguineous unions are between an uncle and niece, a marital arrangement which occurs mainly in South Indian Hindu communities, or between double first cousins in Muslim populations in the Middle East and Pakistan (Chapters 2 and 5). In both of these types of marriage, the partners share one quarter of their genes ($r = 0.25$) and so the coefficient of inbreeding in their progeny is $F = 0.125$. As illustrated in Figure 6.1, double first cousins share both sets of grandparents, whereas in a first-cousin marriage, the couple have two of their four grandparents in common.

Following the same logic, second cousins have inherited $1/32$ of their genes from a common ancestor ($r = 0.0313$), and so the offspring of a second-cousin union would be expected to be homozygous (or more strictly autozygous) at $1/64$ of their gene loci (i.e. $F = 0.0156$). In populations with restricted marriage-partner choice, couples who are not second cousins may be related through multiple pathways involving more remote ancestors. Under these circumstances, the coefficient of inbreeding for an individual is calculated by summing each of the known pathways of inheritance. Thus, for an individual whose parents are third, fourth and fifth cousins ($r = 0.0078, 0.0039$ and 0.00195), the corresponding coefficient of inbreeding is ($F = 0.0039 + 0.00195 + 0.00098$); i.e. a composite coefficient of inbreeding of $F = 0.00683$.

The mean population coefficient of inbreeding

Using information derived from the coefficients of inbreeding of individuals in a community or population, the mean coefficient of inbreeding (α) can be

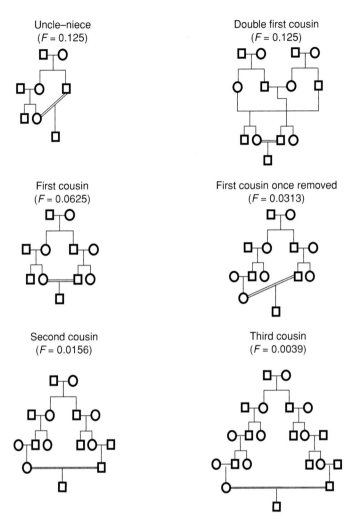

Figure 6.1 Consanguineous pedigrees. Top left: uncle–niece ($F = 0.125$); top right: double first cousin ($F = 0.125$); middle left: first cousin ($F = 0.0625$); middle right: first cousin once removed ($F = 0.0313$); bottom left: second cousin ($F = 0.0156$); bottom right: third cousin ($F = 0.0039$)

calculated by applying the formula:

$$\alpha = \sum p_i F_i$$

where Σ is the sum of the proportion of couples (p_i) in each class of consanguineous relationship (F_i). In practice, the values obtained for α calculated

over a single generation typically range from 0.0001–0.0010 in the present-day indigenous populations of Western countries, to values as high as 0.0414 in Pakistan (Hashmi, 1997) and 0.0449 in South India (Puri *et al.*, 1978).

Cumulative consanguinity and its calculation

In many communities, there is a long tradition of preferential consanguineous marriage, and where this custom is unbroken across generations, the resultant cumulative coefficient of inbreeding will be substantially higher than the value calculated for a single generation only. To at least partially account for this practice, the adjusted coefficient of inbreeding can be calculated as:

$$F = \sum (1/2)^n (1 + F_A)$$

where F_A is the inbreeding coefficient for a known ancestor and measures the probability that the ancestor carried two genes which were identical at a specific locus or loci, n is the number of individuals in the path connecting the parents of the individual under investigation including the parents themselves, and the summation Σ is calculated over each path that goes through a common ancestor (Wright, 1951).

In small endogamous communities with limited numbers of marriage partners, cumulative inbreeding via multiple consanguineous pathways can result in a significant build-up of homozygosity, even within a few generations. Pedigrees often have insufficient detail and depth to trace back more than three to five generations, and even in populations for which there are reliable, comprehensive data, the task can become extremely complex because of multiple pathways of inheritance, necessitating extended computational analysis.

Where this information is available – e.g. via detailed family reconstructions from northern coastal Sweden that included marriages up to sixth cousins once removed ($F = 0.00003$) (Bittles & Egerbladh, 2005; Egerbladh & Bittles, 2011) coefficients of inbreeding which incorporate multiple pathways of consanguinity can substantially exceed those based on a single generation, especially affecting the more remote types of consanguineous unions (Table 6.4). In the Swedish study, only 44.7% of first-cousin marriages had a coefficient of inbreeding of 0.0625 (Table 6.5), with F values across all 'first-cousin' unions ranging from $F = 0.0625$ to 0.1484 (I. Egerbladh & A.H. Bittles, unpublished data).

The Swedish data illustrate the influence of cumulative inbreeding on gene-pool structure at both individual and community levels, and confirm the desirability of a prior understanding among researchers and clinicians of the demographic and social structures of communities, and the legal and customary

Table 6.4 *Sweden 1800–1899*

Expected and calculated mean coefficients of inbreeding ($>F$)			
Marriage	F expected	F pedigree	Mean increase (%)
1st cousin	0.0625	0.06703	7.3
2nd cousin	0.0156	0.01882	20.6
3rd cousin	0.0039	0.00509	30.5
4th cousin	0.0098	0.00140	42.6
5th cousin	0.00024	0.00038	55.3

Table 6.5 *Sweden 1800–1899*

Pedigree-based F values for First-cousin marriages		
Coefficient of inbreeding (F)	Number	Percentage
0.0625	130	44.7
0.0626–0.0781	137	47.1
0.0782–0.0938	17	5.8
0.0939–0.1094	–	
0.1095–0.124	–	
≥ 0.125	7	2.4
Total number	291	227

status of specific types of marital unions. In particular, their preferred marriage patterns, because, as shown in Table 6.3, information of this nature could play a major role in determining gene expression and thus the patterns and frequencies of specific genetic disorders.

Recognition of distant genealogical loops is especially important in homozygosity mapping because failure to adequately allow for their impact on gene-pool composition can lead to false positives being reported. This problem was highlighted in a study of Alzheimer disease in a Dutch religious isolate, where on average 57.7% of the total consanguinity was found to be due to distant genealogical loops in the 16-generation pedigree (Liu *et al.*, 2006b).

Recessive disease incidence and rates of consanguineous marriage

As was indicated in Table 6.2, according to the Hardy–Weinberg principle, the less common the incidence of a mutant gene causing a recessive disorder, the greater the probability that the parents of an affected individual are consanguineous. In 1929, Dahlberg described a method to calculate the probability of consanguineous unions occurring in a population by random chance. However, a number of practical problems with his approach were identified, in particular the difficulty of defining a fixed population of potential spouses in human societies (e.g. with respect to the age range within which males and females actually marry), and its neglect of factors such as age preference in the choice of a partner, with males in virtually all societies older than their female spouses although by varying numbers of years (Hajnal, 1963).

Despite these practical shortcomings, Dahlberg (1947) extended his calculations to determine the relationship between the frequency of first-cousin marriages in the parents of subjects with an autosomal recessive disorder (C^1), the frequency of first-cousin unions in the general population (C), and the frequency of the causative recessive gene (q), according to the formula:

$$C^1 = C(1 + 15q)/(C + 16q - Cq)$$

This formula was further adapted to estimate the incidence of the gene frequency of the autosomal recessive disorders. For first-cousin marriages alone, the formula is:

$$q = C(1 - C^1)/C(1 - C^1) = 16(C^1 - C)$$

where q is derived from q^2, the observed incidence of the disorder, which in studies on cystic fibrosis, phenylketonuria (PKU) and Friedreich ataxia were identified by clinical diagnosis and, in the case of PKU, partially via neonatal screening (Romeo *et al.*, 1981, 1983a, 1983b). Roman Catholic Church dispensation records for consanguineous marriage were accessed to calculate the relevant levels of consanguinity in the three groups of patients and in the general population with, as previously suggested (Barrai *et al.*, 1965), data on first-cousin, first-cousin-once-removed and second-cousin unions used to calculate mean coefficients of inbreeding α (for the general population) and α^1 (for the progeny of couples in consanguineous unions).

The method proved helpful in providing comparative incidence rates for major autosomal recessive disorders in populations where reliable data were available on the prevalence of consanguineous marriage, and it was widely adopted in part due to its ready applicability. However, its application has lessened with the recognition that thousands of different mutations are responsible for disorders such as cystic fibrosis and phenylketonuria, together with the demonstration of population-specific mutation profiles and the recognition of previously undocumented levels of population stratification.

The concept of genetic load and its application in consanguinity studies

Because all humans are heterozygous for numbers of detrimental recessive genes, the term *genetic load* refers to the decrease in the average fitness of a population caused by the expression of genes which reduce survival. Lethal gene equivalents are defined as the numbers of detrimental recessive genes carried by an individual in the heterozygous state which, if homozygous, would result in death. By comparing death rates in the progeny of consanguineous and unrelated couples, it therefore is theoretically possible to estimate the number of lethal gene equivalents in a community or population according to the formula:

$$-\log_e S = A + BF$$

where S is the proportion of survivors in the study population, A measures all deaths that occur under random mating, B represents all deaths caused by the expression of recessive genes via inbreeding, and F is the coefficient of inbreeding (Morton *et al.*, 1956; Crow, 1958). By plotting a weighted regression of the log proportion of survivors (S) at different levels of inbreeding (F), A can be determined from the intercept on the Y-axis at zero inbreeding ($F = 0$), and B (the number of lethal gene equivalents) is given by the slope of the regression.

Because consanguineous individuals have a greater probability of inheriting the same mutant allele(s) from a common ancestor, their progeny will be at a higher risk of expressing one or more recessive disorders. By calculating the number of lethal gene equivalents, it was proposed that the results of inbreeding surveys could be transformed into a reproducible format for use in comparative studies on the health effects of consanguinity in different populations.

The formula devised by Morton *et al.* (1956) was widely applied in consanguinity studies from the 1960s onwards, despite the identification of a number of theoretical and methodological limitations (Li, 1963a, 1963b; Sanghvi, 1963; Bittles & Makov, 1985; Makov & Bittles, 1986). The interpretation of results obtained in terms of lethal gene equivalents also proved problematic when the values derived for A and B were used to calculate B/A ratios, as a means of differentiating between the roles of mutation and segregation in the total genetic load (Morton *et al.*, 1956; Crow, 1958; Crow, 1963).

Through time, estimates for lethal gene equivalents derived from studies on which the genetic-load method originally had been based proved to be significantly higher than in later investigations which had larger sample sizes, were more representative of the study populations, and included better control for non-genetic variables (Table 6.6). The calculation of lethal gene equivalents in human populations has therefore gradually been replaced by more

Table 6.6 *Estimated number of lethal equivalent genes per zygote*

Lethal equivalents	Study
3–5	Morton *et al.* (1956)
4.2	Marçallo *et al.* (1964)
1.8	Freire-Maia *et al.* (1964)
2.2	Bodmer & Cavalli-Sforza (1976)
1.5	Freire-Maia & Takehara (1977)
1.4	Bittles & Neel (1994)
1.2	Cavalli-Sforza *et al.* (2004)
1.2	Bittles & Black (2010a)

direct measures of consanguinity-associated effects – e.g. as presented in Figures 9.1–9.3 for stillbirths, infant deaths and total pre-reproductive mortality, respectively.

Genomic approaches to consanguinity measurement

Direct estimation of an individual's coefficient of inbreeding by reference to genomic homozygosity data offers many potential advantages. Initially, a maximum-likelihood method of analysis was developed using simulated whole genome data, which permitted inference of the IBD status of both alleles of an individual at each marker along the genome (Leutenegger *et al.*, 2003). The method also provided a variance measure for these calculated estimates and, for example, it was shown that whereas the mean value for IBD status for first cousins was 0.0625, theoretically the calculated values could range from 0.03 to 0.12 at individual loci. However, when individual coefficients of inbreeding were compared from genealogies and via genetic polymorphisms based on microsatellite and SNP analysis, the excess levels of homozygosity were discrepant (Carothers *et al.*, 2006).

As expected, direct microsatellite analysis of DNA samples obtained from UK ethnic migrants showed that in the Pakistani Muslim community, for which estimates of consanguineous marriage range from 33% (Qureshi *et al.*, 2003) to 67% (Darr & Modell, 1988), the observed F values were much higher than in a co-resident South Asian community which avoided consanguinity (Overall *et al.*, 2003). Significant genetic sub-structuring also was demonstrated in the Pakistani community which, according to the Wahlund principle, could significantly interfere with frequency estimates of recessive disease genes and lead to the calculation of spuriously high values (Overall, 2009).

It has been proposed that the consanguineous parents of children with an autosomal recessive disorder have a greater proportion of their genomes identical by descent than parents in similar consanguineous relationships but who conceive only unaffected offspring (Teeuw *et al.*, 2010). Using both SNP and microsatellite analysis, a study of UK Pakistani consanguineous individuals with a range of autosomal recessive diseases showed that, on average, persons whose parents were first cousins ($F = 0.0625$) were actually homozygous at 11% of the loci tested, with a range of 5–20% (Woods *et al.*, 2006). This is in quite close agreement with the values derived for some of the Swedish first-cousin couples obtained from the analysis of just three consecutive generations of consanguineous marriage (Table 6.5).

Consanguinity and runs of homozygosity

Estimates of consanguinity and the resultant levels of homozygosity that are based on standard three-generation pedigrees are, by definition, unable to provide information on close kin marriages in more distant generations, and they also fail to account for cases of mistakenly ascribed paternity (Bittles & Black, 2010a). To preclude these issues, and at the same time complement pedigree-based approaches, high-density single nucleotide polymorphism (SNP) scans have been used to directly estimate individual homozygosity levels from uninterrupted runs of homozygosity (ROH) located throughout the genome. According to the results of an initial ROH study, extended tracts of homozygosity over 1 Mb long frequently occurred in individuals regarded as outbred; in one Japanese individual, an interrupted run of 3922 homozygous SNPs spanning 17.9 Mb was identified (Gibson *et al.*, 2006).

Subsequent investigations on the population of the Orkney Isles, Scotland, confirmed these findings and, importantly, demonstrated a good correlation between the results of the ROH studies and genealogical records (McQuillan *et al.*, 2008). The applicability of the ROH approach to assessing historical family relationships was apparent in the Orkney population, since ROHs measuring up to 4 Mb in length were present in individuals whose pedigrees indicated non-consanguineous ancestry extending back for five to more than ten generations.

The application of ROH analysis to individuals in Phase III of the International Haplotype Map Project (HapMap), with genomic data derived from approximately three million SNP loci, confirmed formerly known genetic relationships between test subjects. But the investigators also detected numerous additional and previously unidentified pairs of close relatives who had been included in the HapMap sample (Pemberton *et al.*, 2010). The findings thus provide further evidence that the levels of genetic identicality within human populations are greater than earlier predicted or expected, although they also call into question the recruitment and sampling protocols used in population-based research initiatives.

Dense genome-wide SNP genotypes also have been used to investigate homozygosity in samples from unrelated individuals recruited into the Human Genome Diversity Panel (HGDP-CEPH). Two alternative methods of assessment have been applied: first, as runs of homozygosity examined in 1043 individuals with the proportion of individuals with ROH of lengths 2–4, 4–8, 8–16 and > 16 Mb calculated (Kirin *et al.*, 2010); and second, expressed in terms of homozygosity by descent (HBD); i.e. the proportion of an individual's genome that is IBD (Leutenegger *et al.*, 2011).

As expected from genetic epidemiological and anthropological surveys, in both sets of studies the highest rates of homozygosity were determined in individuals from populations resident in the Middle East, Central South Asia and the Americas. More unexpectedly, low levels of ROH and homozygosity were detected in sub-Saharan populations (Leutenegger *et al.*, 2011), which is at variance with anthropological and medical genetics reports from a number of countries in the region (Chapter 5; www.consang.net). Whether these findings reflect the recruiting frameworks adopted, with individuals from tribes that avoid consanguineous unions over-sampled, and/or are associated with polygny and informal adoptions which are known to occur in many sub-Saharan populations, remains to be determined.

Control for population stratification

From a genetic counselling perspective (Chapter 13), the combined roles of consanguinity and population subdivision are important in many clinical situations (Bittles, 2008). In genome-wide association studies (GWAS), it has been claimed that methods using a set of ancestry-informative markers or genome-wide data can be employed to statistically correct single-SNP association tests for population stratification in unrelated samples (Pritchard *et al.*, 2000; Hoggart *et al.*, 2003; Price *et al.*, 2006; Kimmel *et al.*, 2007; Price *et al.*, 2010; Thornton & McPeek, 2010; Zhang *et al.*, 2011), and a recent simulation-based comparative study indicated that principal component-based logistic regression performed well under most of the experimental circumstances modelled (Wu *et al.*, 2011). However, whether any current method based on these approaches can be successfully and consistently applied to control for population stratification in a society as large and multiply subdivided as India (Indian Genome Variation Consortium, 2008; Reich *et al.*, 2009) remains to be demonstrated.

Commentary

Unexpected complexities can be encountered when dealing with consanguineous relationships in actual human populations, and at least a basic knowledge of the demographic structure of a population is important in differentiating between the outcomes of random inbreeding, brought about by founder effect, drift and endogamy, and by preferential consanguinity. Given the increasing accessibility and decreasing costs of whole-genome and exome analysis, direct estimates of consanguinity based on homozygosity by descent seems to be the most promising and appropriate method for current use at the individual level. But, for wider community and population surveys, genealogical approaches often are sufficiently informative to remain an alternative method of choice into the foreseeable future.

7   *Consanguinity and reproductive behaviour*

Introduction

To a large extent, opinions on the overall effect of consanguinity on reproduction have been influenced by the conclusions drawn from studies conducted in non-human species, where reduced fertility has been ascribed to the generalized phenomenon of inbreeding depression (Amos *et al.*, 2001), mediated in part via poor sperm quality in some species (Table 7.1). A problem in assessing the impact of inbreeding depression in captive animals is that, as in the studies cited in Table 7.1, many investigations were conducted on small numbers of animals and the observed levels of inbreeding depression referred either to sib ($F = 0.25$) or half-sib ($F = 0.25$) matings.

A meta-analysis of zoo populations representing 88 species of mammals, birds, reptiles and amphibians reported evidence of purging (i.e. the removal of deleterious alleles from the gene pool through deliberate inbreeding), which was statistically significant in 14 of the 119 populations examined (Boakes *et al.*, 2007). However, a loss of genetic diversity with inbreeding through time is not inevitable in mammalian populations, as demonstrated by the increased levels of heterozygosity in an island population of mouflon (*Ovis aries*) since the herd was founded by a single pair of animals in 1957 (Kaeuffer *et al.*, 2007).

The relevance of these studies to human populations is limited, with great care needed in extrapolating data derived from captive animals subjected to sib- and half-sib mating to the outcomes of first-cousin marriage ($F = 0.0625$) in humans. Further, although gene-pool purging may have been a salient factor in small early human groups subjected to major population bottlenecks (Gherman *et al.*, 2007; Manica *et al.*, 2007), it probably has been much less important in continental populations during the past 500 years.

Basic shortcomings in studies of human fertility

As with non-human species, where empirical, population-based information has been collected on human populations, the investigations often have relied on

93

Table 7.1 *The influence of inbreeding depression on biological characteristics of mammalian species*

Parameter	Species	Status	Reference
Reproductive capacity			
Reduced sperm quality	Cheetahs	Captive	Wildt *et al.* (1983, 1987a)
	Cheetahs	Wild	Wildt *et al.* (1987a)
	Lions	Wild	Wildt *et al.* (1987b)
	Ungulates	Captive	Roldan *et al.* (1998)
	Shetland ponies	Domestic	van Eldik *et al.* (2006)
Reduced circulating testosterone levels	Lions	Wild	Wildt *et al.* (1987b)
Prenatal growth			
Reduced birthweight	Hereford cattle	Captive	Pariacote *et al.* (1998)
	Seal pups	Wild	Coltman *et al.* (1998)
Survival			
Increased infant mortality	Primates	Captive	Ralls & Ballou (1982a)
Increased juvenile mortality	Big-horn sheep	Wild	Hass (1989)
Increased juvenile mortality	Ungulates	Captive	Ralls *et al.* (1979)
	Small mammals	Captive	Ralls & Ballou (1982b)
Resistance to disease			
Reduced parasitic resistance	Soay sheep	Wild	Coltman *et al.* (1999)
Life expectancy			
Reduced longevity	Deer	Semi-wild	Sternicki *et al.* (2003)

small sample numbers which make the results difficult to assess. This problem was illustrated in a number of studies reported from the South Indian state of Andhra Pradesh, with numbers of live-born children used as the measure of fertility (Bittles *et al.*, 2002). The total sample sizes investigated ranged from just 20 to 927 families, with a median size of 174 families; in all cases, there was a high proportion of consanguineous marriages, predominantly contracted at the levels of first cousin and uncle–niece.

Although greater mean numbers of children were born to consanguineous parents in 12 of the 18 communities investigated, because of the inadequate sample sizes, the differences only attained statistical significance in a minority of cases. To add to the problems of interpretation, often there was no indication why particular families or communities were included, with a suspicion that in many cases the sampling procedure was largely opportunistic, or how representative the sampled families and communities might be in terms of the ethnicity, religion, demography or socioeconomic status of the population in general.

Social and demographic factors that are known to influence reproductive behaviour and, hence, the comparative fertility of consanguineous couples were addressed in Chapter 5. In general terms, because many consanguineous unions are contracted at younger spousal ages, the consequent earlier opportunity of commencing reproduction can place consanguineous couples at a fertility advantage (Bittles *et al.*, 2002; Bittles & Black, 2010c). The fact that communities in which consanguinity is favoured may often be less likely to routinely use reliable methods of contraception also means that the potential maternal reproductive span is optimized (Hussain & Bittles, 1999), and so there are good non-biological reasons to suppose that consanguineous couples may, on average, have larger numbers of live-born children.

Genetically determined factors influencing human mate choice

Two major genetic mechanisms can be identified that govern or contribute to human fertility:

(i)   factors that influence mate choice
(ii) factors that facilitate embryonic implantation and fetal growth.

Olfactory recognition systems in mate choice

Studies on the H-2 immune response system in mice have indicated a significant role for major histocompatibility complex (MHC) haplotypes in the initial stage of mate selection, with a marked preference for MHC-dissimilar partners and hence inbreeding avoidance. The underlying mechanism in mice appears to be olfactory in nature, and it has been interpreted as indicating chemosensory imprinting because it could be reversed by cross-fostering (Penn & Potts, 1998). Subsequent studies have demonstrated the ability of mice to discriminate between some, but not all, naturally occurring allelic variants at classical MHC loci, which suggests that MHC-disassortative mating preferences may operate via small MHC-driven odour differences (Carroll *et al.*, 2002).

An analogous olfactory mate-choice system has been proposed in humans, with the phenomenon initially investigated in a group of female Swiss university students who were asked to conduct blind smell-testing of cotton T-shirts that had been worn for two consecutive nights by male students (Wedekind *et al.*, 1995). The resultant male body odours were scored in terms of pleasantness and 'sexiness', and when the results were assessed with respect to MHC-similarity or dissimilarity, females scored male body odours as more pleasant the greater their MHC-dissimilarity to the test male.

In support of these findings, a lower than expected incidence of human leucocyte antigen (HLA) haplotype matches was reported in the S-leut Hutterites, a highly endogamous Anabaptist sect living in South Dakota, USA, with the results interpreted as evidence for the avoidance of spouses with similar HLA haplotypes (Ober *et al.*, 1997a). However, studies conducted across other populations either failed to confirm non-random mating at HLA loci (Pollack *et al.*, 1982; Jin *et al.*, 1995a; Hedrick & Black, 1997; Ihara *et al.*, 2000), or even indicated a greater likelihood of HLA sharing between couples, possibly indicative of ethnic clustering in mate choice (Rosenberg *et al.*, 1983). In a follow-up of the original T-shirt studies, it was reported that the level of variance in the scoring of odour pleasantness explicable by the degree of MHC-similarity (r^2) ranged from 0% to 23%, dependent on the identity of the T-shirt-wearer (Wedekind & Füri, 1997). Further, the highest r^2 value was obtained by a male subject smelling a male body odour, a finding that presumably fell somewhat outside the scope of the hypothesis under examination.

Subsequent olfactory investigations have also produced anomalous results. For example, in a study of the reactions of females to male body odours, once again based on T-shirts worn by males for two consecutive nights, significantly more HLA allele matches were observed between the 'donor' of the most preferred male body odour and the female test subject. But it appeared that the women's ability to discriminate and choose odours was based on HLA alleles inherited from fathers but not from mothers, suggesting that paternally inherited HLA-associated odours influenced odour preference and acted as social cues (Jacob *et al.*, 2002).

A puzzling feature of the original Swiss study was that the pattern of MHC odour preferences appeared to be reversed in women taking oral contraceptives (Wedekind *et al.*, 1995). To see if contraceptive-pill use could be shown to alter their odour preferences, in the UK women during their follicular phase were requested to rate the odours of MHC-similar and MHC-dissimilar males. Whereas no significant difference was observed for the group as a whole, single women preferred the odours of MHC-similar men, whereas women in relationships preferred the odours of MHC-dissimilar men. This finding was minimally interpreted by the authors as indicating the potential disruptive effect of contraceptive-pill usage on disassortative human-mate preferences (Roberts *et al.*, 2008), but the implications in terms of human-relationship preferences and relationship stability in general are perhaps more noteworthy.

A small-scale study also reported that opposite-sex siblings but not same-sex siblings could recognize each other's body odours, with olfactory aversion reported only for the father–daughter and brother–sister nuclear family relationships (Weisfeld *et al.*, 2003). If these findings can be verified, they would appear to support the Westermarck hypothesis (see Chapter 12), that incest avoidance develops during childhood, possibly via olfactory cues (Schneider & Hendrix, 2000). However, the data appear to have little practical relevance in terms of the more remote, legal levels of consanguinity (e.g. between first cousins) because in communities where consanguineous unions are strongly preferential, marriage-partner choice is usually subject to parental decision making (Bittles, 1994; Bittles *et al.*, 2002), and premarital olfactory sensing of a prospective son-in-law by the bride's parents not a recognized social custom.

Visual recognition cues in mate choice

Clearly, in most modern human societies where there is widespread use of deodorants and perfumes, the concept of straightforward mate choice preferences based solely on body odour is overly simplistic, although it has been suggested that at least some perfumes may function as amplifiers of MHC-related body odours (Milinski, 2006). The emphasis on selection based on body odours also overlooks the obvious importance of visual cues in partner choice, in particular facial preferences, which when tested in females suggested a preference for MHC-similar males (Roberts *et al.*, 2005). But facial cues may need to be considered in a context-specific manner; e.g. with familiar faces judged as trustworthy and possibly worthy of a long-term but not a short-term relationship (DeBruine, 2005). Also, although unstressed individuals showed a preference for visually similar mates, when subjected to physiological stress by cooling, they were more likely to prefer dissimilar mates (Lass-Hennemann *et al.*, 2010).

Even in devout Islamic societies, facial photographs may be exchanged as a preliminary part of marriage-partner selection processes, and in Western societies multiple, often quite obvious, visual body cues are frequently on display. Thus, in overall biological terms, it could be that to achieve an optimal level of genetic variability, olfactory and visual channels operate in an essentially complementary manner; i.e. with odour preferences indicative of MHC-dissimilarity but visual preferences favouring MHC-similarity (Havlicek & Roberts, 2009).

Combined genome-wide and HLA-based studies on mate choice

Genome-wide analysis has been used in conjunction with HLA typing to investigate whether or not couples enrolled in the *HapMap* study exhibited a significant pattern of similarity/dissimilarity in the MHC region. The results indicated that African couples did not show a significant pattern of similarity/dissimilarity in the MHC region, whereas across the remainder of the genome, they were more similar than random pairs of individuals. By comparison, European Americans were significantly more MHC-dissimilar than random pairs of individuals, and this pattern of dissimilarity was more pronounced than in non-MHC regions of the genome (Chaix *et al.*, 2008).

The differences between Africans and European Americans indicated that if humans are capable of discriminating between the MHC types of potential partners via odour cues, either the system is specific to certain ancestries or in some instances it can be over-ridden by other more potent, possibly visual or sociodemographic, cues. However, speculation of this nature may be unnecessary as the results of the MHC data on European-Americans have been queried on the grounds that they were based on a few extreme data points and could not be substantiated in an equivalent but independent set of European-Americans (Derti *et al.*, 2010).

Genetic factors that facilitate embryonic implantation and fetal growth

Two opposing hypotheses relating to pregnancy success have been advanced, proposing either that:

   (i)   Pregnancies initiated by consanguineous partners result in enhanced genetic compatibility between mother and fetus due to the greater proportion of shared maternal and paternal genes, which in turn is beneficial to fetal survival and growth; or

  (ii)   According to the fetal allograft concept, a balanced polymorphism for transplantation antigens is maintained via a selection mechanism in which antigenic disparity between the mother and fetus is beneficial to fetal development.

Studies which support the mother–fetus compatibility hypothesis

Several common genetic mechanisms have been investigated to explain these findings:

   (i)   ABO blood group incompatibility

Both the ABO and Rhesus (Rh) blood group systems have been implicated as potential causes of maternal–fetal incompatibility and hence fetal losses.

The role of maternal–fetal ABO incompatibility as a differential determinant of pregnancy loss is probably limited because the causative anti-A and/or anti-B antibodies are predominantly IgM and so cannot readily cross the placenta (Bittles & Matson, 2000). Nonetheless, a significantly higher frequency of maternal–fetal ABO incompatibility has been reported in cases of spontaneous abortion than in successful pregnancy outcomes (Bandyopadhyay *et al.*, 2011), and levels of ABO incompatibility were comparatively lower in a study of pregnancies involving consanguineous partners (Nair & Murty, 1985).

(ii)   Rhesus incompatibility

Haemolytic disease due to Rhesus (Rh) incompatibility is less common because of the low frequency of Rh −ve alleles in most populations, but it usually is severe because the anti-D antibodies are IgG, which can readily cross the placenta and cause large-scale red cell destruction. Red blood cells have 49 different Rh antigens (Avent *et al.*, 2006), and Rh incompatibility can arise following the escape of Rh+ve red cells from a fetus into the maternal circulation, resulting in the formation of maternal anti-Rh(D) antibody if the mother is Rh −ve. This situation may commence at parturition or, less commonly, following an abortion or some form of prenatal testing (e.g. amniocentesis or chorionic villus sampling). Empirical data on the degree to which consanguinity influences maternal–fetal Rh incompatability are scarce; however, an early study did report a reduced prevalence of Rh incompatibility in consanguineous pregnancies, thus favouring fetal and/or newborn survival (Stern & Charles, 1945).

(iii)   Pre-eclampsia

Information on a possible association between consanguinity and pre-eclampsia involving components of the HLA system, in particular HLA-DR4, is at best mixed (Saftlas *et al.*, 2005). Early studies suggested a reduced prevalence of pre-eclampsia in consanguineous pregnancies (Stevenson *et al.*, 1971, 1976), and in subsequent investigations, no significant effect on prevalence (George *et al.*, 1992; Badria *et al.*, 2001; Sezik *et al.*, 2006), no effect on the severity of symptoms (Badria & Amarin, 2003; Sezik *et al.*, 2006), and a possible positive association with consanguinity (Anvar *et al.*, 2011) were variously reported. A population-based study of 1.7 million births in Norway indicated that both maternal and fetal genes contributed to the risk of pre-eclampsia, with the fetal component also affected by paternal genes (Lie *et al.*, 1998).

Familial aggregation of pre-eclampsia and intra-uterine growth restriction was reported in a genetic isolate in the Netherlands, with women born to consanguineous unions significantly more likely to have experienced previous pre-eclampsia (82%) and intra-uterine growth restriction (78%), thus suggesting the expression of an underlying recessive mutation (Berends *et al.*, 2008). In addition, mutations in complement regulatory proteins and, hence, complement activation have been putatively identified as predisposing to pre-eclampsia (Salmon *et al.*, 2011). But, as has been noted in other contexts, studies into a possible association between consanguinity and pre-eclampsia mostly have been statistically under-powered and heterogeneous in terms of their study design, outcome and exposure assessment, and so the inconclusive overall nature of the results obtained is unsurprising (Saftlas *et al.*, 2005).

The fetal allograft hypothesis

The fetal allograft hypothesis (Clarke & Kirby, 1966; Adinolfi, 1986; Ober, 1998) has been extensively examined in human populations, especially among women who have experienced primary recurrent spontaneous abortions (i.e. with all conceptions lost pre-term). Some studies reported a significant positive association between parental allele sharing at HLA loci and recurrent abortion, variously involving HLA-A, HLA-B, HLA-DR, HLA-DQ alleles (Komlos *et al.*, 1977; Schacter *et al.*, 1984; Reznikoff-Etievant *et al.*, 1984; Thomas *et al.*, 1985; McIntyre *et al.*, 1986; Coulam *et al.*, 1987; Johnson *et al.*, 1988; Ho *et al.*, 1990), and HLA-G (Aldrich *et al.*, 2001). In support of these findings, strong negative selection was claimed against individuals homozygous at HLA-A, HLA-B, HLA-C, HLA-DR and HLA-DQ loci in the S-leut Hutterite community, for whom mean coefficients of inbreeding ranging from $\alpha = 0.0211$ to 0.0396 have been calculated (Kostyu *et al.*, 1993); in a prior study, surviving HLA-DR–compatible infants had been shown to have lower birthweights (Ober *et al.*, 1987).

On the basis of the early findings, women who had experienced recurrent abortion were immunized with purified lymphocytes prepared from their partner's blood, with 70–80% pregnancy success rates claimed for the procedure (Mowbray *et al.*, 1985). However, the occurrence of successful pregnancies among women given their own lymphocytes, rather than those of their partners, raised the possibility of a placebo effect (Adinolfi, 1986; Clark & Daya, 1991). In addition, in other studies, no specific association between HLA antigen-sharing and recurrent infertility was detected (Oksenberg *et al.*, 1984; Laitenen *et al.*, 1993; Martin-Villa *et al.*, 1993; Jin *et al.*, 1995b), nor was any association among HLA-antigens, HLA-linked genes, consanguinity and recurrent fetal loss apparent when both partners were tested at Class 1 HLA, HLA-A, HLA-B and HLA-C or Class II HLA, HLA-DR and HLA-DQ loci (Moghraby *et al.*, 2010).

Because successful pregnancies have been described in which the HLA haplotypes of the mother and offspring were identical at HLA-A, HLA-B, HLA-C and HLA-DR loci (Kilpatrick, 1984; Oksenberg *et al*., 1984), it appears that even if a positive association between infertility and parental HLA allele-sharing can be demonstrated in some couples, the relationship must be far from absolute. Furthermore, Hutterite couples who shared HLA-DR antigens were reported to have a median completed family size of 6.5, as opposed to 9.0 among those with no HLA alleles in common (Ober *et al*., 1988), indicating a limited degree of reduction in fecundity. Intriguingly, maternal–fetal HLA-DR antigen sharing appears to influence the sex ratio of firstborns, resulting in a preponderance of male children at birth, and this trend may continue in higher-order births when parents share both HLA-DR antigens (Radvany *et al*., 1987).

Retrospective data on pregnancies and prenatal losses are unfortunately known to be unreliable, with major problems encountered in their collection and significant levels of recall bias (Wilcox & Horney, 1984; Wilcox *et al*., 1988), and so a degree of caution needs to be exercised in interpreting all studies of this type. In addition, whereas approximately 15% of clinically recognized pregnancies spontaneously abort (Beydoun & Saftlas, 2005), much higher levels of earlier spontaneous pregnancy losses are indicated in studies based on urinary or serum human chorionic gonadotrophin (hCG) assays to detect implantation of the embryo (Miller *et al*., 1980; Wilcox *et al*., 1988).

For example, a study in Bangladesh based on sequential twice-weekly urinary hCG assays showed that 45% of the pregnancies detected among women at 18 years of age spontaneously miscarried; in women aged 38 years, the rate of fetal loss had increased to 92%, with the age-related decline in fecundability primarily due to fetal losses caused by meiotic nondisjunction (O'Connor *et al*., 1998). These results suggest that many data sets on fetal losses refer only to later stages of pregnancy and may be unrepresentative of the entire spectrum of post-implantation reproductive losses.

Comparative fertility in consanguineous and non-consanguineous couples

Consanguinity, infertility, and spontaneous prenatal losses

Reduced levels of primary and/or secondary sterility have been reported for consanguineous couples in different populations (Yanase *et al*., 1973; Yamaguchi *et al*., 1975; Rao & Inbaraj, 1977b; Rao & Inbaraj, 1979; Hann, 1985; Edmond & De Braekeleer, 1993), and an early investigation in a French

Table 7.2 *Prenatal losses in first-cousin versus non-consanguineous pregnancies*

Country	Increased losses	No or reduced losses	Study
Japan (Hiroshima)	+		Schull (1958)
Japan (Nagasaki)		+	Schull (1958)
Japan (Kure)		+	Schull (1958)
Brazil (Bauru, Japanese)	+		Freire-Maia *et al.* (1964)
Canada (Montreal)		+	Warburton & Fraser (1964)
Nigeria (Yoruba)	+		Scott-Emuakpor (1974)
Canada (Montreal)	+		Fraser & Biddle (1976)
Sri Lanka (Goyigama)		+	Reid (1976)
Brazil (Bauru, Japanese)	+		Freire-Maia & Takehara (1977)
India (Tamil Nadu, rural)	+		Rao & Inbaraj (1977b)
India (Tamil Nadu, urban)	+		Rao & Inbaraj (1977b)
India (Tamil Nadu, rural)	+		Rao & Inbaraj (1979a)
India (Tamil Nadu, urban)		+	Rao & Inbaraj (1979a)
India (Karnataka, rural)		+	Hann (1985)
Kuwait (National)		+	Al-Awadi *et al.* (1986)
Lebanon (Beirut)		+	Khlat (1988)
Saudi Arabia (Riyadh)		+	Wong & Anokute (1990)
Sudan (Khartoum)		+	Saha *et al.* (1990)
UAE (Al Ain)		+	Abdulrazzaq *et al.* (1997)
Saudi Arabia (Riyadh)		+	al Husain & al Bunyan (1997)
Israel (Arab)		+	Jaber *et al.* (1997a)
Pakistan (Karachi)	+		Hussain (1998)
Saudi Arabia (Damman)		+	al-Abdulkareem & Ballal (1998)
Jordan (National)		+	Khoury & Massad (2000)
Egypt (Alexandria)	+		Mokhtar & Abdel-Fattah (2001)
Qatar (Doha)		+	Saad & Jauniaux (2002)
Tunisia (Monastir)		+	Kerkeni *et al.* (2007)
Palestinian Territories	+		Assaf *et al.* (2009)
Oman		+	Gowri *et al.* (2011)

Canadian isolate concluded that intra-uterine mortality was lower in consanguineous couples, resulting in greater overall fertility (Philippe, 1974).

As shown in Table 7.2, most empirical investigations of spontaneous prenatal losses have reached the same conclusion, with just 11 of 29 studies conducted in 19 countries indicating greater losses among first-cousin pregnancies when compared with non-consanguineous couples. Indirect indicators of fetal survival, such as multiple birth rates and the secondary sex ratio, also have failed to show any adverse effect of consanguinity on prenatal survival (Bittles *et al.*, 1988). A positive association, however, has been reported between consanguinity and spontaneous preterm birth (Al-Eissa & Ba'Aqeel 1994; Mumtaz *et al.*, 2010), and with apnoea of prematurity (Tamim *et al.*, 2003).

Consanguinity and reproductive success

To overcome the problematic nature of the results obtained with small-scale studies into the comparative fertility of consanguineous and non-consanguineous couples (Bittles *et al.*, 2002), a meta-analysis was conducted to determine whether the increased levels of genetic homozygosity associated with consanguineous unions, arising from the expression of identical genes inherited from a common ancestor through both parents, resulted in reduced levels of fertility. Data were collated from 41 studies conducted in 9 countries (Table 7.3), with acceptance of studies into the meta-analysis contingent on 3 basic pre-conditions: (i) a post-1950s publication date to limit the study to contemporary populations; (ii) a minimum sample size of 750 subjects; and (iii) where feasible, a minimum of 3 consanguinity data points (e.g. first cousin, second cousin, and non-consanguineous).

The data set contained information on approximately one million pregnancies and births covering six categories of consanguinity: uncle–niece/double first cousins ($F = 0.125$); first cousins ($F = 0.0625$); first cousins once removed ($F = 0.0313$); second cousins ($F = 0.0156$); beyond second cousins ($F < 0.0156$); and non-consanguineous ($F = 0$). The levels of consanguinity recorded refer to a single generation only with no allowance made for consanguineous unions in preceding generations, and fertility was measured as the total number of live births which had been borne by a woman at the time of the survey. The fertility rates at different levels of consanguinity are shown in Table 7.3, with mean values ranging from 1.87 live births for non-consanguineous Muslim couples in Israel to 7.48 live births among first cousins in Lucknow, India.

In dichotomous comparisons, a positive association between consanguinity and the number of live-born children was found in 13/21 studies at $F = 0.125$, 34/41 studies at $F = 0.0625$, 16/20 studies at $F = 0.0313$, and 14/19 studies at $F = 0.0156$. A non-parametric sign test for symmetry (Lentner, 1982) was applied to test the associations between each consanguinity category and the numbers of live-born children per family, based on the assumption that there would be equal numbers of positive and negative associations. The differences in the mean number of live-born children between the consanguineous and non-consanguineous couples failed to attain statistical significance at the 0.05 level for $F = 0.125$ and $F = 0.0156$. The differences were, however, highly significant at $F = 0.0625$ ($p < 0.0001$), reflecting the larger numbers of first-cousin subjects and controls examined, and at $F = 0.0313$ ($p < 0.02$).

Table 7.3 *Mean number of live-born children by consanguineous marriage and study population*

Study populations	Coefficient of inbreeding (F)	0.125	0.0625	0.0313	0.0156	0	Authors
	Japan						
1.	Hiroshima		3.10	3.05	3.14	2.95	Neel & Schull (1962)
2.	Nagasaki		3.99	4.08	3.97	3.84	Neel & Schull (1962)
3.	Hirado	4.00	4.58	4.63	4.68	3.95	Schull *et al.* (1970a)
	India						
4.	Delhi		7.05	6.31	6.10	5.83	Basu (1978)
5.	Lucknow		7.48	6.16	6.00	5.41	Basu (1978)
6.	Udaipur		4.83	4.98	5.00	5.63	Basu (1978)
7.	Tamil Nadu, rural	3.18	3.59			3.24	Rao & Inbaraj (1977b)
8.	Tamil Nadu, urban	3.15	3.28			3.09	Rao & Inbaraj (1977b)
9.	Tamil Nadu, rural	3.32	3.72			3.34	Rao & Inbaraj (1979a)
10.	Tamil Nadu, urban	3.40	3.49			3.31	Rao & Inbaraj (1979a)
11.	Pondicherry	5.10	6.69			3.60	Puri *et al.* (1978)
12.	Karnataka, all women	2.73	2.98			3.07	Hann (1985)
13.	Karnataka, completed reproduction	3.07	4.18			4.07	Hann (1985)
14.	Vadde, <40 yr.	3.41	2.94	3.50	2.84	2.89	Reddy (1992)
15.	Vadde, ≥40 yr.	6.28	6.10	5.67	6.43	5.97	Reddy (1992)
16.	Karnataka, Hindu	2.30	2.28		2.27	2.14	Bittles *et al.* (1992)
17.	Karnataka, Muslim	2.47	2.62		2.77	2.58	Bittles *et al.* (1992)
18.	Karnataka, Christian	2.31	2.26		2.23	2.16	Bittles *et al.* (1992)
19.	Andhra Pradesh, tribal	4.36	3.56			3.44	Naidu *et al.* (1995)
20.	Andhra Pradesh		3.64			4.16	Yasmin *et al.* (1997)
	Pakistan						
21.	Lahore	3.74	3.36	3.72	3.17	3.35	Shami & Zahida (1982)
22.	Sheikhupura		4.61	4.78		4.19	Shami & Iqbal (1983)
23.	Gujrat	4.44	4.65	4.65	4.44	4.16	Shami & Hussain (1984)
24.	Jhelum	3.75	4.41	4.00	7.00	4.34	Shami & Minhas (1984)
25.	Rawalpindi	2.00	3.96	4.23	2.75	4.13	Shami & Siddiqui (1984)

Table 7.3 (*cont.*)

Study populations	Coefficient of inbreeding (F)	0.125	0.0625	0.0313	0.0156	0	Authors
26.	Gujranwala		4.15	3.93		3.66	Bittles *et al.* (1993)
27.	Sahiwal	5.40	5.24	4.86	5.60	4.65	Bittles *et al.* (1993)
28.	Faisalabad		4.58	4.23		3.97	Bittles *et al.* (1993)
29.	Sialkot	3.83	4.34	4.13	4.00	3.99	Bittles *et al.* (1993)
30.	Karachi		5.27	3.77	4.50	4.72	Hussain & Bittles (1998)
	Kuwait						
31.	Kuwait		2.27	2.73		2.44	Al-Awadi *et al.* (1986)
	Lebanon						
32.	Beirut, Christian		3.58			3.23	Khlat (1988)
33.	Beirut, Muslim		4.87			4.36	Khlat (1988)
	Israel						
34.	Arab		2.25			1.87	Jaber *et al.* (1997a)
	Saudi Arabia						
35.	Damman		4.90			4.40	al-Abdulkareem & Ballal (1998)
	Turkey						
36.	Ankara		4.50			3.78	Başaran *et al.* (1989)
37.	Diyarbakir		5.84			3.89	Başaran *et al.* (1989)
38.	Eskisehir		2.49			4.31	Başaran *et al.* (1989)
39.	All-Turkey		3.75			3.46	Tunçbilek & Koc (1994)
40.	Konya		2.93			2.88	Demirel *et al.* (1997)
	Nigeria						
41.	Yoruba	3.26	3.43	3.18	2.96	2.57	Scott-Emuakpor (1974)

Because first-cousin unions were by far the largest category of consanguineous marriage with data available on 41 studies, a direct comparison was made of the mean numbers of live births to first cousins versus nonconsanguineous couples in individual studies, with the data plotted as an unweighted linear regression according to the standard equation $y = a + bx$, where y is the predicted mean value, a is the intercept and b is the slope of the regression line. As shown in Figure 7.1, across the study populations, first-cousin couples on average had 0.05 additional births when compared with

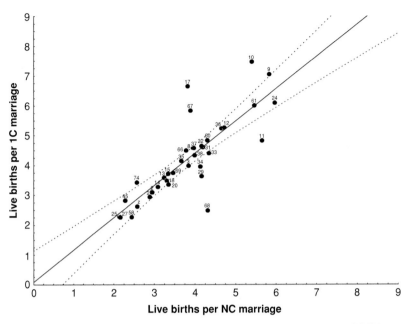

Figure 7.1 Mean numbers of live-born children in first-cousin (1C; $F = 0.0625$: y-axis) versus non-consanguineous marriages (NC; $F = 0$: x-axis) in 41 study populations

co-resident non-consanguineous couples ($r^2 = 0.38$, p $< 10^{-5}$), with the high level of statistical significance reflecting the large numbers of subjects analysed.

The variation in reported fertility partially reflects the individual survey designs, temporal and geographical differentials in fertility, and the small numbers of births recorded in some of the less common consanguinity categories. For example, the fertility rates cited for Lucknow, ranging from 5.41 to 7.48 live births, were for completed family sizes and were derived from data collected during the 1970s (Basu, 1978). By comparison, the live births reported for Rawalpindi were based on retrospective, cross-sectional household interviews, and in the case of double first-cousin unions, they were derived from just eight live births occurring within four marital unions (Shami & Siddiqui, 1984).

The two study populations in the state of Karnataka in South India comprised all women and women who had completed their reproduction, respectively (Hann, 1985). Similarly, in Andhra Pradesh, South India, the two study populations were defined by maternal age: < 40 years and ≥ 40 years (Reddy, 1992). Thus, the women in the two surveys had been subject to different time periods of risk of exposure to pregnancy, which at least in part explains the greater than two-fold difference in the number of children born per woman in the first-cousin category.

An overview of the relationship between consanguinity and fertility

Contrary to perceived opinion, analysis of the published literature drawn from a number of different regions and continents produced no convincing evidence that the increased levels of genetic homozygosity associated with consanguineous unions resulted in reduced fertility, either through failure to conceive and/or because of recurrent spontaneous abortions (Table 7.2). Indeed, in overall terms, reproductive success was positively associated with consanguinity, resulting in increased mean numbers of children born to consanguineous couples (Table 7.3).

All studies of human fertility are subject to uncertainty given the high, and still largely unexplained, losses of human conceptions (Bittles & Matson, 2000), and the varied designs of the studies incorporated in Table 7.3 allowed comparisons of completed family sizes in only two studies. It therefore seems probable that to some degree, the results were influenced by biodemographic and socioeconomic covariates known to affect fertility; for example, age at marriage and the length of the reproductive span (Bittles *et al.*, 1993). In addition, spurious positive associations may have been generated by sampling populations at different points in their economic development (Bittles *et al.*, 2002). That said, in Iceland, where detailed multi-generational genealogical data are available and socioeconomic variation is limited, reproductive success was shown to be greatest at a level of parental consanguinity equivalent to third to fourth cousins ($F = 0.0039$–0.00098) (Helgason *et al.*, 2008).

Consanguinity and proximal and modifying determinants of fertility

A range of proximate determinants exerts a direct influence on fertility (Bongaarts *et al.*, 1984; Wood, 1989), including age at menarche and menopause, the proportion of women who are in a sexual union, sterility/fecundity, the frequency of sexual intercourse, contraceptive use, induced abortions, prenatal losses, lactational amenorrhoea and postpartum abstinence. By comparison, the cultural, environmental and socioeconomic factors discussed in Chapter 5 are mainly regarded as acting in a modifying manner.

An analysis of Muslim women recruited into the Indian National Family and Health Survey (NFHS) of 1992–1993 failed to show any evidence that contraceptive prevalence, postpartum abstinence from intercourse, induced abortion and frequency of intercourse were directly influenced by consanguinity (Bittles *et al.*, 2002). It is, however, possible that these variables were confounded by socioeconomic status and so they were indirectly associated with consanguineous marriage, e.g. the lower contraceptive prevalence reported by consanguineous couples in Pakistan may be due to a negative association between both of these variables and their socioeconomic status (Saha & Ali, 1992; Hussain & Bittles, 1999).

More detailed empirical information is available on the proportion of women in consanguineous and non-consanguineous unions. Thus, in Pakistan, maternal age at marriage was negatively associated with consanguinity, averaging 18.4 years for double first-cousin unions ($F = 0.125$), 18.8 years for first cousins ($F = 0.0625$), and 19.7 years for non-consanguineous couples (Bittles *et al.*, 1993). Further, a younger maternal age at first birth among consanguineous couples in South India (Bittles *et al.*, 1991) and a higher mean age of motherhood in Turkey (Tunçbilek & Koc, 1994) have been reported. On this basis, it appears that early marriage, the earlier commencement of reproduction, and maximization of the maternal reproductive span by consanguineous couples have been critical factors in determining family size in many less-developed countries (Bittles *et al.*, 1991; Bittles *et al.*, 1993).

Consanguinity and reproductive compensation

Reproductive compensation through the replacement of an infant dying at an early age provides another convincing explanation for the positive association between consanguinity and fertility (Philippe, 1974). Reproductive compensation may involve a conscious decision by parents to achieve their desired family size (Scrimshaw, 1978), and this appeared to be the case in comparative studies conducted in Colombia, Kenya, Pakistan and Sri Lanka (Richter & Adlakha, 1989). However, initiation of a further pregnancy closely following the death of a breast-fed infant could mainly be a biological phenomenon due to cessation of lactational amenorrhoea. In studies in Bangladesh, the death of an infant resulted in a significant reduction in the median birth interval which was attributed to the interruption of lactation, the earlier onset of post-partum ovulation, and the consequent vulnerability of females to conception (Chowdhury *et al.*, 1976).

Support for the concept of reproductive compensation with respect to consanguineous unions was reported from the island of Hirado, Japan, where increased mortality in the progeny of consanguineous couples was offset by their greater fertility (Schull *et al.*, 1970a; Schull & Neel, 1972). Similarly, in Pakistan, greater numbers of live births have been reported in families where childhood deaths have occurred, especially when the children who died were male (Rukanuddin, 1982; Yusuf & Rukanuddin, 1989). This relationship may

Table 7.4 *Family size and the probability of an affected birth to carrier parents of an autosomal recessive disorder*

No. of conceptions	Risk of an affected pregnancy (%)
1.	$1 - (0.75)^1 = 25$
2.	$1 - (0.75)^2 = 42$
3.	$1 - (0.75)^3 = 58$
4.	$1 - (0.75)^4 = 68$
5.	$1 - (0.75)^5 = 76$
6.	$1 - (0.75)^6 = 82$
8.	$1 - (0.75)^8 = 90$
10.	$1 - (0.75)^{10} = 94$

not be entirely straightforward in urban communities because it also has been shown in Pakistan that rather than reporting an exclusive son preference, couples preferred one or more sons and at least one surviving daughter (Hussain *et al.*, 2000), an outcome that perhaps coincidentally would facilitate the *watta satta* marriage arrangements discussed in Chapter 5.

From a genetic perspective, the overall effect would be the same whether a primarily social or biological mechanism was involved, with an effective reduction in the rate at which deleterious genes were eliminated from the gene pool (Schull & Neel, 1972; Overall *et al.*, 2002). However, the relationship among consanguinity, fertility and reproductive compensation is complicated because, as shown in Table 7.4, the greater the number of children conceived by parents who are carriers of a detrimental autosomal recessive allele, the higher the mathematical expectation that at least one of their offspring will be affected and may die in early childhood.

The number of surviving children may determine whether a couple opts for reproductive compensation (Rahman, 1998), and an additional contributory factor is whether or not there is spare reproductive capacity within a particular population. Quite simply, if reproduction is at or close to its potential biological and social maxima, replacement births will be very difficult to identify. Thus, in the Hutterite Anabaptist community in South Dakota, no evidence of reproductive compensation was detected in the birth cohorts of the first two decades of the twentieth century when mean family sizes exceeded 10 children. But, as family sizes declined in the subsequent cohorts, compensation for early deaths was observed (Ober *et al.*, 1999).

The problem of undeclared contraception

In early investigations into a possible association between HLA haplotypes and fertility in the Hutterites, it had been assumed that the community prohibited contraception (Ober *et al.*, 1983; Ober *et al.*, 1988; Kostyu *et al.*, 1993; Dawson *et al.*, 1995). But, despite community reluctance to discuss the subject, it became obvious that the decreases in family sizes observed in the Hutterites across generations were due at least in part to increasing use of contraception (Ober *et al.*, 1999). Contraception was subsequently confirmed in a Dariusleut Hutterite colony, with sterilization and tubal ligation the most prevalent methods, but IUDs and oral contraceptives were also accessed by some women (White, 2002).

A significant decline in fertility had been observed among the Dariusleut Hutterites from the 1960s onwards, corresponding to the cohort of women born between 1931 and 1935 (Sato *et al.*, 1994). For example, the total fertility rates (TFR) in women aged 15–49 years decreased from 8.80–9.83 offspring in the years 1946–1954 to 8.13 in 1966–1970, 7.22 in 1971–1975, 6.39 in 1976–1980, and 6.29 in 1981–1985 (Nonaka *et al.*, 1994). The reduction in fertility was particularly marked for the age groups 35–39 years and older, with a $> 50\%$ decrease in the age-specific nuptial fertility rate between 1951–1955 and 1981–1985.

Because the possibility of contraceptive usage by some members of the community but not others was not considered in the earlier fertility studies, the conclusions drawn require careful re-evaluation. In addition to completed family sizes (Ober *et al.*, 1988), a reassessment of the reported HLA-associated odour preferences (Ober *et al.*, 1997) is required because, in some cases, the results may have been influenced by oral-contraceptive usage (Wedekind *et al.*, 1995; Roberts *et al.*, 2008).

Commentary

In assessing the impact of consanguinity on fertility, a range of non-genetic factors may operate, including earlier age at marriage, lower contraceptive uptake associated with disadvantaged socioeconomic status, and shorter birth intervals because of reduced postpartum lactational amenorrhoea following the early death of a child. When the effects of these various factors are adjusted at the multivariate level, there appear to be few differences in mean fertility levels at the levels of consanguinity most commonly contracted in human populations. This situation may change in future generations with, as an example, older maternal ages at marriage and the greater use of reliable methods of contraception combining to result in smaller initial family sizes and a reduced likelihood of reproductive compensation. However, the global epidemiological transition from a communicable to a non-communicable disease profile, and a concomitant improvement in the detection and diagnosis of lethal and sub-lethal genetic diseases, may act as an effective balancing mechanism by diminishing the genetic burden of inbreeding depression. Only time and continued empirical observation will reveal which, if any, of these possible scenarios is correct.

8 *Consanguinity and metrical traits at birth and in childhood*

Introduction

Despite the potential consequences in terms of selection, a limited number of studies have been conducted into the possible impact of consanguinity on continuous variables, such as body dimensions, psychometric testing and blood pressure. As will be indicated in the following sections, part of the difficulty in addressing these and similar topics lies in differentiating between what could reasonably be described as normal variation as opposed to pathological changes, whether the latter are of an inherited nature or are caused by external agencies including infections.

While anthropometric measurements are not so obviously subject to these problems, as will become apparent, the results of psychometric tests may be compromised by the homozygous expression of rare detrimental recessive alleles in a small proportion of cases, which in turn can negatively influence mean values and increase the level of variance. Because visual and hearing measurements may be even more commonly influenced by genetic and non-genetic factors, the relationship between consanguinity and a lack of visual acuity, or with hearing loss will be comprehensively considered in Chapter 10.

Consanguinity and body dimensions

Birth measurements

Physical dimensions at birth, especially birthweight, are an important predictor of infant mortality and childhood morbidity (McCormick, 1985; Beaty, 2007), and at the later end of the age spectrum, low birthweight also has been associated with an increased risk of major chronic diseases in mid to late adulthood (Margetts *et al.*, 2002). Therefore, early indications that infants born to consanguineous parents showed lower intra-uterine growth and on average were smaller at birth raised concerns as to their subsequent well-being. In fact, there is little evidence either to suggest a generalized increase in prenatal losses among first-cousin progeny (Table 7.2) or reduced fertility (Table 7.3),

112

as might reasonably be expected with reduced intra-uterine growth. However, the association between consanguinity and infant deaths (Table 9.2) may in part be indicative of reduced birth measurements in babies born to consanguineous parents and for that reason, comparative data on birthweight and other physical measures at birth have been compiled.

Little published information is available on an association between consanguinity and Apgar scores, i.e. measures of a newborn baby's basic physiological status taken at one and five minutes after birth. This omission is extremely puzzling, especially in countries and communities where consanguineous marriage is highly prevalent, because Apgar measures are simple to assess with no complex instrumentation required and they are routinely recorded in most hospital-based deliveries. Given the absence of Apgar scores, it is not surprising that studies on consanguinity and birth measurements also have been few in number and they were mostly undertaken some 30 to 40 years ago. Nineteen studies are available for study (Table 8.1), although in a majority of cases, birthweights only were recorded.

Gestational age was reduced among first-cousin progeny in just three of nine studies. Significant decreases in birthweight were reported in just over half (10/19) of the studies, and the picture was similarly mixed for other birth measures, including recumbent length, head circumference and chest circumference. Many studies had been conducted on small numbers of babies, which may have influenced the results of the statistical analyses. But even when a significant decrease was observed in these measures, authors often noted that the mean differences in body sizes between babies born to consanguineous and non-consanguineous parents were slight (Morton, 1958; Schork, 1962).

To an extent, the inability to demonstrate a significant adverse effect of consanguinity is unsurprising, given the fact that birthweight and other associated measures are not simply dependent on the expression of fetal genes but additionally involve both maternal genes and many external factors (Beaty, 2007). For example, in Pakistan, gestational age was initially found to exert the largest effect on the incidence of birthweights of < 2500 g, but after controlling for gestational age, significant associations with low birthweight were also demonstrated for maternal age, consanguinity, ethnicity, maternal anaemia, and a poor pregnancy history (Badshah *et al.*, 2008).

An understanding of the complex inter-relationships that result in the new-born phenotype was originally gained through investigations based on familial trends in birthweight within and across generations (Ounsted, 1974; Johnstone & Inglis, 1974; Klebanoff *et al.*, 1984; Magnus *et al.*, 1993). Whereas twin studies have suggested that most of the intra- and inter-familial variation in birthweight, birth length and head circumference could be ascribed to the expression of fetal genes (Magnus, 1984; Lunde *et al.*, 2007), it also has been

Table 8.1 *Comparison of mean birth measurements in first-cousin versus non-consanguineous progeny*

Study populations	Gestational age	Birthweight	Length	Head circum.	Chest circum.	Triceps thickness	Subscapular thickness	Authors
Japan								
Hiroshima, Nagasaki & Kure	↔	→	→	↔	→			Morton (1958)
USA								
Chicago	↔							Slatis & Hoene (1961)
Kuwait								
Kuwait	→							El-Alfi et al. (1969)
India								
Tamil Nadu	→	↔	↔	↔	↔			Sibert et al. (1979)
India								
Tamil Nadu, rural	↔	↔	↔	↔	↔			Rao & Inbaraj (1980)
Tamil Nadu, urban	↔	↔	↔	↔	↔			
India								
Tamil Nadu	↔	↔	↔	↔	↔			Asha Bai & John (1982)
Iraq								
Basrah	→							Ramankutty et al. (1983)
Norway								
All-Norway	→	→	→					Magnus et al. (1985)
UK								
Pakistani	↔	↔	↔	↔				Honeyman et al. (1987)

Region	Location					Reference
Lebanon	Beirut	↔	↔	↔	↔	Khlat (1989)
Saudi Arabia	Riyadh	↔	↔	↔	↔	Saedi-Wong & al-Frayh (1989)
Saudi Arabia	Riyadh	↔				Wong & Anokute (1990)
India	Karnataka	→	→			Kulkarni & Kurian (1990)
Pakistan	Punjab	→	→	→	→	Shami *et al.* (1991)
Saudi Arabia	Riyadh	→				Al Eissa *et al.* (1991)
Turkey	Eskisehir	↔				Başaran *et al.* (1994)
Israel	Muslim	→				Jaber *et al.* (1997a)
India	Aligarh	→	→			Badaruddoza *et al.* (1998)
Saudi Arabia		↔				al-Abdulkareem & Ballal (1998)
Lebanon	National	→				Mumtaz *et al.* (2007)

↑ Significant increase
↓ Significant decrease
↔ No significant effect

claimed that the overall genetic contribution to birthweight is probably small (Langhoff-Roos *et al.*, 1987; Carr-Hill *et al.*, 1987).

In lower-income countries, where mothers themselves often are much smaller in body dimensions and factors such as parity and their nutritional status can significantly influence intra-uterine growth rates, non-genetic variables may exert a proportionately greater influence on birth measurements. Nevertheless, after controlling for medical and sociodemographic variables, a significant negative association between consanguinity and birthweight was demonstrated at each gestational age in Lebanon (Mumtaz *et al.*, 2007). Although from a genetic perspective, the lack of difference in the observed mean decreases in birthweight between babies born to first-cousin ($F = 0.0625$) and second-cousin parents ($F = 0.0156$) in the study is difficult to explain.

The importance of rigorous control for medical and sociodemographic variables was apparent in neighbouring Jordan, where univariate analysis initially indicated a highly significant negative association between consanguinity and low birthweight. But this relationship disappeared when control for age, body mass index, occupation, education, smoking, gravidity, parity, medical problems during pregnancy, and a family history of premature deliveries was introduced (Obeidat *et al.*, 2010). The combined results therefore suggest that although a negative association between consanguinity and birth measurements may exist, the effect is modest in extent and probably would not be detected in most small-scale and less well-controlled studies.

Anthropometric measurements in childhood

Even fewer studies have been reported with respect to the influence of consanguinity on anthropometric measurements in childhood, possibly because in this age group, any effects that may be observed are less likely to be associated with a major disadvantageous influence on health and survival. In the comprehensive post-war studies conducted in the Japanese cities of Hiroshima and Nagasaki, there was a very slight consanguinity-associated reduction in weight gain and in head circumference and chest girth during the first nine months of life (Schork, 1964). But the effects were much less pronounced than those due to a wide range of non-genetic variables, including parental age, residence, sex, month of birth and parity. By an average age of ten years, extended studies in Hiroshima and Nagasaki indicated that in terms of physical dimensions, measured as weight, height, head circumference, chest girth, calf girth, head length, head breadth, head height, sitting height and knee height, first-cousin progeny were on average 0.2–4.7% smaller than their non-consanguineous peers according to the measure assessed, but consanguinity had no statistically significant effect on these mainly skeletal measurements (Schull & Neel, 1965).

Subsequent anthropometric studies on children aged 13–16 years on the island of Hirado, Japan, again failed to show any significant effect of consanguinity on standing height, sitting height, span, head width, head length, head circumference and calf circumference (Neel *et al.*, 1970a). However, among primary schoolchildren in Shizuoka, Japan, a significant consanguinity-associated reduction was observed in stature, weight, biacromial breadth, chest circumference, sitting height, head length, head breadth and head circumference in girls but not in boys (Ito, 1972), and there was no effect on auricular height or skeletal age (Sugiura, 1972; Komai & Tanaka, 1972).

In other countries, no significant reduction in childhood weight or height was found in consanguineous progeny in São Paulo, north-eastern Brazil (Krieger, 1969), whereas a modest reduction in height was reported at first-cousin level among primary-grade schoolchildren in the southern Brazilian state of Paraná (Chautard-Freire-Maia *et al.*, 1983). A reduction also was reported with consanguinity in the overall stature of Egyptian pre-schoolchildren (Abolfotouh *et al.*, 1990), and slight consanguinity-associated declines were observed in the mean values for stature, arm span, sitting height, head length, head circumference, chest girth and calf circumference both of Muslim boys aged 11–16 years in Aligarh, North India (Krishnan, 1986) and in boys and girls aged 6–14 years from the same city (Badaruddoza, 2004a). As with birth measurements, the observed anthropometric effects in each of these studies were limited in scale, despite the substantial numbers of children who were investigated.

Considered collectively, the data suggest a probable but very modest negative effect of consanguinity on growth in childhood. Detailed studies based on large numbers of subjects and incorporating demonstrable rigorous control for socioeconomic and nutritional variation would, however, be needed to confirm the findings reported to date.

Consanguinity, vision and hearing

Whereas there is a substantial literature on inherited disorders of sight and hearing, few studies have been reported on eye and ear accommodation or visual and hearing acuity in individuals without significant eye disease or hearing impairment. These properties were examined on the island of Hirado with schoolchildren examined for visual acuity and accommodation, and auditory acuity. The results indicated a lower mean level of visual acuity in the children of consanguineous parents but the differences did not approach statistical significance, and the influence of consanguinity on visual accommodation also was negligible. Similarly, there was no indication that the mean hearing acuity measured by audiometry, which classified hearing loss by 5-decibel (dB) intervals, was significantly affected by consanguinity (Neel *et al.*, 1970b). If underlying adverse consanguinity effects on sight and hearing did exist, they were small in their effect, and the numbers tested precluded the attainment of statistical significance. A very different picture, however, emerges in Chapter 10 when consanguinity-associated pathological changes in vision and hearing during infancy and childhood are assessed.

Dental characteristics

A majority of children who show numerical anomalies in their primary teeth also exhibit an increased likelihood of anomalies in the corresponding permanent teeth (Whittington & Durward, 1996). However, the mode of inheritance of conditions such as hypodontia (i.e. the congenital absence of one or more teeth) appears to be consistent with an autosomal dominant mode of inheritance with incomplete penetrance (Burzynski & Escobar, 1983; Jursić & Skrinjarić, 1988) and therefore would not be expected to be significantly influenced by consanguinity.

The Hiroshima and Nagasaki studies of the influence of consanguinity on the development and health of Japanese schoolchildren included an investigation of dental occlusion (i.e. the manner in which the upper and lower jaws meet either at rest or during mastication) and dental maturation, in both cases undertaken by an expert in dentition (Schull & Neel, 1965). Significantly lower levels of normal occlusion were found in male and female schoolchildren born to first-cousin parents in both cities, strongly suggesting that consanguinity adversely affected dental occlusion. As noted by the authors, orthodontic correction was virtually unknown in Japan at the time of the study; therefore, there was little chance that the primary dental picture would have been obscured by prior dental treatment. By comparison, there was no evidence of a consanguinity effect with respect to dental maturation, with similar results obtained in both cities for children of both sexes.

More recent studies conducted on isolated village communities in the Croatian Adriatic island of Hvar also reported an association between consanguinity and dental malocclusion in 224 children aged 7 to 14 years, 8.0% of whom had received orthodontic treatment. In the children of endogamous or consanguineous parents, there was a higher proportion of overjet (i.e. extension of the incisal or buccal cusp ridges of the upper teeth to the incisal margins and ridges of the lower teeth when the jaws are closed normally); overbite, which is an extension of the upper incisor teeth over the lower teeth vertically when the opposing posterior teeth are in contact; and vertical bite. Little or no effect, however, was observed in terms of crowding and/or the spacing of teeth (Lauc, 2003; Lauc *et al.*, 2003). From these changes, it was concluded that the genetic basis of some of the observed occlusal traits was polygenic, although significantly influenced by the expression of rare, recessive gene variants.

Psychometric measures

An early small-scale study in the USA of the influence of consanguinity on intellectual performance reported no significant effect with respect to mean intelligence quotient (IQ), with scores of 101.5 in first cousins versus 104.1 in non-consanguineous progeny. But the first-cousin progeny had a significantly higher standard deviation in their IQ scores due to the inclusion of two children, one child with an IQ score of 57 and the other of 154 (Slatis & Hoene, 1961). In the former case, it seems probable that the low IQ was indicative of some form of intellectual disability, whether genetic or non-genetic in origin. With respect to the latter child, it would be difficult to construct a convincing case that consanguinity had been intellectually detrimental.

Psychometric testing was included in a number of much larger studies into the possible effects of consanguinity on intellectual achievement. After allowance for differential socioeconomic status, a range of neuromuscular tests administered to children at an average age of ten years in Hiroshima and Nagasaki indicated a decline in the performance of first-cousin progeny ranging from 0.20–0.68% by comparison with non-consanguineous children (Schull & Neel, 1965). In terms of school performance, the first-cousin progeny were 0.09–0.13% less successful than their non-consanguineous counterparts according to the subject assessed, whereas WISC (Weschler Intelligence Scale of Children) psychometric tests indicated a mean reduction in verbal scores of 2.76% and in performance scores of 2.06% (Schull & Neel, 1965).

On the island of Hirado, no significant effect of consanguinity on tapping rate, IQ or school performance was detected, as measured by the Tanaka-Binet Test Method, which was used extensively at that time in Japanese schools. The findings possibly were due to the smaller numbers enrolled than in Hiroshima and Nagasaki, but they also reflected the difficulty of adequately controlling the many non-genetic variables that potentially could influence students' performance in such tests (Neel *et al.*, 1970a). Once again, the Hirado data contrasted with the larger Shizuoka studies which did report a negative association between increasing levels of consanguinity and both school performance and IQ, also measured by the Tanaka-Binet Test (Nakazawa *et al.*, 1972). However, as with the study on childhood anthropometric measurements, the Shizuoka data attained statistical significance only with girls and not boys (Kudo *et al.*, 1972).

Studies conducted in North India among Muslim children aged 12–15 years (Agarwal *et al.*, 1984), 9–12 years (Afzal, 1988), 6–11 years (Badaruddoza & Afzal, 1993), and 8–12 years (Pandey *et al.*, 1994) all reported lower mean IQ scores in first-cousin offspring, with similar findings in a group of 6- to 14-year-old children who had a mean coefficient of inbreeding (α) of 0.0460 (Badaruddoza, 2004a, 2004b). By comparison with the test protocols adopted by Schull, Neel and their co-workers in Japan to closely match consanguineous and non-consanguineous groups with respect to socioeconomic status and other variables, in four of the five Indian studies the attempts made to ensure comparability of the consanguineous and non-consanguineous subjects were quite rudimentary. This methodological shortcoming had previously been strongly criticized when other studies into the claimed negative impact of consanguinity on intelligence were reviewed (Kamin, 1980).

Probably the most convincing investigation into the adverse influence of consanguinity on cognitive performance was conducted in the Israeli Arab community, with children born to double first-cousin parents, first-cousin parents and non-consanguineous parents compared (Bashi, 1977). The performance of the children was measured by three mental ability and four achievement tests in school grades 4 and 6. At both stages in their school education, the highest mean scores were obtained by the children of non-consanguineous parents and the lowest by the children of double first cousins ($F = 0.125$), suggesting a negative genetic influence associated with consanguinity.

As in the US study by Slatis & Hoene (1961) on first-cousin progeny, the group of children born to double first-cousin parents showed higher variance in some of their tests (Bashi, 1977), which would be consistent with the homozygous expression of specific recessive alleles that include mild to moderate intellectual disability in their phenotype. With this possibility in mind, it has been proposed that the secular rise in IQ of approximately three IQ points per decade over recent decades (Flynn, 1987) may reflect not only favourable environmental changes but also an increase in heterosis following the reduction in consanguineous marriage in many countries during the course of the twentieth century (Mingroni, 2004).

Although certainly feasible, the small decreases that have been reported in the mean IQ scores of first-cousin progeny, and the often imperfect control for non-genetic variables, suggest that all such hypotheses need to be addressed with caution. Support for this contention was provided in a study that examined the negative association between mean consanguineous marriage and mean national IQ scores across 72 countries. When the data were subjected to multiple regression analysis, with control for an economic variable measured as gross domestic product (GDP) per capita and an education index, GDP was the most significant variable examined, whereas consanguinity had the least impact on mean IQ (Woodley, 2009).

A further important factor to be considered in assessing consanguinity and cognitive performance is the possible adverse effect of infectious disease on IQ scores. A recent review involving 184 countries demonstrated a zero-order correlation between average national IQ scores and parasite stress, which ranged from $r = -0.76$ to $r = -0.82$ and was significant at p < 0.0001 (Eppig *et al.*, 2010). Given the strong negative correlation between consanguinity and socioeconomic status in many societies (see Chapter 5), it would appear that in at least some cases the poorer performance by the children of consanguineous parents in IQ tests may reflect a trade-off in energetic terms between brain development and the metabolically expensive task of counteracting infection, to the detriment of their intellectual performance.

Consanguinity and blood pressure

Studies conducted on schoolchildren in Hiroshima and Nagasaki failed to detect any effect of consanguinity on pulse rate or diastolic blood pressure (Schull & Neel, 1965). Similar non-significant results were subsequently obtained on Hirado where systolic and diastolic blood pressure were measured (Neel *et al.*, 1970a), and for pulse rate, systolic and diastolic blood pressure among Brazilian schoolchildren (Chautard-Freire-Maia *et al.*, 1983). Conversely, modest increases in consanguinity-associated diastolic blood pressure had been reported among children in north-eastern Brazil (Krieger, 1969) and in North India (Badaruddoza, 2004a).

Higher heritability for systolic but not diastolic blood pressure was reported in consanguineous families in the South Indian state of Andhra Pradesh (Rice *et al.*, 1992). In Oman, the 9- to 10-year-old sons of hypertensive consanguineous parents showed significantly higher systolic and diastolic blood pressure levels than the offspring of normotensive first cousins, possibly caused by familial aggregation of blood-pressure determinants (Hassan *et al.*, 2001). Consanguinity was also associated with increased systolic and diastolic blood pressure in a stratified, random sample of male and female Kuwaiti schoolchildren aged 6–10 years (Saleh *et al.*, 2000).

As previously discussed in terms of comparative anthropometric and psychometric measurements on consanguineous and non-consanguineous infants and children, the results of all studies on metrical traits are critically dependent on the adequacy of control for non-genetic variables, which in practice is often more approximate than guaranteed. Nevertheless, in village-based studies on the Dalmatian islands, Croatia, a strong linear relationship was identified between the coefficient of inbreeding (F) and blood pressure, with an increase in the coefficient of inbreeding (F) of 0.01 calculated to result in increases of 3 mm in systolic and 2 mm in diastolic blood pressure. The study further estimated that recessive or partially recessive quantitative trait locus (QTL) alleles contributed to 10–15% of the total variation in blood pressure in the study population, with at least 300–600 QTL involved (Rudan *et al.*, 2003b). Given these numbers of contributory alleles, and multiple environmental and dietary influences, the inability of earlier smaller-scale studies to detect any statistically significant effect of consanguinity on mean blood-pressure measures is readily understandable.

Commentary

Although many of the studies of consanguinity and normal variation were undertaken up to 50 years ago, they remain unique in the comprehensive nature of their approach and in their quite sophisticated use of statistical methodology to control for biases that may be associated with factors such as age, sex and socioeconomic status. With respect to the anthropometric measurements, psychometric testing and blood-pressure variation, there was no indication that consanguinity exerted a significant influence on the test results. However, in a number of studies, there was a suggestion of a limited consanguinity-associated decline in the measurements, and the consistent results obtained with dental occlusions did indicate an adverse consanguinity-associated effect. On the basis of the findings, there appears to be little reason to implicate consanguinity per se as a major compromising factor in normal fetal, infant and childhood development. Although it remains possible that the increases in childhood and adult stature measured in many human populations during the past 50 years could in part reflect the breakdown of population isolates and a resultant increase in heterosis (Schreider, 1967; Wolański *et al.*, 1970; Altukhov *et al.*, 2000).

Plate 2.1 Table of Consanguinity, artist unknown, Paris, about 1170–80.
Reproduced by permission of the J. Paul Getty Museum, Los Angeles, CA,
Ms. Ludwig XIV 2, fol. 227v.

Plate 2.2 Table of Affinity, artist unknown, Paris, about 1170–80.
Reproduced by permission of the J. Paul Getty Museum, Los Angeles, CA,
Ms. Ludwig XIV 2, fol. 228.

Plate 12.1 Pharaoh Akhenaten
Source: Wikimedia Commons, photograph of a bust of the Pharaoh Akhenaten held
in the National Museum of Alexandria. Licensed under the Creative Commons
Attribution–Share Alike 3.0 Unported license.

Plate 12.2 Pharaoh Akhenaten in profile, illustrating his thin face, long nose, long
hanging chin and the forward curve of this neck.
Reproduced with permission of the Ägyptisches Museum und Papyrussammlung,
Staatliche Museen zu Berlin SPK, ÄM 14152.

Plate 12.3 A family shrine of the Pharaoh Akhenaten with his principal wife Nefertiti and their eldest three daughters, bathed in the rays of the Aten (sun disc). The royal couple are both portrayed with wide hips, thin legs and forward curved necks, in the artistic style typical of the Amarna period.
Reproduced with permission of the Ägyptisches Museum und Papyrussammlung, Staatliche Museen zu Berlin SPK, ÄM 14145.

9 Consanguinity and pre-reproductive mortality

Introduction

Given adequate sample sizes, almost all studies that have examined postnatal mortality and morbidity have confirmed that the progeny of consanguineous unions are to some degree disadvantaged in health terms. However, estimates of the overall adverse effects of consanguinity have been highly variable with, as previously indicated in Table 6.6, marked downward revisions through time that probably reflected better sampling techniques and improved control for non-genetic variables. Despite these changes, much of the clinical data on morbidity, and to a lesser extent on mortality, remain disappointingly weak. However regrettable and traumatic for relatives and the wider community, death represents a clearly defined end-point, and in critically assessing the relationship between consanguinity and health, a meta-analysis study of mortality at different early life-stages stages was first undertaken.

Meta-analysis of consanguinity and early deaths

For the meta-analyses, data were collated from a total of 64 studies conducted in 14 countries located across 4 continents (Tables 9.1– 9.3). As in Chapter 7, the acceptance of studies into the meta-analysis was contingent on three basic pre-conditions: (i) a post-1950s publication date to limit the study to contemporary populations; (ii) a minimum sample size of 750 subjects; and (iii) where feasible, a minimum of three consanguinity data points (e.g. first cousin, second cousin, and non-consanguineous). In some cases, however, the latter pre-condition was relaxed because the definition of a number of the reported consanguineous relationships was unacceptably vague to allow inclusion for analysis (e.g. 'closer than first cousin' or 'beyond second cousin').

The total data set contained information on some 5.0 million births and included the progeny of six categories of consanguineous parentage: uncle–niece/double first cousin ($F = 0.125$); first cousin ($F = 0.0625$); first cousin once removed ($F = 0.0313$); second cousin ($F = 0.0156$); beyond second cousin ($F < 0.0156$); and non-consanguineous ($F = 0$). The levels of consanguinity

Table 9.1 *Mean proportions of stillbirths by consanguineous marriage and study population*

Study populations	Coefficient of inbreeding (F)	0.125	0.0625	0.0313	0.0156	0	Authors
	Japan						
1.	Kure		0.016	0.009	0.029	0.019	Schull (1958)
2.	Hiroshima		0.024	0.014	0.031	0.020	Neel & Schull (1962)
3.	Nagasaki		0.019	0.009	0.017	0.019	Neel & Schull (1962)
4.	Hirado	0.055	0.034	0.046	0.029	0.036	Schull *et al.* (1970b)
5.	Shizuoka, rural		0.007	0.031	0.006	0.016	Tanaka (1973)
6.	Shizuoka, peri-urban		0.009	0.017	0.003	0.007	Tanaka (1973)
7.	Shizuoka, urban		0.004	0.000	0.000	0.009	Tanaka (1973)
	India						
8.	Udaipur		0.023	0.009	0.018	0	Basu (1978)
9.	Tamil Nadu, rural	0.019	0.020		0.031	0.019	Rao & Inbaraj (1977b)
10.	Tamil Nadu, urban	0.011	0.030		0.027	0.023	Rao & Inbaraj (1977b)
11.	Tamil Nadu	0.015	0.003	0		0.021	Asha Bai *et al.* (1981)
12.	Andhra Pradesh	0	0.008	0		0.007	Chengal Reddy (1983)
13.	Pondicherry	0.042	0.036	0.021		0.037	Jain *et al.* (1993)
14.	Andhra Pradesh		0.018			0.018	Yasmin *et al.* (1997)
	Pakistan						
15.	Lahore	0.195	0.083	0.043	0.048	0.037	Shami & Zahida (1982)
16.	Sheikhupura		0.026	0.013		0.010	Shami & Iqbal (1983)
17.	Gujrat	0.025	0.030	0.027	0.044	0.019	Shami & Hussain (1984)
18.	Jhelum	0.250	0.059	0.108	0	0.034	Shami & Minhas (1984)
19.	Rawalpindi	0	0.042	0.027	0.083	0.026	Shami & Siddiqui (1984)
20.	Gujranwala		0.046	0.062		0.035	Bittles *et al.* (1993)
21.	Sahiwal	0.065	0.036	0.042	0	0.020	Bittles *et al.* (1993)
22.	Faisalabad		0.024	0.017		0.038	Bittles *et al.* (1993)
23.	Sialkot	0	0.008	0.009	0	0.002	Bittles *et al.* (1993)
	France						
24.	Morbihan		0.054	0.013	0.052	0.021	Sutter & Tabah (1953)
25.	Loir-et-Cher		0.028	0.029	0.021	0.013	Sutter & Tabah 1953

Table 9.1 (*cont.*)

Study populations	Coefficient of inbreeding (F)	0.125	0.0625	0.0313	0.0156	0	Authors
Norway							
26.	Norwegian		0.017			0.012	Stoltenberg et al. (1998)
27.	Pakistani		0.012			0.012	Stoltenberg et al. (1998)
Israel							
28.	Arab, Muslim		0.015			0.009	Jaber et al. (1997a)
Egypt							
29.	Alexandria		0.096			0.079	Abul-Einem & Toppozada (1966)
Kuwait							
30.	Kuwait	0.008	0.021			0.023	Al-Awadi et al. (1986)
UAE							
31.	Al Ain	0.014	0.014	0.011	0	0.014	Abdulrazzaq et al. (1997)
Saudi Arabia							
32.	Riyadh		0.015	0.021		0.016	al Husain & al Bunyan (1997)
33.	Damman	0.035	0.036	0.045	0.005	0.023	al-Abdulkareem & Ballal (1998)
Turkey							
34.	Ankara		0.033			0.009	Başaran et al. (1989)
35.	Diyarbakir		0.014			0.011	Başaran et al. (1989)
36.	Eskisehir		0.033			0.029	Başaran et al. (1989)
37.	Antalya		0.027			0.029	Alper et al. (2004)
38.	Antalya		0.014			0.008	Güz et al. (1989)
Brazil							
39.	Minas Gerais, Black		0.071	0.087	0.125	0.038	Freire-Maia et al. (1963)
40.	Minas Gerais, Mixed	0.067	0.020	0.053	0.045	0.040	Freire-Maia et al. (1963)
41.	Minas Gerais, White		0.064		0.029	0.038	Freire-Maia et al. (1963)
42.	Minas Gerais, Black		0.047	0.039	0.047	0.046	Freire-Maia (1963)
43.	Minas Gerais, Mixed		0.050		0.008	0.033	Freire-Maia (1963)
44.	Minas Gerais, White	0.037	0.011	0.058	0.022	0.026	Freire-Maia (1963)
45.	Saõ Paulo, Japanese		0.037			0.027	Freire-Maia et al. (1964)
Sudan							
46.	Khartoum		0.055			0.056	Saha et al. (1990)

recorded refer to a single generation only, with no allowance made for consanguineous unions that may have been contracted in preceding generations or for multiple consanguineous relationships.

To complement the data on prenatal losses summarized in Table 7.2, deaths were assessed at three definable life-stages: stillbirths, deaths in the first year of life, and all deaths from approximately the 28th week of pregnancy to 10–12 years of age. Additional information was available on deaths at other life-stages; e.g. perinatal mortality, neonatal and post-neonatal deaths, deaths from years one to four, and <five-year mortality. However, at all of these ages, the information was restricted to small numbers of studies which were of variable reproducibility.

As described in Chapter 7, dichotomous comparisons were drawn between mean deaths among the progeny of specific consanguinity categories and the progeny of non-consanguineous parents, with a non-parametric sign test for symmetry applied to test for statistical significance on the assumption that there would be equal numbers of positive and negative associations (Lentner, 1982). To quantify the difference between stillbirths in the progeny of first cousins versus non-consanguineous couples, the comparative data were plotted as an unweighted linear regression according to the standard equation $y = a + bx$.

Consanguinity and stillbirths

Data were available for 46 studies on stillbirths, collected in 13 countries across 4 continents, with a total sample size of 1.87 million (Table 9.1). According to the non-parametric sign test, there was no significant association between consanguinity and stillbirths at $F = 0.0156$, $F = 0.0313$ or $F = 0.125$, but there was a positive difference in 29/46 studies at $F = 0.0625$ (p < 0.02). As shown in Figure 9.1, mean stillbirths were 0.5% higher in the progeny of first cousins than non-consanguineous couples ($r^2 = 0.58$, p $< 10^{-5}$). When considered in combination with the results obtained with the non-parametric sign tests, it appears that any positive association between consanguinity and stillbirths is quite restricted in scale. Further, in some studies there may have been bias in the recording of deaths, with the possibility that a proportion of severely affected neonates who died immediately after birth had been erroneously reported as stillborn (Dorsten *et al.*, 1999).

Consanguinity and infant deaths

A similar analytical procedure was followed for deaths in infancy, with data available for 1.81 million subjects enrolled in 45 studies conducted in

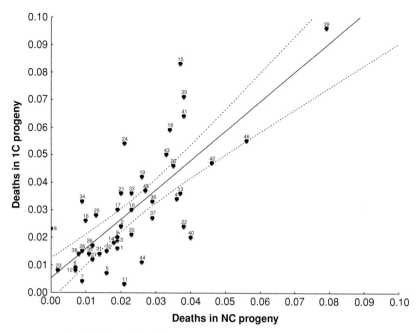

Figure 9.1 Proportion of stillbirths in first-cousin (1C; $F = 0.0625$: *y*-axis) versus non-consanguineous marriages (NC; $F = 0$: *x*-axis) in 46 study populations

12 countries across 3 continents (Table 9.2). The non-parametric sign test indicated a positive association between consanguinity and infant deaths in 15/17 studies at $F = 0.125$ (p < 0.01), 41/45 studies at $F = 0.0625$ (p < 0.0001), and 16/20 studies at $F = 0.0156$ (p < 0.02), but a non-significant result for the 16/22 studies at $F = 0.0313$.

The regression plot of infant deaths in first-cousin versus non-consanguineous progeny is shown in Figure 9.2. Mean infant deaths were 1.3% higher in the progeny of first cousins than those born to non-consanguineous couples ($r^2 = 0.75$, p < 10^{-5}), confirming that in this quantitative measure, a significantly higher numbers of deaths had occurred in infancy in the consanguineous offspring at $F = 0.0625$.

Consanguinity and total pre-reproductive mortality

Data on pre-reproductive mortality from approximately 28 weeks gestation to a mean of 10–12 years of age were collated from 64 studies conducted in 13 countries across 4 continents, with 2.37 million subjects studied

Table 9.2 *Mean proportions of infant deaths by consanguineous marriage and study population*

Study populations	Coefficient of inbreeding (F)	0.125	0.0625	0.0313	0.0156	0	Authors
	Japan						
1.	Hiroshima		0.063	0.043	0.053	0.045	Neel & Schull (1962)
2.	Nagasaki		0.067	0.020	0.073	0.052	Neel & Schull (1962)
3.	Fukuoka		0.053	0.053	0.030	0.048	Yamaguchi *et al.* (1970)
4.	Shizuoka, rural		0.065	0.117	0.082	0.062	Tanaka (1973)
5.	Shizuoka, peri-urban		0.064	0.063	0.077	0.050	Tanaka (1973)
6.	Shizuoka, urban		0.060	0.047	0.015	0.040	Tanaka (1973)
	India						
7.	Udaipur		0.173	0.192	0.205	0.097	Basu (1978)
8.	Tamil Nadu	0.069	0.133			0.084	Jacob John & Jayabal (1971)
9.	Tamil Nadu, rural	0.102	0.113			0.094	Rao & Inbaraj (1977b)
10.	Tamil Nadu, urban	0.133	0.102			0.093	Rao & Inbaraj (1977b)
11.	Tamil Nadu, rural	0.139	0.137		0.128	0.114	Rao & Inbaraj (1979)
12.	Tamil Nadu, urban	0.168	0.171		0.152	0.146	Rao & Inbaraj (1979)
13.	Tamil Nadu	0.031	0.025	0.023		0.019	Asha Bai *et al.* (1981)
14.	Andhra Pradesh	0.057	0.053	0		0.044	Chengal Reddy (1983)
15.	Tamil Nadu	0.160	0.109			0.093	Sivakumaran & Karthikeyan (1997)
16.	Andhra Pradesh		0.128			0.134	Yasmin *et al.* (1997)
	Pakistan						
17.	Lahore	0.140	0.097	0.095	0.095	0.040	Shami & Zahida (1982)
18.	Sheikhupura		0.106	0.096		0.089	Shami & Iqbal (1983)
19.	Gujrat	0.179	0.107	0.166		0.076	Shami & Hussain (1984)
20.	Jhelum	0.200	0.054	0.046		0.023	Shami & Minhas (1984)
21.	Rawalpindi		0.075	0.108	0.273	0.056	Shami & Siddiqui (1984)
22.	All-Pakistan, rural		0.114		0.093	0.083	Ahmed *et al.* (1992)
23.	All-Pakistan, urban		0.086		0.075	0.064	Ahmed *et al.* (1992)
24.	Gujranwala		0.082	0.067		0.058	Bittles *et al.* (1993)
25.	Sahiwal	0.099	0.097	0.122	0.071	0.070	Bittles *et al.* (1993)

Table 9.2 (*cont.*)

Study populations	Coefficient of inbreeding (*F*)	0.125	0.0625	0.0313	0.0156	0	Authors
26.	Faisalabad		0.090	0.084		0.083	Bittles *et al.* (1993)
27.	Sialkot	0.174	0.096	0.078	0.125	0.081	Bittles *et al.* (1993)
28.	Lahore, rural		0.105			0.070	Yaqoob *et al.* (1998)
29.	Lahore, peri-urban		0.208			0.109	Yaqoob *et al.* (1998)
30.	Lahore, urban		0.107			0.041	Yaqoob *et al.* (1998)
	France						
31.	Morbihan		0.099	0.065	0.065	0.054	Sutter & Tabah (1953)
32.	Loir-et-Cher		0.069	0.029	0.051	0.036	Sutter & Tabah (1953)
	Norway						
33.	Norwegian		0.031			0.010	Stoltenberg *et al.* (1998)
34.	Pakistani		0.018			0.009	Stoltenberg *et al.* (1998)
	Israel						
35.	Arab, Muslim		0.063			0.013	Jaber *et al.* (1997a)
	Egypt						
36.	Alexandria		0.007			0.008	Abul-Einem & Toppozada (1966)
	Kuwait						
37.	Kuwait	0.145	0.033			0.025	Al-Awadi *et al.* (1986)
	Saudi Arabia						
38.	Riyadh		0.054			0.044	Wong & Anokute (1990)
39.	Riyadh	0.057	0.011		0.010	0.012	al Husain & al Bunyan (1997)
40.	Damman	0.029	0.029	0.030	0.030	0.036	al-Abdulkareem & Ballal (1998)
	Jordan						
41.	National		0.071			0.049	Khoury & Massad (2000)
	Turkey						
42.	All-Turkey		0.088		0.078	0.074	Tunçbílek & Koc (1994)
43.	Konya		0.052			0.038	Demirel *et al.* (1997)
44.	Antalya		0.067			0.038	Güz *et al.* (1989)
	Brazil						
45.	Parana	0.333	0.221	0.148		0.118	Freire-Maia *et al.* (1983)

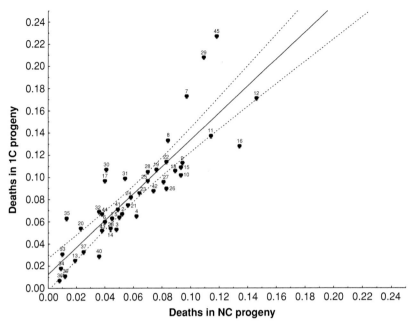

Figure 9.2 Proportion of infant mortality in first-cousin (1C; $F = 0.0625$: y-axis) versus non-consanguineous marriages (NC; $F = 0$: x-axis) in 45 study populations

(Table 9.3). Once again, the data were assessed by a non-parametric sign test with a positive association between consanguinity and infant deaths indicated for 11/12 studies at $F = 0.125$ (p < 0.02), 57/64 studies at $F = 0.0625$ (p < 0.0001), and 17/23 studies at $F = 0.0313$ (p < 0.05), but the results failed to attain significance for the 16/23 studies at $F = 0.0156$.

The comparative regression plot of mean pre-reproductive deaths in first-cousin versus non-consanguineous progeny showed a mean excess of 3.7% mortality in the consanguineous offspring ($r^2 = 0.84$, p $< 10^{-5}$), which lies between the earlier estimates of 4.4% excess mortality calculated on the basis of 38 studies (Bittles & Neel, 1994) and the 3.5% excess deaths at first-cousin level reported by Cavalli-Sforza *et al.* (2004) and Bittles & Black (2010a). The variation in mean excess mortality between these three studies is indicative of the powerful effect that a small number of even quite mildly outlier data sets can exert.

Overall assessment of the mortality findings

As discussed in Chapter 6, the variation in reported death rates partially reflects the individual survey designs, temporal and geographical differentials, and the

Table 9.3 *Mean proportions of pre-reproductive deaths by consanguineous marriage and study population*

Study populations	Coefficient of inbreeding (F)	0.125	0.0625	0.0313	0.0156	0	Authors
	Japan						
1.	Kure		0.095	0.045	0.125	0.087	Schull & Neel (1966)
2.	Hiroshima		0.114	0.090	0.108	0.082	Neel & Schull (1962)
3.	Nagasaki		0.104	0.056	0.112	0.094	Neel & Schull (1962)
4.	Hirado		0.122	0.115	0.091	0.143	Schull *et al.* (1970b)
5.	Fukuoka		0.093	0.066	0.089	0.062	Yamaguchi *et al.* (1970)
6.	Shizuoka, rural		0.147	0.215	0.131	0.129	Tanaka (1973)
7.	Shizuoka, peri-urban		0.147	0.173	0.139	0.109	Tanaka (1973)
8.	Shizuoka, urban		0.130	0.091	0.100	0.087	Tanaka (1973)
	India						
9.	Udaipur		0.249			0.147	Basu (1978)
10.	Tamil Nadu	0.213	0.240			0.205	Jacob John & Jayabal (1971)
11.	Tamil Nadu, rural		0.331			0.305	Rao & Inbaraj (1977b)
12.	Tamil Nadu, urban		0.323			0.300	Rao & Inbaraj (1977b)
13.	Tamil Nadu, rural		0.171			0.144	Rao & Inbaraj (1979b)
14.	Tamil Nadu, urban		0.224			0.207	Rao & Inbaraj (1979b)
15.	Tamil Nadu		0.082			0.102	Asha Bai *et al.* (1981)
16.	Andhra Pradesh		0.054			0.051	Chengal Reddy (1983)
17.	Karnataka, Hindu	0.043	0.044		0.044	0.037	Bittles *et al.* (1992)
18.	Karnataka, Muslim	0.036	0.034		0.047	0.035	Bittles *et al.* (1992)
19.	Karnataka, Christian	0.035	0.049		0.058	0.042	Bittles *et al.* (1992)
20.	Andhra Pradesh, tribal	0.283	0.274			0.271	Naidu *et al.* (1995)
21.	Tamil Nadu	0.464	0.297			0.268	Sivakumaran & Karthikeyan (1997)
22.	Andhra Pradesh		0.210			0.188	Yasmin *et al.* (1997)
	Pakistan						
23.	Lahore	0.420	0.300	0.220	0.210	0.200	Shami & Zahida (1982)
24.	Mianchannu		0.310			0.210	Shami (1983)
25.	Muridke		0.190	0.130		0.120	Shami (1983)
26.	Sheikhupura		0.190	0.140		0.130	Shami & Iqbal (1983)
27.	Gujrat	0.300	0.200	0.500	0.130	0.140	Shami & Hussain (1984)

(cont.)

Table 9.3 (*cont.*)

Study populations	Coefficient of inbreeding (*F*)	0.125	0.0625	0.0313	0.0156	0	Authors
28.	Jhelum	0.400	0.180	0.150	0	0.090	Shami & Minhas (1984)
29.	Rawalpindi	0.330	0.190	0.210	0.330	0.090	Shami & Siddiqui (1984)
30.	Gujranwala		0.197			0.146	Bittles *et al.* (1993)
31.	Sahiwal		0.196			0.174	Bittles *et al.* (1993)
32.	Faisalabad		0.164			0.159	Bittles *et al.* (1993)
33.	Sialkot		0.179			0.130	Bittles *et al.* (1993)
34.	Lahore, rural		0.105			0.070	Yaqoob *et al.* (1998)
35.	Lahore, peri-urban		0.208			0.109	Yaqoob *et al.* (1998)
36.	Lahore, urban		0.107			0.041	Yaqoob *et al.* (1998)
	France						
37.	Morbihan		0.148			0.074	Sutter & Tabah (1953)
38.	Loir-et-Cher		0.096			0.048	Sutter & Tabah (1953)
	Norway						
39.	Norwegian		0.047			0.021	Stoltenberg *et al.* (1998)
40.	Pakistani		0.030			0.021	Stoltenberg *et al.* (1998)
	Israel						
41.	Arab, Muslim		0.079			0.021	Jaber *et al.* (1997a)
	Egypt						
42.	Alexandria		0.102			0.086	Abul-Einem & Toppozada (1966)
	Lebanon						
43.	Beirut, Christian		0.300			0.288	Khlat (1988)
44.	Beirut, Muslim		0.244			0.253	Khlat (1988)
45.	Bekaa Valley		0.079			0.049	Joseph (2007)
	Kuwait						
46.	Kuwait		0.049			0.045	Al-Awadi *et al.* (1986)
	UAE						
	Saudi Arabia						
47.	Riyadh		0.286			0.218	Wong & Anokute (1990)
48.	Damman		0.150			0.134	al-Abdulkareem & Ballal (1998)
	Turkey						
49.	All-Turkey		0.252			0.281	Tunçbílek & Koc (1994)
50.	Antalya		0.143			0.080	Güz *et al.* (1989)
51.	Antalya		0.144			0.177	Alper *et al.* (2004)

Table 9.3 (*cont.*)

Study populations	Coefficient of inbreeding (F)	0.125	0.0625	0.0313	0.0156	0	Authors
52.	Konya		0.160			0.149	Demirel *et al.* (1997)
	Brazil						
53.	Indians	0.467	0.494	0.333	0.377	0.417	Salzano *et al.* (1962)
54.	Mestizos		0.385	0.800	0.364	0.333	Salzano *et al.* (1962)
55.	Minas Gerais, Black		0.474	0.435	0.375	0.306	Freire-Maia *et al.* (1963)
56.	Minas Gerais, Mixed		0.283	0.472	0.174	0.299	Freire-Maia *et al.* (1963)
57.	Minas Gerais, White		0.313	0.342	0.236	0.311	Freire-Maia *et al.* (1963)
58.	Minas Gerais, Black		0.460	0.450	0.372	0.312	Freire-Maia (1963)
59.	Minas Gerais, Mixed		0.368	0.307	0.263	0.325	Freire-Maia (1963)
60.	Minas Gerais, White		0.280	0.286	0.350	0.289	Freire-Maia (1963)
61.	Minas Gerais, Japanese		0.110			0.057	Freire-Maia & Takehara (1977)
62.	Sao Paulo, Japanese		0.186			0.135	Freire-Maia *et al.* (1964)
63.	Parana		0.227			0.118	Freire-Maia *et al.* (1983)
	Nigeria						
64.	Yoruba	0.200	0.253	0.182		0.097	Scott-Emuakpor (1974)

small numbers of deaths recorded in certain of the less frequently reported consanguinity categories. However, in these studies, control for non-genetic variables also is an ongoing problem and when comparing mortality rates in consanguineous and non-consanguineous couples, it is critical that, for example, differentials in socioeconomic status are carefully controlled.

Furthermore, because the lifestyles and socioeconomic circumstances of individual communities can differ significantly and alter across time, it is not sufficient to assume that the findings of a study conducted in a specific region or community will apply to all regions or communities in the same national population. Thus, in Japan, whereas first-cousin marriages were found to be associated with lower socioeconomic status in the cities of Hiroshima and Nagasaki (Schull & Neel, 1965) and in urban families on the island of Hirado, among the rural farming families on Hirado, the frequency of consanguineous marriage was positively correlated with economic status (Schull & Neel, 1972).

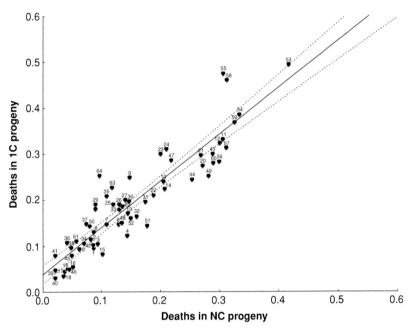

Figure 9.3 Comparative pre-reproductive mortality in first-cousin (1C; $F = 0.0625$: *y*-axis) versus non-consanguineous marriages (NC; $F = 0$: *x*-axis) in 64 study populations

To examine the comparative influence of socioeconomic status versus consanguineous parentage on infant mortality, current data from 11 Middle Eastern countries were collated, together with information on female literacy and female life expectancy as more general indicators of national social advancement. As illustrated in Table 9.4, no major positive or negative relationship was apparent between the reported prevalence of consanguineous marriages ($F \geq 0.0156$) in each country (range: 31.7% to 44.5%) and the widely variant infant mortality rates (range: 7/1000 to 51/1000).

By comparison, the inverse relationship between infant mortality and socioeconomic status, measured in terms of GNP per person, was strikingly obvious, ranging from 8/1000 infant deaths in Qatar with a mean GNP of US$179 000 to 51/1000 infant deaths in Yemen, which has an estimated GNP of just US$2330. The inverse relationships between infant mortality and female literacy, which likewise has a strong influence on childhood mortality (Bittles, 1994), and female life expectancy which is indicative both of problems arising in pregnancy and childbirth and the loss of maternal care experienced by other children in the family following a mother's death, were equally apparent.

Table 9.4 *Infant mortality by consanguinity, female literacy, female life expectancy, and income per person*

Country	Infant mortality (per 1000 births)	Consanguinity (%)	Female literacy (%)	Female life expectancy (year)	Income per person* (US$)
Qatar	8	44.5	88.6	78	179,000
Kuwait	9	34.3	91.0	75	52,610
Bahrain	7	31.8	83.6	78	33,690
Oman	12	35.9	73.5	76	24,530
Saudi Arabia	18	40.6	70.8	75	24,020
Lebanon	21	35.5	82.2	75	13,400
UAE	7	34.7	81.7	77	11,200
Jordan	23	39.7	84.7	74	5,730
Syria	17	31.7	73.6	76	4,620
Iraq	34	33.0	64.2	72	3,330
Yemen	51	33.9	30.0	66	2,330

* Income per person is expressed as Gross National Income (GNI) per capita converted to US$ using purchasing power parity (PPP) rates
Sources: Global Consanguinity Website www.consang.net; PRB (2011); The World Bank (2011); CIA (2011)

Commentary

With the probable exception of stillbirths, the data confirm a significant positive relationship between consanguineous parentage and increased mortality. However, the mean increase in infant mortality of 1.3% and a 3.7% increase in total pre-reproductive mortality at first-cousin level sit at the lower end of estimates reported in most earlier investigations. The ancillary data presented in Table 9.4 clearly illustrate the critical role of parental socioeconomic status in infant mortality, and any inefficiency in control for parental income would therefore lead to significantly biased upward estimates of the adverse effect of consanguinity on postnatal survival (Bittles, 2001; Bittles & Black, 2010b). Similar findings have been reported for other major demographic variables, including reduced birth interval, maternal illiteracy and young maternal age, each of which were shown to be more important independent determinants of infant and childhood mortality in Pakistan than first-cousin marriage (Grant & Bittles, 1997; Bittles, 2005).

10 Consanguinity and morbidity in early life

Introduction

On average, deaths in infancy are 1.3% higher in first-cousin progeny than in the progeny of non-consanguineous couples (Figure 9.2), and accompanying these findings there is a very extensive literature to indicate a significant increase in infant and childhood morbidity among the progeny of consanguineous parents. Any situation in which the expression of detrimental autosomal recessive genes is implicated will potentially be associated with parental consanguinity. Rather than attempting to compile an exhaustive summary of all such possibilities, the intention of this chapter is to provide an overview of the more prevalent causes of childhood morbidity in which an association with consanguinity has been demonstrated or appears probable. However, it should be emphasized that an observed association alone does not provide proof of a causal relationship (Sheehan *et al.*, 2008).

Single gene defects

The significant breakthrough in understanding the nature of many single gene disorders and their association with consanguinity came with the studies of Sir Archibald Garrod into alkaptonuria, a generally quite benign condition that is recognized by characteristic darkening of an individual's urine on exposure to the air, with homogentisic acid identified as the chemical in the urine causing this change (Garrod, 1902). Garrod further recognized that alkaptonuria had a familial distribution, and that approximately 60% of the cases he had investigated had been born to parents who were first cousins. At the same time, he was careful to note that: 'On the other hand, it is equally clear that only a minute proportion of the children of such unions are alkaptonuric'.

Citing William Bateson's recently published book *Mendel's Principles of Heredity* (Bateson, 1902), Garrod observed that the distribution of alkaptonuria within families appeared to mirror the recessive mode of inheritance originally described by Mendel in his experimental work on peas, and raised the possibility that the condition might represent an alternative mode of metabolism and not

136

be detrimental. Finally, in this remarkably insightful paper, Garrod queried whether albinism and cystinuria might similarly be ascribed to alternative modes of metabolism. In presenting a series of four Croonian Lectures to the Royal College of Physicians of London in July 1908, he further refined his ideas and developed the concept of inborn errors of metabolism, a term which continues in worldwide medical and scientific usage (Garrod, 1908).

Not all autosomal recessive disorders are categorized as inborn errors of metabolism but an association with consanguinity is a common finding (Tchen *et al.*, 1977), examples being the skin disorders xeroderma pigmentosa, Mal de Meleda and ichthyoses (Khatri *et al.*, 1999; Charfeddine *et al.*, 2003; Eckl *et al.*, 2003; Al-Zayir & Al-Amro Alakloby, 2004) and primary ciliary dyskinesia (Jeganathan *et al.*, 2004; O'Callaghan *et al.*, 2010). As previously illustrated in Table 6.2, according to the Hardy–Weinberg principle, the less common the incidence of the mutant gene causing a recessive disorder, the greater the probability that the parents of an affected individual are consanguineous, especially where substantial population stratification also exists.

The inheritance of β-thalassaemia major in Pakistan (Ahmed *et al.*, 2002) and South India (Bashyam *et al.*, 2004), both areas in which consanguineous marriage is strongly favoured (Figure 1.1), provide examples of this relationship with a majority of affected individuals homozygous for the causative mutation. The opposite situation is observed in North India where consanguineous marriage is avoided and, despite caste endogamy, most cases of β-thalassaemia are compound heterozygotes (Sinha *et al.*, 2009). The theoretical probability that an affected child will be born to parents who are carriers of a recessive disorder is, however, also dependent on the number of conceptions; e.g. increasing from 25% in a singleton family to 76% in families with five children (Table 7.4).

These quite basic numerical relationships are often overlooked when associations between consanguinity and rare recessive gene disorders are claimed, and it is also usually assumed that where parents are biological relatives, affected offspring will have inherited identical mutations from each parent. Both assumptions may be wrong, as illustrated by a Dutch family in which the parents were members of a religious isolate and multiply related as second cousins, third cousins, and third cousins once removed. Despite a calculated coefficient of inbreeding $F = 0.0215$ in two sons diagnosed with cystic fibrosis, both boys were found to be compound heterozygotes for the disease (ten Kate, *et al.*, 1991). Similarly, in a child diagnosed with both ocular albinism and neurofibromatosis 2 (NF2) born to UK Pakistani first-cousin parents, although both parents were carriers of ocular albinism, there had been no evidence of NF2 in the previous four generations of the family pedigree and the causative mutation almost certainly had arisen de novo (Buchan *et al.*, 2003).

Table 10.1 *Consanguinity and co-inheritance of recessive disorders*

UK (Jewish)	Retinitis pigmentosa and cystinuria	Brooks *et al.* (1949)
Israel (Arab)	Mucolipidosis III and Bardet-Biedl syndrome	Gordon *et al.* (1990)
Germany (Turkish)	Cystic fibrosis and Ehlers-Danlos syndrome type VI	Jarisch *et al.* (1998)
UK (South Asian)	Fabry disease and aspartyl-glucosaminuria	Guy *et al.* (2001)
UK (Arab)	Amelogenesis imperfecta and cone-rod dystrophy	Downey *et al.* (2002)
Brazil	Hereditary deafness (*GJB2* gene) and oculotaneous albinism	Lezirovitz *et al.* (2006)
Tunisia	Retinitis pigmentosa (ARRP) and Usher syndrome type II	Hmani-Aifa *et al.* (2009)

The co-expression of different disease alleles within individual families had previously been reported in the UK Pakistani community, e.g. involving β-thalassaemia and cystic fibrosis, and β-thalassaemia and maple syrup disease (Darr & Modell, 1988); and different branches of a large consanguineous Tunisian kindred were affected with either limb-girdle muscular dystrophy or congenital muscular dystrophy (Hadj Salem *et al.*, 2011). Other examples of this phenomenon of co-expressed recessive disorders which segregate independently in consanguineous families are shown in Table 10.1.

As illustrated in Table 10.2, a further complexity that increasingly has been reported in consanguineous kindreds is the co-existence of multiple mutations for a single disease phenotype (Zlotogora, 2007), which has led several groups of researchers to draw attention to the potentially misleading findings of homozygosity mapping in highly endogamous communities (Pannain *et al.*, 1999; Ducroq *et al.*, 2006; Laurier *et al.*, 2006; Frishberg *et al.*, 2007). It follows that in these families and communities, genetic counselling can become equivalently complex (Bittles & Hamamy, 2010).

Childhood deafness

A recent systematic search into the causes of permanent childhood hearing impairment (PCHI) indicated that 30.4% of cases were hereditary, 19.2% were acquired and the cause was unknown in 48.3% of affected persons (Korver *et al.*, 2011). From a genetic perspective, the aetiology of hearing loss is immensely complex, with more than 400 genetic syndromes and 140 genetic loci associated with non-syndromic hereditary hearing loss (NSHL) mapped to date, and more than 60 genes identified (Alford, 2011). The complexity of NSHL is increased by the fact that the modes of inheritance of hereditary hearing loss can be

Table 10.2 *The co-occurrence of multiple mutations encoding single gene disorders in consanguineous families and communities*

Country	Disorder	Mutations	Authors
USA	Tay-Sachs disease	Two different mutations in the Louisiana Cajun community	McDowell *et al.* (1992)
Israel	Metachromatic leucodystrophy	Five different mutations clustering in Arab communities in a single geographical locality	Heinisch *et al.* (1995)
USA	Congenital hypothyroidism	Two different mutations in the thyroid peroxidise gene of an Amish kindred	Pannain *et al.* (1999)
Israel	Primary hyperoxaluria type I	Seven different mutations in eight families, seven Arab and one Jewish-Libyan, with four novel mutations, and three mutations in single-clan villages	Rinat *et al.* (1999)
Israel	Familial Mediterranean fever	Four different FMF mutations in 13 individuals from a single highly consanguineous Arab family	Gershoni-Baruch *et al.* (2002)
Israel	β-thalassaemia	Three different mutations in a highly endogamous Arab village of 8600 persons	Zlotogora *et al.* (2005)
Israel	Cone-rod dystrophy	Three different *ABCA4* mutations in the same extended Christian Arab family, with multiple consanguineous loops	Ducroq *et al.* (2006)
Lebanon	Bardet-Biedel syndrome	Two mutant genes and three mutations in a single extended Sunni Muslim consanguineous family	Laurier *et al.* (2006)
Israel	Congenital nephrotic syndrome	Three novel mutations identified in 12 children from a highly consanguineous Arab Muslim village established ~250 years ago by a single founder family	Frishberg *et al.* (2007)
Turkey	Familial Mediterranean fever	Three common FMF mutations encoding compound heterozygotes in a single extended family with multiple consanguineous relationships	Seidel & Steinlein (2008)

(cont.)

Table 10.2 (*cont.*)

Country	Disorder	Mutations	Authors
Saudi Arabia	Alström syndrome	Four novel mutations identified in the *ALMS1* gene in five Saudi patients, illustrating the level of genetic heterogeneity even in this rare disorder	Aldahmesh *et al.* (2009a)
Lebanon	Familial Mediterranean fever	Five different FMF mutations and phenotypic variability in a single family with 31 affected individuals	Medlej-Hashim *et al.* (2011)

autosomal dominant, autosomal recessive, X-linked or mitochondrial, with differing ages of symptom onset, and variable penetrance adding to the phenotypic mix.

To date, the most widely reported cause of non-syndromic hereditary hearing loss is a mutation in the *GJB2* gene coding for connexin 26; however, other than testing for connexin 26 mutations, the extreme genetic heterogeneity of NHSL makes genetic diagnosis expensive and time-consuming. This situation can be ameliorated in communities where strict endogamy and a high prevalence of consanguineous marriage exists, since in many such situations a single founder mutation may be identified thereby simplifying subsequent prognostic and diagnostic testing. As an example, in a small Israeli Arab community, all persons with prelingual non-syndromic hearing impairment originated from the same village and bore the same family name. A c.406C>T (p.R136X) nonsense mutation was identified in the *DFNB59* gene of all affected individuals, and the carrier frequency of the mutation in the community as a whole was 8.2% (Borck *et al.*, 2011).

Massive parallel sequencing for causative NSHL mutations has been described in consanguineous and non-consanguineous societies (Shearer *et al.*, 2010; Brownstein *et al.*, 2011), and it has been suggested that in silico prioritization could usefully be applied to the putative identification of NSHL candidate genes (Accetturo *et al.*, 2010). Even within large consanguineous pedigrees there is, however, no guarantee that deafness will be caused by a single mutation, as in a Brazilian pedigree with 26 subjects affected by prelingual deafness (Lezirovitz *et al.*, 2008). Although 15 individuals were homozygous for the novel c.1057delA mutation in the *MYO15A* gene, five were compound heterozygotes for c.1057delA and the novel deletion c.9957_9960delTGAC, and although one person inherited a single copy of the mutant c.1057delA allele in *MYO15A*, the second mutation could not be identified, raising the possibility of a third causative locus for NSHL segregating in the family.

Given the large number of autosomal recessive gene loci known to be associated with hearing loss and the rarity of some of the disease alleles, it is unsurprising that consanguinity has been identified as a significant predisposing factor in studies conducted across many different countries and communities (Costeff & Dar, 1980; Kabarity *et al.*, 1981; Feinmesser *et al.*, 1989; Zakzouk *et al.*, 1993; Liu *et al.*, 1994; Zakzouk *et al.*, 1995; Al-Gazali, 1998; Dereköy, 2000; Shahin *et al.*, 2002; Egeli *et al.*, 2003; Maheshwari *et al.*, 2003; RamShankar *et al.*, 2003; Attias *et al.*, 2006; Olusanya & Okolo, 2006; Khabori & Patton, 2008; Sajjad *et al.*, 2008; Bajaj *et al.*, 2009; Mahdieh *et al.*, 2011; Musani *et al.*, 2011).

A majority of the causative mutations are homozygous rather than compound heterozygotes as in communities where consanguineous marriage is common, which has led some commentators to urge that consanguineous marriage is discouraged (Zakzouk, 2002). Although superficially understandable, given the social and economic benefits perceived to flow from consanguineous unions (Table 5.3), it is questionable whether local families would pay much attention to such calls, especially in communities such as the Al-Sayyid Bedouin in the Negev desert in Israel, who collectively have socially accommodated high levels of deafness in the community by developing their own indigenous sign language (Kisch, 2008).

Where marriages have been customarily contracted within small clans, tribes or equivalent groupings, all members of the marriage pool are genetically related to a significant extent. Under such circumstances, founder effect is probable, resulting in novel mutations that are unique to different sub-communities and which can rise to high frequency via genetic drift, as in the Moroccan Jewish community in Israel, where 34% of genetic deafness was caused by the TMC1 p.5647P allele inherited as a founder mutation (Brownstein *et al.*, 2011).

Although proscribing first- or second-cousin marriages may initially reduce the numbers of persons with congenital hearing loss, the only real solution would be to ban marriage within the clan or tribe, a solution that justifiably would be anathema within traditional communities. An alternative solution adopted in an isolated Israeli Arab community was that in recent generations, most women who were deaf did not marry (Zlotogora & Barges, 2003). Rather than resorting to such a socially and personally drastic approach, it would have been more efficacious to arrange routine hearing screening for infants at three months of age, so that cases of hearing loss could be quickly identified and subsequent language-intervention programmes implemented if requested (Bennett *et al.*, 2002).

Visual defects

As with hearing loss, a wide range of genetic and environmental factors have been implicated in visual defects, with the overall reported prevalence and types of defect largely dependent on the economic status and disease profile of the study population. Thus, during the 1950s, and early 1960s, measles was reported to be responsible for 81.3% of cases of blindness in pre-independence Zambia (Phillips, 1961), and in neighbouring Malawi measles was responsible for 43.7% of all childhood blindness during the 1970s (Chirambo & Benezra, 1976).

The causative role of genetic disorders in visual defect may frequently be overlooked against the background of high rates of blindness due to commom childhood infectious diseases. Genetic visual disorders can, however, rapidly become apparent once the predominant disease profile changes from a communicable to a non-communicable disease pattern, as in Saudi Arabia, where prior to 1962 acquired diseases led to blindness in 75% of patients, but from 1962 it was claimed that 84% of cases of childhood blindness were genetic in origin (Tabbara & Badr, 1985). In neighbouring Jordan, the reported change in the cause of blindness was in the same direction but less dramatic, from 67% genetic causes of childhood blindness prior to 1970 to 78% thereafter (Al-Salem & Rawashdeh, 1992).

Consanguinity has been identified as a contributory factor in cases of single gene visual defects (Tabbara & Badr, 1985; Al-Idrissi *et al.*, 1992; Al-Salem & Rawashdeh, 1992; Elder & De Cock, 1993; Rahi *et al.*, 1995), with the autosomal recessive form of retinitis pigmentosa (RP) a common finding in many different communities with high rates of consanguineous unions (Baghdassarian & Tabbara, 1975; Bundey & Crews, 1984; van Soest *et al.*, 1994; Kar *et al.*, 1995; Bonneau *et al.*, 1992; Atmaca *et al.*, 1995; Gu *et al.*, 1999; Kumaramanickavel *et al.*, 2002; Aldamesh *et al.*, 2009b). In Japan, where consanguineous marriage has significantly reduced in prevalence since the 1950s (Chapter 5), there has been an accompanying shift in the profile of RP probands, with a decline in the proportion of patients showing an autosomal recessive mode of inheritance and an increase in simplex cases (Hayakawa *et al.*, 1997a). However, because of greater longevity and past population growth in Japan, the total number of autosomal recessive retinitis pigmentosa cases actually increased by 21% from 1970 to 1990 (Hayakawa *et al.*, 1997b).

Congenital glaucoma has been identified in consanguineous kindreds and in a range of communities with a high prevalence of consanguineous unions (Ferák *et al.*, 1982; Bejjani *et al.*, 1998; Martin *et al.*, 2000; Panicker *et al.*, 2002; Khan, A.O., *et al.*, 2011), and an apparent association between consanguinity and bilateral but not unilateral retinoblastoma was noted in Saudi Arabia (Bahakim

& El-Idrissy, 1989). The autosomal recessive form of congenital cataract also has been associated with parental consanguinity in Lebanon (Baghdassarian & Tabbara, 1975) and Uzbekhistan (Rogers *et al.*, 1999), as has nonsyndromal congenital retinal non-attachment (NCRNA) in a Kurdish founder population in Iran (Ghiasvand *et al.*, 1998), non-syndromic coloboma in South India (Hornby *et al.*, 2003), and keratoconus in a UK Pakistani community (Georgiou *et al.*, 2004).

However, from a wider population perspective, when the causes of severe visual impairment and blindness in schoolchildren in nine Indian states were compared with the state levels of consanguineous marriage and the equivalent mean coefficients of inbreeding (α), the overall disease profiles did not reveal any obvious causative association between known major single gene visual disorders and consanguinity (Table 10.3). For example, while retinal dystrophies and albinism caused 25.5% of severe visual impairment and blindness in Harayana which had the lowest mean level of consanguinity (1%, $\alpha = 0.0004$), the same disorders accounted for 23.7% and 23.5%, respectively, of blindness in the two southern states of Karnataka and Tamil Nadu, where consanguinity was 29.7% ($\alpha = 0.0180$) and 38.2% ($\alpha = 0.0266$) (Rahi *et al.*, 1995).

The fact that many mutations reported in consanguineous populations had previously been unknown strongly suggests the presence of founder mutations. The influence of family, sub-community and community endogamy on the expression of autosomal recessive disorders was apparent in studies of primary congenital glaucoma in Slovakia, where the prevalence of the disorder in the general population was estimated to be 1/22 000, but 1/1250 in the Roma community among whom 45.6% of the parents of affected individuals were consanguineous (Genčik *et al.*, 1982; Ferák *et al.*, 1982). Similarly, the high rate of autosomal recessive retinitis pigmentosa in the Marranos, a formerly Jewish community that fled from Spain to Portugal during the late-15th-century Inquisition and subsequently converted to Christianity, has been ascribed to a founder mutation in the photoreceptor cell-specific nuclear receptor gene that occurred after their settlement in Portugal and thereafter increased in frequency because of strict community endogamy (Gerber *et al.*, 2000).

As was discussed in terms of hearing loss, the expression of genes located on the X-chromosome could be influenced by the specific type of consanguineous union, and in Pakistan all four daughters born to a first-cousin marriage between a man and his mother's brother's daughter ($Fx = 0.125$) were diagnosed with X-linked retinoschisis (Ali *et al.*, 2003). Detailed pedigree analyses can readily clarify findings of this unexpected nature and, in conjunction with genome-wide screening and homozygosity mapping, it should be possible in the near future to successfully identify the causative mutations responsible for inherited

Table 10.3 *Consanguinity and causes of childhood severe visual impairment and blindness in India by state*

Region/State	Vitamin A deficiency (%)	Retinal dystrophies and albinism (%)	Cataracts and aphakia/amblyopia (%)	Anophthalmos, microphthalmos and coloboma (%)	Consanguineous marriage (%)	Mean coefficient of inbreeding (α)
North						
Haryana	15.6	25.5	5.7	20.5	1.0	0.0004
Central						
Madya Pradesh	26.7	11.9	3.9	18.8	4.1	0.0025
Uttar Pradesh	21.6	6.7	2.9	32.8	7.5	0.0044
East						
West Bengal	24.7	11.2	16.8	19.1	5.0	0.0030
West						
Gujarat	21.6	9.2	11.3	17.5	4.9	0.0029
Maharashtra	20.4	13.8	12.7	17.2	21.0	0.0131
South						
Karnataka	11.5	23.7	6.5	28.4	29.7	0.0180
Kerala	7.5	33.2	23.6	6.5	7.5	0.0042
Tamil Nadu	18.5	23.5	9.8	20.6	38.2	0.0266

Source: Rahi *et al.* (1995)

forms of blindness, despite the multiple genes and loci known to be involved in their pathogenesis (Aldamesh *et al.*, 2009b).

Dental anomalies

There is a limited body of evidence to suggest that dental anomalies are more common among first-cousin progeny (Chapter 8), although in most cases the control for possible non-genetic variables was imperfect. As yet, only a few studies have been reported on major consanguinity-associated dental abnormalities, with a higher rate of structural defects and malocclusions (Maatouk *et al.*, 1995); talon cusp affecting permanent maxillary lateral incisors (Segura & Jiménez-Rubio, 1999); prepubertal and juvenile periodontitis (López, 1992; Hart *et al.*, 2000; Rapp *et al.*, 2011); and oligodontia and other dental anomalies (Chishti *et al.*, 2006; Noor *et al.*, 2009) ascribed to single gene defects. Despite the diagnosis of hypodontia in two families of Arab descent, both with parents who were first cousins, as noted in Chapter 8, the disorder has been identified as having an autosomal dominant mode of inheritance (Phillip & Caurdy, 1985), and so its occurrence in these families may simply have been coincidental.

Dental anomalies have been reported in ectodermal dysplasias and other rare syndromic conditions affecting specific consanguineous families, which has been accepted as presumptive evidence for the involvement of rare recessive genes (Rushton & Genel, 1981; Mégarbané *et al.*, 1998; Paula *et al.*, 2005; Micali *et al.*, 2005; Basel-Vanagaite *et al.*, 2007a; Adaimy *et al.*, 2007; Martelli-Junior *et al.*, 2008; Nanda *et al.*, 2010). Under such circumstances, individual founder effects may prove to be the primary driving factor.

Congenital defects

It has been estimated that in developed countries, 4–5% of children are born with some form of congenital defect (Christianson *et al.*, 2006), although it has also been claimed that the actual prevalence could increase to 8% of births if active rather than passive surveillance systems were routinely employed (Queisser-Luft *et al.*, 2002). Although there is virtual unanimity that congenital defects are more prevalent in the progeny of consanguineous marriages (Temtamy *et al.*, 1998; Stoll *et al.*, 1999), with the underlying implication that detrimental recessive gene expression is involved, the reported rates of congenital defects in first-cousin and non-consanguineous progeny have varied widely among studies (Table 10.4).

Table 10.4 *The prevalence of major congenital abnormalities (%) in first cousin versus non-consanguineous infants*

Country	First cousin ($F = 0.0625$)	Non-consanguineous ($F = 0$)	Authors
Japan			
Hiroshima	9.6	8.2	Neel & Schull (1962)
Nagasaki	12.8	8.7	
Norway			
National	4.6	2.2	Magnus *et al.* (1985)
Turkey			
Antalya	9.1	1.0	Guz *et al.* (1989)
Israel			
Muslim	15.8	5.8	Jaber *et al.* (1992)
Norway			
Norwegian	3.4	1.5	Stoltenberg *et al.* (1997)
Pakistani	4.5	2.1	
Turkey			
Konya	2.7	0.8	Demirel *et al.* (1997)
Israel			
Muslim	6.4	2.5	Zlotogora (2002a)
Israel			
Arab	8.7	2.6	Bromiker *et al.* (2004)
Iran			
Kashan	7.0	2.0	Mosayebi & Movahedian (2007)
Israel			
Non-Jewish	9.3	3.3	Harlap *et al.* (2008)
Jewish, West Asian	6.4	3.5	
Jewish, North African	3.7	3.4	
Jewish, European	12.0	2.7	
Israel			
Muslim	5.5	2.2	Zlotogora & Shalev (2010)
Kuwait	3.5	2.3	Al-Kandari & Crews (2011)

Of the 17 studies comparing the outcomes of first-cousin and non-consanguineous marriages, excess rates of congenital defects among first-cousin progeny ranged from 0.3% in Jews of North African origin in Jerusalem (Harlap *et al.*, 2008) to 10.0% in Israeli Arabs (Jaber *et al.*, 1992), with mean and median excess levels of congenital defects of 4.1% and 3.3%, respectively. Inadequate periconceptual maternal nutrition and trans-placental infections – e.g. by syphilis; common viruses including rubella, cytomegalovirus, varicella-zoster, herpes simplex and human parvovirus B19; and intracellular parasitic agents such as toxoplasmosis, in particular *Toxoplasma gondii* – are common causes of congenital defects and may have contributed significantly to the variation in the rates of defect reported (Christianson *et al.*, 2006). With respect to both nutritional inadequacy and congenital infections, there is a strong probability that impoverished communities are adversely affected to the greatest extent, which makes rigorous control for socioeconomic variation all the more important in comparative studies.

The reported outcomes of studies into the prevalence of birth defects also are dependent on the the nature of the diagnostic criteria employed; the diagnostic expertise of the medical, nursing and paramedical support staff; and the methods and investigative capacity of the unit. Additionally, in low-income countries, recognition of the symptoms of congenital disorders may have been obscured by high rates of neonatal mortality. Given the level of variability in the reported rates of congenital defects (Table 10.4), it is difficult to identify categories of disease that are specifically over-represented in consanguineous progeny, although an association with consanguinity has been claimed for many non-syndromic congenital defects, e.g. ranging from bone and cartilage disorders including scoliosis and syndactyly (Shahcheraghi & Hobbi, 1999; Al-Gazali *et al.*, 2003; Malik *et al.*, 2004), to hypospadias (Neto *et al.*, 1981; Manson & Carr, 2003), and small intestinal and anal atresias (Martínez-Frías *et al.*, 2000; Stoll *et al.*, 1997).

Very rare disorders with a complex aetiology and a higher rate of recurrence do seem to be more prevalent in consanguineous progeny, but the lack of adequate control for non-genetic variables, including birth order (Sheiner *et al.*, 1999), renders many such claims questionable. Perhaps for this reason, in the multi-national Latin-American Collaborative Study of Congenital Malformations (ECLAMC), based on 34 102 newborns and with 47 different congenital anomalies assessed, only hydrocephalus, postaxial hand polydactyly, and bilateral cleft lip with or without cleft palate showed a significant association with consanguinity (Rittler *et al.*, 2001). Another branch of the ECLAMC study showed that the progeny of couples living at altitudes higher than 2000 m above sea level had significantly increased levels of four types of craniofacial defect but lower frequencies of two types of neural tube defect (Castilla *et al.*, 1999), illustrating the importance of control for basic environmental variables in assessing the occurrence of specific types of congenital defects.

Congenital heart defects

A registry-based study in Denmark has shown that the relative risk of recurrence for all types of congential heart defect (CHD) is ≈3 when a first-degree relative has CHD. However, the authors also reported that the contribution of a family history of CHD to the overall prevalence of these defects was only 2–4%, indicating that most cases in the Danish population, which has a prevalence of consanguineous marriage of < 1%, did not have a family history of CHD (Øyen *et al.*, 2009). There also was no control for possible dietary influences associated with CHD (e.g. vitamin E ingested as a pre-conception dietary supplement) (Smedts *et al.*, 2009).

With a few exceptions (Roguin *et al.*, 1995; Roodpeyma *et al.*, 2002), a consistent positive association has been reported between consanguinity and specific types of congenital heart disorders (CHD), including ventricular septal defect and atrial septal defect, indicating the probable contribution of genetic variants with a recessive mode of expression. However, both positive and negative associations between consanguinity and CHD such as patent ductus arteriosus, atrioventricular septal defect, pulmonary atresia and tetralogy of Fallot have been reported in different populations (Becker *et al.*, 2001; Nabulsi *et al.*, 2003; Yunis *et al.*, 2006; Chéhab *et al.*, 2007). It is possible that these differences either reflect non-standardized diagnostic protocols in different study centres or alternatively they may be indicative of community-specific founder mutations (Bittles, 2011a).

A question that merits urgent attention relates to the actual incidence of CHD in human populations. The widely accepted estimate has been that CHD occur in some 4–8/1000 live births and that 90% of cases are multifactorial in nature (Christianson *et al.*, 2006). But, in a review prepared by the American Heart Association Congenital Defects Committee and endorsed by the American Academy of Pediatrics, a CHD incidence of 50/1000 births was described as conservative (Pierpont *et al.*, 2007), with an incidence for ventricular septal defects (VSD) alone of 53/1000 in a study of Israeli neonates (Roguin *et al.*, 1995). In the Israeli study, no association between consanguinity and the incidence of VSD was apparent, the defects were asymptomatic, and 89% closed spontaneously within the first ten months of life, suggesting either the action of environmental factors or, with minor muscular VSD, some form of delayed physiological development (Roguin *et al.*, 1995). A subsequent study of muscular VSD in Israeli preterm neonates closely mirrored these findings with 87.5% spontaneous closure by 6–11 months of age (Du *et al.*, 1996).

The scale of the apparent discrepancy in total CHD incidence between different studies raises significant concerns. Further, while a single paper which reported a significant association between consanguinity and CHD was conducted solely on neonates (Yunis *et al.*, 2006), other studies recruited symptomatic infants, including patients with chromosomal anomalies such as Down syndrome, children of differing ages, and even adults aged in their 20s and 30s as cases (Gev *et al.*, 1986; Badaruddoza *et al.*, 1994; Gnanalingam *et al.*, 1999; Becker *et al.*, 2001; Nabulsi *et al.*, 2003; Ramegowda & Ramachandra, 2006; Chéhab *et al.*, 2007).

It seems probable that age-independent case selection could largely negate the validity of inter-study comparisons because defects that could have been diagnosed at birth may have spontaneously corrected in children assessed at a later age. Conversely, in developing countries, many children with CHD die prior to diagnosis, while individuals not diagnosed until early adulthood presumably have a more mild disorder that is compatible with life (Bittles, 2011a). Therefore, from an ongoing health perspective, it is important, and disappointing, that few details of the relative clinical severity of individual CHDs have been evaluated or in many cases discussed.

With these reservations in mind, and knowing the highly subdivided nature of the Indian population (Chapter 2), the inability of a consanguinity mapping study in Bangalore, the capital of the South Indian state of Karnataka, to identify a single gene of major effect in a study group composed of clinically heterogeneous cases of CHD was somewhat predictable (McGregor *et al.*, 2010). If the cases and controls had been matched with respect to their community of origin, the probability of successfully identifying one or more major genes could have been substantially increased.

Neural tube defects

A positive relationship between consanguinity and the prevalence of neural tube defects (NTD) has been reported in a majority of studies in the Middle East and South Asia (Table 10.5), with the only neutral or negative associations to date in cases of spina bifida in Senegal, West Africa (Kabré *et al.*, 1994) and infantile hydrocephalus in Saudi Arabia (Murshid *et al.*, 2000).

Diet, in particular the B-vitamin folate, is known to exert a major influence on the incidence of NTD, with low maternal levels of other nutrients including vitamin B12, iron, magnesium and niacin also associated with the birth of spina bifida offspring (Vujkovic *et al.*, 2009). In a majority of the studies which purported to show a positive association of NTD and consanguinity, it was unclear how successfully the investigators had matched the cases and controls in terms of their diet and socioeconomic status or, in later papers, the level of access and compliance to preconceptual folate supplementation.

Table 10.5 *Consanguinity and neural tube defects by country*

Country	Increased	No effect/reduced	Reference
Egypt	+		Stevenson *et al*. (1966)
Algeria	+		Benallègue & Kedji (1984)
Tunisia	+		Khrouf *et al*. (1986)
Iran	+		Naderi (1979)
India, South	+		Kulkarni *et al*. (1989)
India, North	+		Agarwal *et al*. (1991)
Senegal		+	Kabré *et al*. (1994)
Israel, Iranian Jews	+		Zlotogora (1995)
Israel, Arabs	+		Zlotogora (1997)
Oman	+		Rajab *et al*. (1998)
UAE	+		Al-Gazali *et al*. (1999)
Saudi Arabia	+		Murshid (2000)
Saudi Arabia		+	Murshid *et al*. (2000)
Saudi Arabia	+		Asindi & Shehri (2001)
India, South	+		Mahadevan & Bhat (2005)
Pakistan	+		Perveen & Tyyab (2007)
Iraq	+		Al-Ani *et al*. (2010)

Matching for consanguinity was also mostly rudimentary and dependent on a simple 'consanguineous' versus 'non-consanguineous' dichotomy. Therefore, although at first sight there is overwhelming evidence for a causative association between consanguinity and NTD, larger and better balanced investigations are needed to confirm this finding. Perhaps instructively, the most convincing association with consanguinity was observed in rare syndromic disorders that included NTDs in the disease profile and therefore were strongly suggestive of an autosomal recessive mode of inheritance (Al-Gazali *et al*., 1999).

Oral and facial clefts

Published data on the possible role of consanguinity in the aetiology of non-syndromic oral and facial clefts (OFC) have been very mixed, with approximately equal numbers of studies indicating positive versus neutral or negative associations, and with divergent national results in Israel, Iran, India and Brazil (Table 10.6). The reported prevalence of OFC among newborns ranged from 0.97/1000 (Golalipour *et al*., 2007) to 2.14/1000 (Jamilian *et al*., 2007), which is consistent with the widely cited prevalence of between 1 and 2 per 1000 livebirths (Rittler *et al*., 2008), with general agreement that males had a higher prevalence of oral cleft and/or palate than females (Milan *et al*., 1994; al-Bustan *et al*., 2002; Elahi *et al*., 2004; Aljohar *et al*., 2008).

Table 10.6 *Consanguinity and non-syndromic oral and facial clefts by country*

Country	Increased	No effect/decreased	Reference
Italy	+		Milan *et al.* (1994)
Indonesia	+		Glinka *et al.* (1996)
Israel, Arab	+		Zlotogora (1997)
Kuwait		+	al-Bustan *et al.* (2002)
Israel, Arab		+	Jaber *et al.* (2002)
Pakistan	+		Elahi *et al.* (2004)
Iran		+	Golalipour *et al.* (2007)
Iran	+		Jamilian *et al.* (2007)
Saudi Arabia		+	Aljohar *et al.* (2008)
Lebanon	+		Kanaan *et al.* (2008)
India, South		+	Sridhar (2009)
India, South	+		Reddy *et al.* (2010)
Brazil		+	Paranaiba *et al.* (2010)
Brazil	+		de Aquino *et al.* (2011)
Brazil		+	Brito *et al.* (2011)

Thereafter there was a notable lack of consensus whether females were more likely to be diagnosed with cleft lip only or cleft palate only and if unilateral clefts were more commonly left-side or right-side. Based on extensive epidemiological data from the Norwegian Medical Birth Registry, it was shown that first-degree relatives had a relative risk of recurrence for any form of cleft lip of 32, whereas the risk was 56 for cleft palate only, suggesting a larger genetic component (Sivertsen *et al.*, 2008). The study also indicated that the cross-over risk between any form of cleft lip and cleft palate was low, and the severity of the primary case was unrelated to the risk of recurrence in this population.

Possible differences in the prevalence rates between co-resident ethnic communities were noted in Iran (Golalipour *et al.*, 2007), and a seasonal effect appeared to operate in South India with a peak incidence of facial cleft births occurring between March and June (Sridhar, 2009). But also in South India, it was noted that 3.4% of the mothers of affected children had taken anti-convulsant medication during their first trimester of pregnancy (Sridhar, 2009) which, as discussed in Chapter 15, is a subject that often has been omitted in assessments of the role of consanguinity in the aetiology of congenital anomalies.

A genome-scan on affected consanguineous individuals in Turkey born with cleft lip, with or without cleft palate, suggested a possible oligogenic model of inheritance (Marazita *et al.*, 2004). In an alternative approach, complex segregation analysis of the extended pedigrees of First Nation (Amerindian) Canadians with cleft lip and/or palate indicated that the most likely mode of inheritance was a mixed model, incorporating a major autosomal gene with a polygenic component (Lowry *et al.*, 2009). Given the significantly different study protocols used by investigators, and the wide range of potential genetic, environmental, lifestyle and medication variables involved, the fact that there has been no compelling evidence for a positive association between consanguinity and oral and facial clefts in major populations is unsurprising.

Intellectual and developmental disability

As described in Chapter 8, cognitive impairment generally is more common in the progeny of consanguineous parents, suggesting the expression of detrimental recessive genes in some individuals. Intellectual and developmental disability (IDD), previously termed *mental retardation* or *learning disability*, is determined by performance in cognition (IQ) and adaptive behaviour (life skills) assessments, usually after five years of age when standard psychometric testing becomes more valid and reliable. Prior to this age, the term *global developmental delay* is generally applied to describe children who fail to meet standard developmental norms. Although more males than females are diagnosed with mild IDD, the proportion of excess males declines with decreasing IQ scores, and some studies have suggested that severe IDD may be more prevalent among females (Kaufman *et al.*, 2010).

People with a mild level of intellectual disability diagnosed before the age of 18 years (i.e. with IQ scores ranging from 55 to 69) form the largest sub-group and represent some 85% of the overall population with IDD. By comparison, about 10% of cases have moderate disability (IQ score 40–54) and 5% have severe disability (IQ score < 40 points) (Bittles *et al.*, 2002). Globally, 0.2% to 4.0% of individuals have mild IDD, with 0.3% to 0.5% of children diagnosed with severe IDD (Roeleveld *et al.*, 1997).

The contributory role of consanguinity in childhood IDD has been apparent for some considerable time (Shields & Slater, 1956; Larson, 1957; Dewey *et al.*, 1965; Costeff *et al.*, 1977; Morton, 1978), and in both low- and high-income countries, elevated levels of global developmental delay, and mild and severe IDD, have been associated with consanguinity (Janson *et al.*, 1990; Madhavan & Narayan, 1991; al-Ansari, 1993; Farag *et al.*, 1993; Girimaji *et al.*, 1994; Temtamy *et al.*, 1994; Abdulrazzaq *et al.*, 1997; Durkin *et al.*, 1998, 2000; Durkin, 2002; Rudan *et al.*, 2002; Bener & Hussain, 2006; Jazayeri *et al.*, 2010; Masri *et al.*, 2010; Sharkia *et al.*, 2010).

Mild IDD was estimated to affect 6.5% of children and severe IDD 1.9% of children in the two- to nine-year age group in Pakistan (Durkin *et al.*, 1998), where the national rate of consanguineous marriage was estimated to be 61.2% ($\alpha = 0.0332$) (Ahmed *et al.*, 1992). As with cognitive impairment, very poor social conditions, inadequate diet including a lack of essential micronutrients, and infection may play significant causative roles in such cases. The extent to which these factors could influence the prevalence of IDD was demonstrated in a separate study in Pakistan, where mild IDD among six- to ten-year-old children was 10.5% in peri-urban slums but just 1.3% in the children of upper middle-class families (Gustavson, 2005).

The association between cognitive ability and infectious disease has been investigated on a global scale (Chapter 8), with a highly significant negative correlation between average IQ scores and parasite stress (Eppig *et al.*, 2010). Because consanguineous marriage is most common in lower socioeconomic communities (Chapter 5), it seems clear that to ensure minimization of the influence of non-genetic variables, rigorous control for socioeconomic factors should be incorporated as an obligatory pre-condition in all association studies involving consanguinity and IDD.

Autosomal recessive non-syndromic intellectual disability

Autosomal recessive forms of intellectual and developmental disability (ARNSID) account for almost 25% of all non-syndromic cases, but the extreme heterogeneity of the disorder has greatly hampered identification of specific mutations (Philippe *et al.*, 2009). In an early theoretical study, it was predicted that rare recessive alleles at some 325 gene loci might be involved in IDD (Morton, 1978). To date, 13 genomic regions, MRT (mental retardation) 1–13, have been positively implicated in ARNSID (Table 10.7). The reports have mainly originated from consanguineous kindreds that also are members of highly endogamous communities in which founder effects would predictably operate, e.g. Israeli Arabs (Basel-Vanagaite *et al.*, 2003; Basel-Vanagaite *et al.*, 2006; Basel-Vanagaite *et al.*, 2007b; Mochida *et al.*, 2009), Tunisian (Philipe *et al.*, 2009), Iranian (Najmabadi *et al.*, 2007; Motazacker *et al.*, 2007; Garshasbi *et al.*, 2008), and Pakistan families (Mir *et al.*, 2009; Khan, A.K., *et al.*, 2011), and in a US religious isolate (Higgins *et al.*, 2004; Nolan *et al.*, 2008; Çalişkan *et al.*, 2011).

In addition, a further three loci, MRT14, 15 and 16, have been putatively identified in highly endogamous and/or consanguineous Pakistani families with linkage to chromosomes 2p25.3-p25.2, 9q34.3 and 9p23-p13.3, respectively, and with two positive loci on chromosome 1q23.2-q23.3 and 8q24.21-q24.23

Table 10.7 *Genomic and phenotypic aspects of autosomal recessive non-syndromic intellectual disability (ARNSID)*

Locus	Chromosomal location	ARNSID genes	IDD phenotype	Authors
MRT1	4q26	*PRSS12*	Moderate to severe	Molinari *et al.* (2002) Basel-Vanagaite *et al.* (2006)
MRT2	3p26.3	*CRBN*	Mild	Higgins *et al.* (2004)
MRT3	19p13.12	*CC2D1A*	Severe	Basel-Vanagaite *et al.* (2003, 2006, 2007b)
MRT4	1p33-p34.3	–	Mild	Najmabadi *et al.* (2007)
MRT5	5p15.2-p15.32	*GRIK2*	Moderate to severe	Najmabadi *et al.* (2007)
MRT6	6q16.3-q21	–	Moderate to severe	Motazacker *et al.* (2007)
MRT7	8p12-p21.1	*TUSC3*	Mild to severe	Molinari *et al.* (2002) Garshasbi *et al.* (2008) Khan, A.K. *et al.* (2011)
MRT8	10q21.3-q22.3	–	Moderate	Najmabadi *et al.* (2007)
MRT9	14q12-q13.1	–	Mild	Najmabadi *et al.* (2007)
MRT10	16p12.1-q12.1	–	Moderate	Najmabadi *et al.* (2007)
MRT11	19q13	*TECR*	Moderate	Nolan *et al.* (2008) Çalişkan *et al.* (2011)
MRT12	1p21.1-p13.3	*ALX3* *RBM15*	Severe	Najmabadi *et al.* (2007) Uyguner *et al.* (2007)
MRT13	8q24.3	*TRAPPC9*	Mild to moderate	Mir *et al.* (2009) Philippe *et al.* (2009) Mochida *et al.* (2009)

(Rafiq *et al.*, 2010), whereas in populations from Iran, homozygosity mapping has indicated ARNSID loci on chromosomes 9q34, 11p11-q13 and 19q13 (Kuss *et al.*, 2011).

Nine genes causing ARNSID have been positively identified (Table 10.7). Each of these genes and the wider chromosomal regions so far associated with ARNSID are involved in neuronal enzymatic and/or signalling functions, but the exact nature of their roles in higher brain function remains speculative (Molinari *et al.*, 2002; Higgins *et al.*, 2004; Basel-Vanagaite *et al.*, 2007b; Najmabadi *et al.*, 2007; Motazacker *et al.*, 2007; Uyguner *et al.*, 2007; Garshasbi *et al.*, 2008; Nolan *et al.*, 2008; Mir *et al.*, 2009; Mochida *et al.*, 2009; Philippe *et al.*, 2009; Kaufman *et al.*, 2010; Çalişkan *et al.*, 2011). The mutations range in their phenotypic expression from mild in MRT2, MRT4 and MRT9, to severe in MRT3 and MRT12.

Other genes and gene variants potentially associated with ARNSID have been identified by bioinformatic scanning, which also revealed associations with other neural and mental disorders. For example, genes and variants associated with Alzheimer disease (ATP13 and APOE in 19q13) and autism (CNVs in 16p12.2 and 16p11.2) (Marshall *et al.*, 2008) were also found to be associated with ARNSID genes MRT11 and MRT10, respectively (Moolhuijzen *et al.*, 2009). In addition, a mutation in *MED23*, a sub-unit of the evolutionarily conserved Mammalian Mediator complex located in 6q23.2, has been shown to co-segregate with ARNSID in a multiply affected consanguineous Algerian family, leading to the conclusion that their cognitive deficit may be due to dysregulation of immediate early gene expression in brain development (Hashimoto *et al.*, 2011).

Severe autosomal dominant forms of intellectual disability usually manifest as sporadic cases because affected individuals rarely reproduce, and in developed countries, most patients with autosomal recessive forms of IDD now also appear as isolated cases because of small family sizes. Because half of the estimated 25 000 human genes are expressed in the brain, the total number of genes associated with ARNSID could potentially run into the thousands (Najmabadi *et al.*, 2007). But even if a disease gene for ARNSID is identified in a specific community, it may well have arisen as a founder mutation, with its subsequent expression and extent dependent on the degree of community marital endogamy and consanguinity. Therefore, there is a high likelihood that many causative genes for ARNSID will prove to be restricted in their distribution and expression (Nolan *et al.*, 2008).

It has, however, been claimed that in Iran some ARNSID mutations may account for several per cent of the patients diagnosed (Kuss *et al.*, 2011). Using a combined approach of homozygosity mapping and next-generation sequencing in consanguineous families, the same group of investigators has reported additional mutations in 23 genes already implicated in IDD or related neurological disorders, and probable IDD-causing gene variants in a further 50 novel candidate genes (Najmabadi *et al.*, 2011).

Neurodevelopmental disorders in childhood

An increased prevalence of speech disorders (Jaber *et al.*, 1997b), including delayed language acquisition despite normal audiometric thresholds (Fuess *et al.*, 2002), learning and reading difficulties (Eapen *et al.*, 1998; Abu-Rabia & Maroun, 2005), attention deficit hyperactivity disorder (ADHD) with aggression and misbehaviour (Al-Sharbati *et al.*, 2003), recurrent febrile seizures (al-Eissa, 1995) and infantile convulsions (Demir *et al.*, 2004) have been reported in first-cousin and other consanguineous progeny, although conversely no association was found between consanguinity and ADHD in Qatar

(Bener *et al.*, 2008), or with speech disorders among schoolchildren in Iran (Akhavan Karbasi *et al.*, 2011). As previously discussed in other contexts, the level of control for variables other than consanguinity was variable, and the association with consanguinity was principally assumed rather than demonstrated to be causal.

The prevalence of paediatric neurological disorders, including developmental delay with or without seizures, and neurometabolic degenerative disorders in particular, was significantly elevated in consanguineous progeny in Syria (Al-Rifai & Woody, 2007). A similar pattern was reported with progressive encephalopathy in Norway, where the risk was seven-fold higher in the endogamous Pakistani migrant community than among persons of Norwegian origin, and eleven-fold higher in the progeny of consanguineous Pakistani parents (Strømme *et al.*, 2010). This association with consanguinity was to some extent predictable because many cases of progressive encephalopathy, both due to individually rare metabolic diseases and neurodegenerative disorders, have an autosomal recessive mode of inheritance.

A wide range of generally rare but often quite severe childhood behavioural and psychiatric disorders has been reported in consanguineous families. In the earlier studies, the findings were largely based on clinical observation but more recently genomic analysis has been incorporated. A nationwide study of neurodegenerative disease in the UK revealed a strong association with the migrant Pakistani community in which consanguinity is favoured (Devereux *et al.*, 2004), and several genes for autosomal recessive primary microcephaly (MCPH) have been identified with *MCPH5*, the most prevalent mutation in consanguineous Pakistani families (Roberts *et al.*, 2002; Bond *et al.*, 2003).

To date, seven genetic loci (*MCPH1–7*) have been identified in MCPH, with the smaller number of cerebral cortical neurons generated during embryonic development in affected individuals causing a reduction in their brain sizes to approximately one-third of the expected volume and resulting in variable levels of intellectual disability (Mahmood *et al.*, 2011). Although displaying an autosomal recessive mode of inheritance, many other microcephalic disorders are complex in nature; e.g. hereditary spastic paraplegia with epileptic myoclonus (Sommerfelt *et al.*, 1991), and microcephaly with early onset seizures and variable spasticity (Silengo *et al.*, 1992; Straussberg *et al.*, 1998; Shen *et al.*, 2010).

Epilepsy

Twin studies have convincingly demonstrated a strong genetic component in the aetiology of generalized epilepsies (Berkovic *et al.*, 1998; Vadlamudi *et al.*,

2004), and seizure disorders and epilepsies were found to be more prevalent than febrile convulsions in a total population survey of neurological disease conducted in eastern Saudi Arabia, where first-cousin marriages formed 54.6% of all marital unions (al Rajeh *et al.*, 1993).

Although a family history of epilepsy, head trauma and febrile convulsions appeared to predispose to epilepsy, no association with consanguinity was apparent in Jordan (Daoud *et al.*, 2003). Likewise, a recent retrospective Israeli study of children with cerebral palsy who developed epilepsy, using as controls children with cerebral palsy but who did not become epileptic, showed that although neonatal seizures were strong predictors of epilepsy, there was no association with consanguinity (Zelnik *et al.*, 2010). However, among children in Iran (Asadi-Pooya, 2004), adolescents and young adults of Indian origin in Malaysia (Ramasundrum & Tan, 2004), and subjects enrolled in the Kerala Registry of Epilepsy and Pregnancy in South India (Nair & Thomas, 2004), positive associations between consanguinity and epilepsy were observed.

Homozygosity mapping in families with epilepsy has enabled putative linkage with specific genomic regions, but to date no specific mutations have been identified (Abouda *et al.*, 2010; Duru *et al.*, 2010; Layouni *et al.*, 2010). These findings are probably indicative both of the heterogeneity of the disorder and the presence of family- and community-specific founder mutations, with heritability estimates higher for juvenile myoclonic and all generalized epilepsies than for localization-related epilepsy (Nair & Thomas, 2004).

Autism

Homozygosity mapping has been successfully employed in the identification of a number of mutations coding for autism, mostly large homozygous deletions and implicating genes such as *PCDH10* (protocadherin 10) and *DIA1* (deleted in autism 1), whose level of expression changes in response to neuronal activity (Morrow *et al.*, 2008). Several other potential mutations causing autism were identified in subjects with unrelated parents, and a small-scale study in Saudi Arabia did not indicate a positive association between autism and consanguinity (Al-Salehi *et al.*, 2009).

Given the heterogeneous nature of autism and autism spectrum disorders, and the huge recent increase in the apparent prevalence of the condition in Western societies where consanguinity generally is rare, it would appear improbable that autosomal recessive mutations account for a significant proportion of all cases. Indeed, an investigation into the expression of neurologically relevant genes in cell lines cultured from monozygotic twins with autism, who were discordant either with respect to their diagnostic features or in the severity of

Table 10.8 *Consanguineous marriage and the incidence of Down syndrome*

Country	Increased	No effect	Reference
Kuwait	+		Alfi *et al.* (1980)
Italy, national		+	Devoto *et al.* (1985)
Kuwait, Bedouin	+		Farag & Teebi (1988)
Iraq, Baghdad		+	Hamamy *et al.* (1990)
UK, Shetlands	+		Roberts *et al.* (1991)
Turkey, national		+	Başaran *et al.* (1992)
Spain, national		+	Cereijo & Martínez-Frías (1993)
Venezuela, Maracaibo	+		Moreno-Fuenmayor *et al.* (1993)
Canada, Québec		+	de Braekeleer *et al.* (1994)
Israel, Arab		+	Zlotogora (1997)
France	+		Stoll *et al.* (1998)
India, Karnataka		+	Sayee & Thomas (1998)
Saudi Arabia, national		+	El Mouzan *et al.* (2008)

their symptoms, hypothesized that epigenetic factors such as DNA methylation or microRNAs could be involved in the disease phenotype (Sarachana *et al.*, 2010).

Down syndrome

Advanced maternal age operating via maternal meiotic non-disjunction has been consistently shown to positively influence the incidence of Down syndrome in human populations. The initial report suggesting an association between consanguinity and Down syndrome implicated the action of one or more recessive genes involved in the control of mitotic non-disjunction in the homozygous fertilized ovum, or the action of an autosomal recessive gene in homozygous parents that resulted in meiotic non-disjunction (Alfi *et al.*, 1980). Five of the subsequent 12 studies conducted in 11 other countries have confirmed a positive association between parental consanguinity and Down syndrome (Table 10.8).

In general, studies which were unable to identify a link between Down syndrome and consanguinity were based on larger sample sizes and had incorporated better control for maternal age, and in the large multinational South American ECLAMC study, the association that appeared to exist between consanguinity and Down syndrome was shown mainly to be due to the confounding effect of maternal age (Rittler *et al.*, 2001). Two studies which indicated a link between consanguinity and trisomy 21 suggested that rather than non-disjunction being the cause of the higher prevalence of Down syndrome births, there may have been lower rates of prenatal losses of trisomy 21 conceptuses in the pregnancies of consanguineous parents (Roberts *et al.*, 1991; de Braekeleer *et al.*, 1994), which matches the overall findings on prenatal losses presented in Table 7.2.

It also was proposed that advanced age in the maternal grandmothers of Down syndrome pregnancies may have been a contributory causative factor (Malini & Ramachandra, 2006). However, in support of an earlier study in Italy (Devoto *et al.*, 1985), a large national registry-based investigation in Norway found no evidence to support the hypothesis that the increased risk of a trisomic pregnancy was transmitted to the offspring of their non-affected daughters (Kazaura *et al.*, 2006), although a weak paternal age effect was observed (Kazaura & Lie, 2002).

As demonstrated in a recent study of a highly endogamous and consanguineous Arab community in Israel, some lingering doubt remains as to whether an as-yet-unidentified causative role may exist for consanguinity in the aetiology of Down syndrome in certain populations (Zlotogora & Shalev, 2010). To some extent, this suspicion was heightened by results from the city of Mysore, South India (Malini & Ramachandra, 2006), where some 24% of marriages are between uncles and their nieces or between first cousins (Bittles *et al.*, 1993). Although the authors of the Indian study concentrated on the possible effects of grandmother's age as a contributory factor in Down syndrome, their data had clearly identified consanguineous marriage as a significant explanatory variable in univariate regression analysis. Multivariate analysis based on specific levels of consanguinity (e.g. uncle–niece ($F = 0.125$) and first cousin ($F = 0.0625$)) would have been instructive.

Mitochondrial disorders

The oxidative phosphorylation (OXPHOS) system comprising five enzyme complexes located in the mitochondrial inner membrane is responsible for the synthesis of ATP in the respiratory chain. OXPHOS is functional before birth and in postnatal life, it serves as the primary energy source for most human tissues and organs. The coordinated expression of both the mitochondrial and nuclear genomes is involved, with maternally inherited mitochondrial DNA (mtDNA) coding for 13 sub-units of the OXPHOS complexes I, III, IV and V, and the tRNAs and rRNAs required for the translation of these transcripts. All other mitochondrial proteins, including more than 70 OXPHOS sub-units in complexes I, II, III, IV and V, are encoded by nuclear genes (Moslemi & Darin, 2007; Smits *et al.*, 2010). Mutations in any of the more than 100 genes involved in the respiratory chain can disrupt its function and at least theoretically cause a wide range of symptoms affecting any organ or tissue, at any age, and with any mode of inheritance (Rötig & Munnich, 2003).

The mitochondrial genome is small, with just 16 659 bases, it lacks introns and it is dependent on nuclear-encoded proteins for repair (van Adel & Tarnopolsky, 2009). Initially, mutations in the mtDNA were held to be solely responsible for respiratory chain disorders, following identification of the causative mtDNA mutations in such phenotypically diverse disorders as Leigh disease, mitochondrial encephalopathy (MELAS), myoclonic epilepsy and ragged red fibres (MERFF), Leber's hereditary optic neuropathy, chronic progressive external ophthalmoplegia (CPEO) and Kearns-Sayre syndrome (Thorburn & Dahl, 2001).

The list of nuclear DNA (nDNA) mutations causing functional abnormalities of OXPHOS has, however, been growing quite substantially, accompanied by increasing numbers of reports of consanguineous families whose offspring present with a mitochondrial cytopathy or myopathy (von Kleist-Retzow *et al.*, 1998; Nevo *et al.*, 2004; Thorburn, 2004; Ferreiro-Barros *et al.*, 2008; Rötig & Poulton, 2009; Smits *et al.*, 2011). In the absence of a genomic diagnosis, it is tempting to accept such cases as inferring the expression of an autosomal recessive defect. However, this may not necessarily be the case, as has been demonstrated by the presence of pathogenic mtDNA mutations in the progeny of first-cousin parents (Alston *et al.*, 2011).

Genetic counselling on recurrence risks and reproductive options in mitochondrial disorders is made additionally complex by the phenomenon of heteroplasmy, the co-existence of variable proportions of normal and mutant mtDNA within the same cells and tissues of an individual. However, in families with a mitochondrial disorder, the transmission of mtDNA mutations can be limited, if not prevented, by pre-implantation genetic diagnosis (PGD) of embryos produced by in vitro fertilization (IVF) (Poulton *et al.*, 2010).

Commentary

At the outset, it is important to acknowledge that consanguinity principally influences the incidence of rare recessive disorders. Therefore, with prevalent disorders, such as cystic fibrosis in populations of northern European origin or haemoglobinopathies in Africa and Asia, the causative alleles may be too common in most gene pools for their expression to be significantly influenced by intra-familial marriage. Further, in many instances, the reports of an association between consanguinity and a specific disorder originated from small endogamous communities, circumstances under which founder effect and drift could be predicted. For these reasons, no single disorder or group of disorders affecting infancy and childhood have been consistently reported in consanguineous offspring, although the empirical evidence implicating consanguinity and non-syndromic deafness, intellectual and developmental disability, and some categories of congenital heart defects is convincing. More speculatively, in the future, heterogeneous inherited disorders such as primary immunodeficiency (Movahedi *et al.*, 2004; Rezaei *et al.*, 2006; Al-Herz *et al.*, 2011; Azarsiz *et al.*, 2011; Subbarayan *et al.*, 2011) and inherited kidney disorders (Madani *et al.*, 2001; Hamed, 2002; Warady & Chadha, 2007; Topaloglu *et al.*, 2011) may warrant addition to the list of diseases showing a significant association with consanguinity and endogamy.

11 Consanguinity and disorders of adulthood

Introduction

Given the rapidly ageing profile of many countries, including those in which consanguineous marriage is most prevalent, disorders of adulthood represent the most challenging current issue in consanguinity-associated health research. Yet the information available with respect to the impact of close kin marriage on health in later life is at best fragmentary, and as a topic, it has been significantly under-investigated. As an introduction to this topic, a brief overview of the findings reported for possible associations between consanguinity and common disorders of adulthood, including cardiovascular disease, cancers and diabetes, is listed in Table 11.1.

As in other aspects of consanguinity research, a major difficulty in assessing most of the findings that have been reported with adult-onset diseases is that they were derived from composite studies combining discrete breeding sub-populations and included little control for sociodemographic variables. Because of the lack of precise information on the comparative composition and structure of the consanguineous and non-consanguineous study groups, and an absence of matching for non-genetic factors, the findings claimed have very often proved to be irreproducible in follow-up studies.

With complex adult diseases, another major factor to be considered is the nature of the genetic contribution, whether in common diseases with a sizeable genetic component some predisposing alleles occur in the gene pool at quite high frequency and lead to the disorder being expressed in the presence of adverse environmental factors, i.e. the common disease–common variant hypothesis. Alternatively, as its description suggests, according to the common disease–rare variant concept, it is the expression of rare mutations which is largely responsible for the genetic contribution to common diseases (Schork et al., 2009; Shields, 2011).

Consanguinity would be expected to exert a greater influence on the aetiology of complex diseases if rare autosomal recessive alleles were causally implicated, whereas if disease alleles that are common in the gene pool are involved, then intra-familial marriage would have a proportionately lesser effect (Bittles & Black, 2010a). Despite extensive genome-wide association studies, the fraction

162

Table 11.1 *Consanguinity and the prevalence of common disorders of adulthood*

Country	Increased	No effect/ Decreased	Reference
Italy	+		Milan *et al.* (1994)
Indonesia	+		Glinka *et al.* (1996)
Israel, Arab	+		Zlotogora (1997)
Kuwait		+	al-Bustan *et al.* (2002)
Israel, Arab		+	Jaber *et al.* (2002)
Croatia, Dalmatian Islands	+		Rudan *et al.* (2002a, 2003a, 2003b, 2004, 2006)
Pakistan	+		Elahi *et al.* (2004)
Iran		+	Golalipour *et al.* (2007)
Iran	+		Jamilian *et al.* (2007)
Saudi Arabia		+	Aljohar *et al.* (2008)
Lebanon	+		Kanaan *et al.* (2008)
India, South		+	Sridhar (2009)
India, South	+		Reddy *et al.* (2010)
Brazil		+	Paranaiba *et al.* (2010)
Brazil	+		de Aquino *et al.* (2011)

of the heritable variance in complex human disorders that can be explained by identified loci remains low, leading to the issue termed *missing heritability* (Maniolo *et al.*, 2009; Shields, 2011). Various technical and theoretical reasons have been advanced for the existence of this problem, including rare variants not being captured by available genotyping arrays, structural variants that are undetected by existing technology, small sample sizes, insufficient power to identify multigene interactions, and environmental and epigenetic effects that influence the disease phenotype (Parker & Palmer, 2011).

Consanguinity and age-related diseases states

Consanguinity and cardiovascular disease

Increases in blood pressure commonly occur with advancing age but, as indicated in Chapter 8, the evidence for an association between consanguinity and diastolic and systolic blood pressure in childhood was quite equivocal. Long-term studies conducted on the Dalmatian Islands, Croatia, have suggested that consanguinity is a strong predictor for a wide range of later-onset disorders, including hypertension, coronary heart disease and stroke (Rudan *et al.*, 2003a, 2003b), a conclusion confirmed in Qatar (Bener *et al.*, 2007) and with a positive association between consanguinity and hypertension also claimed in Saudi Arabia (Wahid Saeed *et al.*, 1996). In Pakistan, consanguinity was identified as a significant risk factor for non-fatal myocardial infarction among males aged 15–45 years (Ismail *et al.*, 2004), but conversely in both Pakistan and Israel, no significant association was observed between consanguinity and a range of adult diseases, including heart disease (Shami *et al.*, 1991; Jaber *et al.*, 1997c).

It has been suggested that familial aggregation of blood pressure may be more pronounced among the progeny of hypertensive as opposed to normotensive consanguineous parents (Hassan *et al.*, 2001). At least in the short term, studies which concentrate on well-characterized families and sub-communities, such as the highly consanguineous Arab village of Abu Ghosh in Israel (Hurwich *et al.*, 1982), or Icelandic subjects for whom a good genealogical record is available (Eldon *et al.*, 2001), are more likely to provide information on predisposing alleles for heart disease than ethnically mixed populations with poor control for non-genetic variables. But the effect of individual risk factors may be small, which necessitates substantial sample sizes, and even in population isolates with extensive pedigree, data failure to allow for the influence of distant genealogical loops can result in false positives in homozygosity mapping (Liu *et al.*, 2006b).

As contrasting examples, the problems of mixed ethnicity and variable lifestyles have largely been overcome in the Oman Family Study, in which five highly consanguineous and geographically isolated Arab pedigrees with a limited number of founders, genetically verified genealogical records, very similar lifestyle conditions, and stringently stipulated phenotypic criteria were recruited for the multidisciplinary investigation of complex diseases (Hassan *et al.*, 2005). In the study group, more than 50% of marriages are consanguineous and tribal endogamy is strictly observed, 21% of individuals have hypertension, 41% have dyslipidaemias and 5% are obese.

By comparison, in a small-scale study of Gypsies (Roma) in the USA with 62% consanguineous marriage ($\alpha = 0.017$) and 39% occlusive vascular disease, 84% of the subjects were obese and 86% smoked (Thomas *et al.*, 1987). Both genetic and lifestyle factors were therefore implicated in the 73% of Gypsy individuals who had hypertension, the 80% with hypertriglyceridaemia, and the 67% with hypercholesterolaemia. Therefore, in this study sample, the likelihood of disentangling the influence of consanguinity on cardiovascular performance and disease was remote, if not impossible.

Consanguinity and diabetes

Parental consanguinity has been associated with a number of rare mutations causing neonatal and juvenile-onset diabetes (Soliman *et al.*, 1999; Sellick *et al.*,

2003; Rubio-Cabezas *et al.*, 2008; Zalloua *et al.*, 2008). But in a community-based case-control study in Saudi Arabia, no association was found between consanguinity and Type 1 diabetes (El Mouzan *et al.*, 2008), while in Algeria investigators were unable to relate *HLA-DR* frequencies with Type 1 diabetes because of the high overall level of consanguinity in the study community (Aribi *et al.*, 2004). Once again, the sheer complexity of the disease represents a major problem, because besides the association with the HLA region on chromosome 6p21, more than 50 non-HLA genomic regions have been identified that are said to significantly affect the risk of Type 1 diabetes (Pociot *et al.*, 2010; Bradfield *et al.*, 2011).

Although a slightly higher risk of Type 2 diabetes among adult consanguineous progeny was reported in Qatar (Bener *et al.*, 2007), no such association was detected among Israeli Arabs (Jaber *et al.*, 1997c), and in many studies, major confounding effects, principally obesity (Thomas *et al.*, 1987; Bener *et al.*, 2005; Bayoumi *et al.*, 2007; Badii *et al.*, 2008) and smoking (Thomas *et al.*, 1987; Bener *et al.*, 2005) effectively masked any associations that might have existed between consanguinity and adult-onset diabetes. In the UK, it has been suggested that ethnic differences in Type 2 diabetes precursors, including glycated haemoglobin, fasting insulin, triglyceride, C-reactive protein and HDL-cholesterol, are already apparent in 9- to 10-year-old children (Whincup *et al.*, 2010). If these findings are confirmed, a major global association between consanguinity and Type 2 diabetes appears unlikely.

It may be possible to determine whether there are significant associations between consanguinity and glucose tolerance, insulin resistance and insulin secretion by reference to the 1969–1973 Vellore Birth Cohort Study in South India, for which robust data are available on the prevalence and types of consanguineous marriage, fetal losses, congenital defects and infant deaths (Rao & Inbaraj, 1977a, 1977b, 1980; Antonisamy *et al.*, 2009). However, as yet, follow-up studies have concentrated on physical growth during childhood and adolescence and cardiovascular risk factors in young adulthood, with the primary focus on factors such as parental size, neonatal size and childhood growth rather than parental consanguinity.

Consanguinity and cancers

Consanguinity has been linked with various cancers, occasionally in families with multiple types of malignancies ascribed to mismatch repair gene mutations (Bandipalliam, 2005). Childhood cancers were mostly implicated in the earlier studies and, where a genetic explanation was advanced, it was almost entirely based on either the parents of affected individuals being consanguineous or an unexpectedly high recurrence in first-degree relatives. For example, in a US sibship where five of eight siblings were diagnosed with congenital leukaemia, all at similar ages (Anderson, 1951); a UK family where a child also diagnosed with congenital leukaemia had been born to first-cousin parents (Bouton *et al.*, 1961); and in two Israeli Arab kindreds with multiple cases of acute lymphoblastic leukaemia (Kende *et al.*, 1994). Although the association between consanguinity and acute lymphocytic leukaemia was supported by a study in Qatar, a negative association was reported for Hodgkin and non-Hodgkin lymphomas (Bener *et al.*, 2001), and in neighbouring UAE, both acute lymphocytic leukaemia and non-Hodgkin lymphoma failed to show any association with consanguinity (Denic *et al.*, 2007a).

Comparable epidemiological reports on positive associations between consanguinity and breast cancer in the USA among women under but not over 45 years of age (Simpson *et al.*, 1981); a large pedigree also in the USA in which 16 of 18 cases of adenocarcinoma had consanguineous parentage (Lebel & Gallagher, 1989); and testicular cancer in two non-twin Japanese brothers (Yonemitsu *et al.*, 1988) added to the speculation that the expression of rare recessive genes could predispose to specific malignancies. Additional support for this hypothesis was drawn from familial aggregates and the first-cousin parentage of hereditary nonpolyposis colorectal cancer cases in Egypt (Soliman *et al.*, 1998), and a generalized increase in the incidence of cancers in population isolates on the Dalmatian Islands, Croatia (Rudan, 1999; Rudan *et al.*, 2003b), with increased glycosylation proposed as a possible mechanism to explain the apparent association between consanguinity and the incidence of cancers (Polašek *et al.*, 2011).

Data on an association between consanguinity and the prevalence of adult cancers are largely inconclusive (Table 11.2). For example, among individuals with breast cancer, studies have variously reported a positive association (Shami *et al.*, 1991; Liede *et al.*, 2002; Gilani *et al.*, 2006); no effect (Denic *et al.*, 2005, 2007a; Bener *et al.*, 2009; Bener *et al.*, 2010a); and a negative association (Denic & Bener, 2001). Each of these findings may be valid and indicative of inter-ethnic differences in the profiles of predisposing or causal recessive breast cancer genes within the different study populations, e.g. as reported in Saudi Arabia and Switzerland (Al-Kuraya *et al.*, 2005). However, the lack of unanimity in the results equally could reflect small numbers and failure to control for significant non-genetic variables, such as the effect of birth size on breast-cancer risk (dos Santos Silva *et al.*, 2008; Lagiou & Trichopoulos, 2008), a reservation that also applies to the claim that women with locally advanced breast cancer whose parents were consanguineous showed a better response to chemotherapy (Saadat *et al.*, 2011).

In a number of studies, patients with different levels of parental consanguinity were coalesced and compared with non-consanguineous progeny. For example,

Table 11.2 *Consanguinity and the prevalence of adult cancers by country*

Country	Type of cancer	Increased	No effect	Decreased	Authors
Japan	Testicular	+			Yonemitsu *et al.* (1988)
USA	Colorectal	+			Lebel & Gallagher (1989)
Pakistan	Breast	+			Shami *et al.* (1991)
UAE	Breast			+	Denic & Bener (2001)
UAE	Cervix		+		Denic & Bener (2001)
Pakistan	Breast	+			Liede *et al.* (2002)
Pakistan	Ovarian	+			Liede *et al.* (2002)
Pakistan	Breast	+			Gilani *et al.* (2006)
UAE	Breast		+		Denic *et al.* (2005, 2007a)
UAE	Colorectal		+		Denic *et al.* (2007a)
UAE	Lung		+		Denic *et al.* (2007a)
UAE	Thyroid		+		Denic *et al.* (2007a)
Qatar	Breast		+		Bener *et al.* (2009)
Qatar	Breast		+		Bener *et al.* (2010a)
Qatar	Colorectal		+		Bener *et al.* (2010b)

comparing subjects who were defined as being more or less consanguineous, $F \geq 0.0625$ versus $F < 0.0625$ (Denic *et al.*, 2005; Denic *et al.*, 2007a); or combining and then comparing all cases who were the progeny of double first cousins ($F = 0.125$), first cousins ($F = 0.0625$), first cousins once removed ($F = 0.0313$), second cousins ($F = 0.0156$) and second cousins once removed ($F = 0.0078$) with non-consanguineous subjects (Denic & Bener, 2001; Bener *et al.*, 2009; Bener *et al.*, 2010b).

Although this type of dichotomous comparison increases the total number of consanguineous cases and simplifies the resultant statistical analyses, second-cousin and second-cousin-once-removed progeny are actually closer to non-consanguineous ($F = 0$) offspring in terms of their coefficient of inbreeding (F) values and thus the assumed levels of genomic homozygosity. The biological rationale behind these approaches is therefore unclear, and acceptance of their conclusions merits appropriate sceptical caution.

Behavioural and psychiatric disorders

In contrast to the mid-nineteenth-century study by George Darwin that failed to show an increased incidence of first-cousin parents among the inmates of British lunatic asylums (Chapter 5), in societies that favour consanguinity, the adult progeny of consanguineous unions often are over-represented in institutions caring for persons with intellectual disability (Farag *et al.*, 1993). Associations between consanguinity and a range of common behavioural and psychiatric illnesses have also been reported in specific communities, e.g. depression in South India (Sathyanarayana Rao *et al.*, 2009), Tourette syndrome in Iran (Motlagh *et al.*, 2008), and early-onset Parkinson disease due to a *PARK2* mutation in Lebanon (Verlaan *et al.*, 2007).

Although the pattern of genetic inheritance seems to be clear in the latter example, in other cases of behavioural disorders, the association with consanguinity may be more social in origin, as in depressive illness following arranged marriages in which the bride or groom has not been consulted (El-Islam, 1976; Chaleby, 1988). However, genomic studies have increasingly demonstrated associations between consanguinity and a range of common behavioural and psychiatric illnesses with onset in adulthood, which raises the real possibility of causative autosomal recessive gene expression, albeit possibly restricted to specific sub-populations.

Schizophrenia

The evidence for a genetic involvement in adult-onset psychiatric illness is currently strongest in schizophrenia. Genome-wide linkage analysis using microsatellite markers and sib-pair analysis based on short tandem repeat polymorphisms were used to detect genomic regions showing an apparent susceptibility to schizophrenia in different populations (Gurling *et al.*, 2001; Irmansyah *et al.*, 2008), and both large recurrent microdeletions and rare chromosomal deletions and duplications were reported by multinational research groups investigating copy-number variants (Stefansson *et al.*, 2008; Stone *et al.*, 2008). Investigations using runs of homozygosity (ROH) also have been used to identify genetic risk loci for schizophrenia, and it was found that in outbred populations, ROHs were significantly more common in individuals diagnosed with schizophrenia (Lencz *et al.*, 2007). Because several of these ROHs contained or were adjacent to previously described schizophrenia risk loci, it was suggested that recessive alleles with high penetrance could be responsible for some of the genetic susceptibility to the disorder.

Earlier epidemiological studies in Sudan, Norway and Saudi Arabia had failed to reveal elevated rates of schizophrenia in the progeny of consanguineous parents (Ahmed, 1979; Saugstad & Ødegård, 1986; Chaleby & Tuma, 1987). However, schizophrenia spectrum psychosis has been associated with consanguinity in genealogy-based studies in Dagestan (Bulayeva *et al.*, 2000; Bulayeva, 2006) and the Dalmatian Islands, Croatia (Britvić *et al.*, 2010), and genomic studies have identified similar associations in Israeli Bedouins

(Dobrusin *et al.*, 2009), South Indian Tamil communities (Holliday *et al.*, 2009) and in Egypt (Mansour *et al.*, 2010). The degree to which schizophrenia risk loci identified in genetic isolates are likely also to be found in outbred populations is questionable, but the identification of gene products that may contribute to the phenotypic expression of the disorder could provide useful clues towards successful treatment regimens.

The phenotypic complexity of schizophrenia and other psychiatric disorders is a major problem. A case-control genome-wide association study provided evidence for a substantial polygenic component to the risk of schizophrenia involving thousands of common alleles of very small effect (Purcell *et al.*, 2009). In conjunction with whole genome studies based on rare copy-number variants (CNVs) and common single nucleotide polymorphisms (SNPs), it also indicated overlap between the genetic loci and alleles previously identified in autism, schizophrenia and bipolar disease. Taken at face value, this seems to indicate that in conjunction with environmental factors, a very substantial number of loci can increase an individual's risk of developing any of these disorders (Carroll & Owen, 2009; Lichenstein *et al.*, 2009).

Consanguinity and bipolar disease

There have been few detailed studies into the possible influence of consanguinity on bipolar disease, other than a case-control study in the Nile delta region of Egypt based on 64 DNA polymorphisms and self-reported parental relationships, with bipolar I disorder more prevalent among the progeny of consanguineous parents (Mansour *et al.*, 2009). But, as previously noted, diagnosis of these complex disorders can be difficult, control for non-genetic variables is seldom perfect, and there appears to be significant overlap between the multiple underlying predisposing genes identified as causative in different disorders (Carroll & Owen, 2009; Lichenstein *et al.*, 2009; Purcell *et al.*, 2009).

Consanguinity and Alzheimer disease

Alzheimer disease (AD) is a genetically complex and heterogeneous disorder in which cognitive deficits may exist prior to the onset of clinical symptoms (Hom *et al.*, 1994). Three genes with an autosomal dominant mode of inheritance that cause early-onset AD (*APP*, *PSEN1* and *PSEN2*) have been described (Bertram *et al.*, 2008), and epidemiological studies and ultra-high density SNP genotyping have positively identified APOE as the major susceptibility locus for sporadic late-onset AD (Coon *et al.*, 2007). The *APOEε4* allele is associated with an increased risk of late-onset AD in most populations, whereas the *APOEε2* allele is protective, although in a North Indian study, it was claimed that in combination with consanguinity, *APOEε2* no longer played a protective role for AD (Kaur & Balgir, 2005). An additional 12 loci (AD5–AD16) have been associated with AD, variously located on chromosomes 1, 3, 7, 8, 9, 10, 12, 19, 20 and the X-chromosome (OMIM, 2011), and preliminary mapping of extended runs of homozygosity have suggested that the method will be applicable to locating rare recessive risk variants contributing to AD (Nalls *et al.*, 2009a).

An association between consanguinity and AD was demonstrated in a genealogical study in the Saguenay region of Québec, Canada (Vézina *et al.*, 1999), and multiple loci for AD were identified in a highly endogamous and consanguineous Arab community in Israel, with more than one-third of cases clustered within a single *hamula* (tribal group) (Farrer *et al.*, 2003a, 2003b). Given the demographic history of both communities, recessive founder mutations would seem a probable explanation. However, high-density SNP genotyping was unable to identify the causative mutation in homozygous regions that were identical by descent in two siblings with early onset AD who were born to neurologically healthy Israeli first-cousin parents (Clarimón *et al.*, 2009), which serves to emphasize how difficult it will be to fully catalogue all of the causative mutations in AD and other complex behavioural disorders.

Consanguinity and infectious disease

In evolutionary terms, optimal MHC heterozygosity should offer increased resistance to infectious diseases and thus convey an important selectional advantage to individuals and communities under pathogen pressure (Doherty & Zinkernagel, 1975; Penn *et al.*, 2002; Frodsham & Hill, 2004; Lyons *et al.*, 2009a, 2009b). The progeny of consanguineous parents would be expected to be at a disadvantage because, predictably, they would have higher levels of homozygosity at MHC loci than their non-consanguineous counterparts and so be less capable of counteracting invading microorganisms (Potts & Wakeland, 1993; Penn *et al.*, 2002; Lipsitch *et al.*, 2003).

From a different perspective, it also has been proposed that due to host-parasite coevolution, consanguinity would initially have been advantageous in regions with high levels of disease-causing parasites. This initial protection only became disadvantageous following increased host population numbers and the geographical dispersal of individuals, families and communities, which in turn resulted in their exposure and susceptibility to novel pathogens (Hoben *et al.*, 2010).

Tuberculosis

Studies in the South Indian state of Tamil Nadu demonstrated differential susceptibility to mycobacterial diseases, with co-resident Hindu caste communities showing greater or lesser resistance to pulmonary tuberculosis (TB) according to their HLA profiles (Brahmajothi *et al.*, 1991; Ravikumar *et al.*, 1999). Besides the influence of caste endogamy, parental consanguinity was shown to adversely affect a number of children who had been inoculated with BCG vaccine (Bacille Calmette-Guérin) to prevent infection with TB, but instead they had developed disseminated BCG infections because of inherited immunodeficiency (Casanova *et al.*, 1995). It is unclear how common this adverse health outcome may be, but it is potentially important because there is a high prevalence of consanguineous marriage in many low-income countries with high rates of TB infection, with BCG vaccination given to infants at birth with or without subsequent booster vaccinations (Zwerling *et al.*, 2011).

In communities where consanguineous marriage is known to be favoured, individuals with higher levels of homozygosity were found to be over-represented among cases of both TB and hepatitis (Lyons *et al.*, 2009a). However, studies also have highlighted the diversity and very large number of genes that may be implicated as susceptibility alleles in infectious diseases affecting humans (Fumagalli *et al.*, 2010), together with the role of age as a variable in determining predisposition to childhood viral infections (Abel *et al.*, 2010).

Leprosy

Although there was no apparent association between consanguinity and leprosy in India (Lyons *et al.*, 2009a), in Libya, conjugal leprosy occurred among 32.0% of consanguineous couples by comparison with just 2.9% of non-consanguineous couples, suggesting a greatly enhanced, shared genetic susceptibility (El-Orfi *et al.*, 1998). Significantly increased shared susceptibility to infection by *Mycobacterium leprae* was also observed in a multidisciplinary study of leprosy co-prevalence in Brazil (Sales *et al.*, 2011). The disease subtype is dependent on the nature of the host immune response, whether Th1 (cell-mediated) or Th2 (humoral), and genome-wide linkage and association investigations, and candidate gene studies, have suggested that there is independent genetic control over susceptibility to leprosy and the development of the clinical subtype (Alter *et al.*, 2011), both of which may be influenced by recessive gene expression and hence consanguinity.

Multiple sclerosis

As a chronic immune-mediated, demyelinating disease of the central nervous system with a well-defined pattern of geographical prevalence, multiple sclerosis (MS) has variously been considered to be an autoimmune disease, a latent or persistent viral infection, or a neurodegenerative condition (Milo & Kahana, 2010). An early study revealed the presence of lymphocytotoxic antibodies in patients with MS (Schocket & Weiner, 1978) and Epstein Barr virus and infectious mononucleosis have been implicated as possible environmental agents (Milo & Kahana, 2010), with vitamin D proposed as a natural inhibitory agent to progression of the illness (Hayes, 2000; VanAmerongen *et al.*, 2004).

MS has been shown to be significantly more common in females than males in a number of high prevalence populations (Sundström *et al.*, 2001; Grytten *et al.*, 2006; Boström *et al.*, 2009), and there is pronounced spatial clustering of cases within local regions that appears to correlate with ethnic differences (Pugliatti *et al.*, 2002; Callander & Landtblom, 2004). Twin, sibling and other family-based studies have indicated a strong genetic predisposition to MS, including specific HLA associations (Prokopenko *et al.*, 2003), and founder effects have been claimed in Sweden and Finland (Callander & Landtblom, 2004, Tienari *et al.*, 2004). A recent large-scale, multi-national genome-wide association study that compared cases and controls of European descent confirmed and refined both the identity of the *HLA-DRB1* risk alleles and *HLA-A* protective alleles for MS (IMSGC & WTCCC, 2011).

Significantly higher recurrence rates of MS were observed in the sibs of index cases whose parents were consanguineous (Sadovnick *et al.*, 2001), and there was evidence of familial clustering and increased consanguinity in MS patients in a Dutch genetic isolate (Hoppenbrouwers *et al.*, 2007). Despite these indications that recessive genes may be involved in the disease aetiology, initial genome-wide linkage screens of kinships and communities with a high rate of consanguineous marriage were unable to locate specific genes contributing to the disease (Eraksoy *et al.*, 2003; Modin *et al.*, 2003). In addition, a recent genome-wide case-control homozygosity study conducted in the Orkney and Shetland Isles, Scotland, which have one of the world's highest rates of MS, was unable to identify any association between MS and consanguinity (McWhirter *et al.*, 2012). Because the genome-wide association study that confirmed the existence of *HLA DRB1* risk alleles predisposing to MS also located 95 distinct genomic regions outside the MHC in which there was an association with MS (IMSGC & WTCCC, 2011), the contribution of multiple common variants of small effect to individual MS susceptibility would appear to reduce the probability of a strong overall association with consanguinity.

Consanguinity and infertility

An increased rate of sperm abnormalities has been reported in adult males born to consanguineous parents (Baccetti *et al.*, 2001), and it appears that consanguinity may be associated with the expression of rare recessive alleles causing infertility in a small number of individuals and families, with sub-communities occasionally affected. For example, steroid 5α-reductase-2 deficiency (pseudovaginal perineoscrotal hypospadias) is an autosomal recessive disorder affecting the steroid 5α-reductase-2 gene, which in normal males is responsible for the conversion of testosterone to dihydrotestosterone, the intracellular mediator of many androgenic actions. Mutation prevents normal male genital development and instead permits the development of a female phenotype, which in humans acts as the 'default' phenotype. Affected individuals are 46,XY males with an external female phenotype at birth and so they are often raised as females. Although they have histologically normal testes with normal testosterone and oestrogen levels and can produce semen, they have a small prostate.

Several case studies of steroid 5α-reductase-2 deficiency indicated a strong family history of consanguinity with, for example, three affected brothers raised as females who were born to double first-cousin parents (Simpson *et al.*, 1971). In these individuals, no breast development or menstruation occurred at puberty and instead normal masculinization was observed. Similarly, three brothers initially identified as girls were born to a family in which the parents and grandparents on one side were first cousins and the great-grandparents also were biological relatives (De Vaal, 1955). There appears to have been a founder effect in both families, as was the case in three large clusters of the disorder occurring as geographical isolates in the Dominican Republic, Turkey and Papua New Guinea (Griffin *et al.*, 1995). Parental consanguinity was also present in a family with four male offspring, all of whom were 46,XY but with gynaecomastia, bifid scrotum and azoospermia. All four of these individuals were classified as dysgenetic male pseudohermaphrodites, and with their agreement and the approval of their parents, they were assigned male gender (Meijer & Groeneveld, 2007).

Rare inherited disorders that adversely affect reproduction can also affect females, e.g. as reported in multiple members of a consanguineous Iranian kindred with repeated familial molar pregnancies and gestational trophoblastic disease causing early abortion and secondary infertility (Fallahian, 2003). Recurrent molar pregnancies additionally occurred in a Lebanese family with extensive inter-marriage: two sisters each married to a first cousin had three and five pathologically confirmed molar pregnancies, respectively, and a second cousin of these women who also was married to a first cousin had five consecutive molar pregnancies and no living children (Seoud *et al.*, 1995).

Where a complete hydatidiform mole is of biparental origin, a majority of subsequent pregnancies also are likely to be complete hydatidiform moles, with only 5% normal pregnancies recorded (Fisher *et al.*, 2004). The relationship between the causative biallelic *NALP7* (nucleotide-binding, leucine-rich repeat, pyrin domains) mutations, familial biparental hydatidiform moles and maternal genomic imprinting has, however, yet to be fully elucidated (Hayward *et al.*, 2009).

Hypogonadotropic hypogonadism, resulting in absent or incomplete puberty, was reported in five of eight children born to first cousins in France, with low plasma testosterone levels in the males and low plasma oestradiol in a female associated with reduced plasma gonadotropin levels (de Roux *et al.*, 2003), and the disorder was also described in a consanguineous Israeli Arab kindred showing cryptorchidism in males and primary amenorrhoea in a female (Nimri *et al.*, 2011). In 50–60% of hypogonadotropic hypogonadism cases, the disorder co-occurs with anosmia (i.e. absent sense of smell) and on this basis is defined as Kallmann syndrome (Mitchell *et al.*, 2011). Common copy-number variants were detected at five loci across the genome in a consanguineous Han Chinese kindred with three family members diagnosed with Kallmann syndrome, including microdeletions in the intron of *KAL1*, which causes X-linked Kallmann syndrome (Zhang *et al.*, 2011).

With the increasing application of whole genome sequencing, it is highly probable that further studies will identify members of consanguineous kindreds with unique inherited infertility profiles caused by founder mutations (e.g. male infertility caused by mutations in the *CATSPER1* gene) (Avenarius *et al.*, 2009). Where there is evidence of both consanguinity and familial clustering of cases, the associations drawn become significantly more persuasive, as in a study of male-factor infertility in Lebanon involving diagnoses of azoospermia and severe oligospermia (Kobeissi & Inhorn, 2007; Inhorn *et al.*, 2009). At the same time, it is important that rare examples of this nature are not confused with the general influence of consanguinity on reproduction, which as indicated in Chapter 7 (Table 7.3 and Figure 7.3) resulted in larger completed family sizes in a majority of studies.

Consanguinity and coagulation disorders

Inherited coagulation disorders often are identified in early life, but in the present context because of co-occurrence with infertility, they have been considered under the heading of diseases of adulthood. Other than coagulation deficiencies due to factor VIII (haemophilia A) and factor IX (haemophilia B), both of which are inherited as X-linked recessive conditions, a majority of genetically encoded bleeding disorders are inherited as autosomal recessive traits, including deficiencies of fibrinogen, prothrombin, and factors V, VII, X, XI and XIII (Peyvandi *et al.*, 2001, 2002). In most populations, these deficiencies are rare, but during the late 1970s and 1980s, consanguinity and community endogamy were associated with clusters of cases of recessive coagulation disorders, including congenital afibrinogenaemia, e.g. with 10 of 27 sibs born into two uncle–niece families in which the fitness of affected individuals was close to zero (Fried & Kaufman, 1980). Factor XIII deficiency also was shown to cause high fetal losses and in some cases resulted in male infertility due to oligospermia and small testes (Kitchens & Newcomb, 1979; Frydman *et al.*, 1986).

The severe Type III form of von Willebrand disease (VWD), which can lead to fatal bleeding episodes, is also inherited as an autosomal recessive disorder and it has been described in homozygous individuals born to consanguineous parents (Berliner *et al.*, 1986). There are at least 50 causative mutations known to be distributed across the entire von Willebrand sequence and in affected homozygotes and compound heterozygotes, a null mutation is predominantly involved, resulting in negligible levels of von Willebrand factor (Baronciani *et al.*, 2003; Gupta *et al.*, 2008; Gadisseur *et al.*, 2009). By the late 1990s, the genomic basis of the rare recessive bleeding disorder, Glanzmann thrombasthenia, had been identified, e.g. with two separate founder mutations characterized in Iraqi Jews (Rosenberg *et al.*, 1997).

The global prevalence of the autosomal recessive forms of bleeding disorders has been described as rare, with prevalences ranging from 1/500 000 to 1/million (Peyvandi *et al.*, 2001, 2002). However, increased numbers of cases may arise due to founder effect in specific families and communities where endogamous marriage is the rule and consanguinity is preferential, despite the high associated mortality in more severely affected individuals. An extreme example of this situation has been reported in a seven-generation family of single tribal origin in Pakistan examined for bleeding disorders (Borhany *et al.*, 2010). Of the 533 individuals identified in the pedigree, 470 were alive, and of the 144 individuals investigated, 98 persons were diagnosed with a bleeding disorder: 51% with von Willebrand disease Types I, II and III, and 49% with platelet functional disorders, including epinephrine receptor defect, Glanzmann thrombasthenia, Bernard Soulier syndrome, adenosine diphosphate (ADP) receptor defect and collagen receptor defect.

A complementary study in Karachi, Pakistan, on patients with a history of congenital bleeding disorders found that 21.3% had von Willebrand disease, and of these individuals, 51.4% had the severe autosomal recessive Type III form of the disorder (Borhany *et al.*, 2011). Likewise in Mumbai, Western India, where consanguineous marriage also is prevalent, 59.5% of von Willebrand disease cases were Type III with severe clinical manifestations (Trasi *et al.*, 2005). These findings are in sharp contrast to autochthonous Western populations where Type I von Willebrand disease is the most common form of the disease and Type III is rare.

In Australia, 6.2% of the costs associated with adult hospitalization for single gene or chromosomal disorders were ascribed to von Willebrand disease (Dye *et al.*, 2011b). Ongoing migration to Western countries from societies where consanguineous marriage is prevalent has meant that larger numbers of previously rare recessive bleeding disorders, such as Type III von Willebrand disease, are now being diagnosed with treatment successfully instituted (Mannucci *et al.*, 2004). However, besides autosomal recessive coagulation disorders, a topic that seems not to have been substantially addressed is the degree to which specific types of consanguineous marriage – e.g. between a man and his mother's sister's daughter with a coefficient of inbreeding at X-chromosome loci, $Fx = 0.1875$, and a mother's brother's daughter marriage ($Fx = 0.125$) (Table 6.3) – might influence the expression of the X-linked recessive disorders haemophilia A and haemophilia B in females.

Commentary

Many of the data on consanguinity and common diseases of adulthood are confused and confusing, as in Table 11.2 where positive, neutral and negative associations with various cancers are listed. Similar contradictory patterns are reported for other disorders in which a genetic component has been suspected (e.g. duodenal or peptic ulcer) (Jaber *et al.*, 1997c; Rudan *et al.*, 2003b), whereas in conditions such as osteoporosis (Rudan *et al.*, 2004) and kidney disease (Barbari *et al.*, 2003; Karnib *et al.*, 2010), only positive associations with consanguinity have so far been published. In almost all cases, the investigations undertaken lacked adequate control for the multiple non-genetic variables to which an adult would have been exposed throughout pre- and post-natal life, and in many instances, it is difficult to differentiate between changes that may principally typify biological ageing from underlying pathological processes. Given the rapidly ageing profile of most of the world's population, the possible influence of consanguinity on common diseases of adulthood must be considered a subject that merits urgent, multidisciplinary investigation.

12 *Incest*

Introduction

Incest occupies a specific niche within the gamut of human consanguineous relationships because in most societies incestuous unions are both illegal and subject to general opprobrium. Furthermore, with some notable exceptions, incest avoidance seems to have been common to almost all societies. At the same time, judging by its regular inclusion as a topic in contemporary fiction (McEwan, 1980; Swift, 1984; Byatt, 1992; Proulx, 1993; Eugenides, 2003), incest also remains a subject of considerable, ongoing public interest.

It has been suggested that there may be an underlying genetic rationale for incest avoidance (Ember, 1975), with the rules prohibiting incestuous unions interpreted as an attempt to produce an optimal fitness balance between the detrimental genetic effects of inbreeding and the adverse influence of out-breeding on kin altruism (van den Berghe, 1980, 1983; Lieberman *et al.*, 2003). According to this scenario, marriage between first cousins is sufficiently distant in genetic terms to obviate severe inbreeding effects while maintaining the benefits that could accrue from kin selection. Alternatively, and less persuasively, a specific mutant 'incestuous allele' has been hypothesized, with carriers of the putative allele more likely to regard incest and other forms of consanguineous relationships as acceptable (Denic & Nicholls, 2006).

Some early human societies were certainly aware of the existence of specific familial disorders and hence problems that could arise if these disorders were ignored. In Judaism, there is a Talmudic dispensation for male circumcision, variously dated to the second to the fifth centuries AD and ascribed to Rabbi Judah the Patriarch and Rabbi Nathan, which appears to recognize the X-linked recessive mode of inheritance of haemophilia (Rosner, 1969). Two specific preconditions for this dispensation apply:

(i) Where two previous sons in a family died following circumcision;
(ii) In cases where two sisters each had a son who had died, the sons of other sisters are exempted from circumcision.

In general, however, the major explanations given for incest avoidance are mainly sociological rather than biological in origin and include Freudian guilt, and theories ascribed to Lévi-Strauss and others which propose that the incest taboo serves simultaneously to maintain the cohesion of the family and encourage the establishment of affinal relations with other kin groups (van den Berghe, 1983; Durham, 1991; Thornhill, 1991). Gene-culture coevolutionary models for brother–sister mating also have been advanced, with the premise that sib mating is avoided due to the recognition of associated genetic defects, possibly operating via cultural pressures (Lumsden & Wilson, 1980; Aoki & Feldman, 1997).

According to the Westermarck hypothesis, incest avoidance can best be explained in terms of negative imprinting against close associates of early childhood (Westermarck, 1937), possibly through the action of olfactory cues (Schneider & Hendrix, 2000). Two sets of studies have been widely cited in support of the Westermarck hypothesis. First, it was reported that individuals raised under mixed-sex child-rearing regimes in Israeli kibbutzim rarely married or entered into sexual relationships (Talmon, 1965; Spiro, 1969), and of 2769 marriages recorded from 211 kibbutzim, only 14 had been contracted between couples raised in the same communal children's house (Shepher, 1971).

The second area is the Chinese practice of adopting a girl into a family, usually during infancy or early childhood, with the expectation that she will marry a son of the adoptive family (Meijer, 1978). Studies of these *sim-pua* marriages in Taiwan showed that the couples were less likely to consummate their marriage, and they also had significantly lower mean fertility and lesser marital stability than their peers in 'major' marriages (i.e. non-*sim-pua* relationships) (Wolf & Huang, 1980). A follow-up study showed that the age difference between spouses exerted a significant effect on fertility, with differences of eight years or greater less likely to be associated with sexual aversion (Wolf, 2002).

A refinement of the Westermarck hypothesis of negative imprinting proposed that there should be some differentiation between types or degrees of early kin bonding. According to this theory, incest avoidance occurs when early kin bonding has been secure. By comparison, if the bonding has been impaired or has totally failed, then incest may occur, the degree of risk being proportional to the level of bonding failure (Erickson, 1993). There also is evidence from Canadian surveys that siblings separated during the first six years of life were more likely to subsequently engage in sexual acts of a potentially procreative nature with their opposite-sex siblings but not other forms of sexual activity (Bevc & Silverman, 1993, 2000). From a more general perspective, it has been argued that incest and inbreeding avoidance are diverse practices which may vary according to environmental circumstances and, more controversially, that incest may not be biologically detrimental (Leavitt, 2007). The evidence presented in Chapters 9–11 would, however, suggest that this latter suggestion is somewhat over-stated.

The legal boundaries of incest

At least some of the opposition in Great Britain to the Marriage with a Deceased Wife's Sister Bill (Chapter 3) seems to have been provoked by an unexplained fear that it would encourage incest. Perhaps for that reason, after its eventual acceptance in 1907, Parliament then passed the Punishment of Incest Act 1908, which effectively removed incest from the purview of Ecclesiastical courts and made it a criminal offence under civil law, with perpetrators defined as: 'Any male person who has carnal knowledge of a person who is to his knowledge his granddaughter, daughter, sister or mother . . . whether or not it was with her consent'; and 'Any female person of or above the age of sixteen years who with consent permits such carnal knowledge'.

This legislation was eventually repealed and its provisions were effectively incorporated within the 1956 Sexual Offences Act, which in Section 10 stipulates that: 'It is an offence for a man to have sexual intercourse with a woman whom he knows to be his granddaughter, daughter, sister' (including half-sister and whether or not the relationship in question is within wedlock); and in Section 11, 'It is an offence for a woman over the age of 16 years to permit a man whom she knows to be her grandfather, father, brother, or son to have sexual intercourse with her by her consent' (including a half-brother and whether or not the relationship is within wedlock) (Bluglass, 1979).

Because of differences in national legislation, an incestuous relationship is most simply defined as any form of union between biological relatives who are genetically closer than permissible under prevailing legislation. Most commonly, this means sexual intercourse between persons defined as first-degree relatives (i.e. father–daughter, mother–son, or brother–sister). In some countries, including Scotland where the Incest and Related Offences (Scotland) Act 1986 applies, the prohibited degrees of relationship have been widened to include half-sib and uncle–niece unions, in which the partners have 25% of their genes in common (Noble & Mason, 1978; Gane & Stoddart, 1988). By comparison, under the terms of the Swedish Marriage Law of 1987, half-sibs can marry, subject to special approval by the Government or the appropriate Authority designated by the Government, whereas in the Netherlands, special permission may be granted by the Ministry of Justice for a male or female to marry their adopted sibling (Chapter 14).

Dynastic incestuous marriage

In contravention of the concept of a universal incest taboo, there have been societies in which dynastic incest was practised over multiple generations. The best known examples are Egypt during the 18th and 19th dynasties (prior to

332 BC) and the Ptolemaic (323–30 BC) and Roman (30 BC to AD 324) periods (Middleton, 1962). Institutionalized royal incest between siblings or parents and offspring has also been described in Zoroastrian Iran, especially by the Achaemenids from the sixth to the fourth centuries BC (Scheidel, 2002); among the Incas and Mixtecs in the Americas; in Hawaii; Thailand; and in various African monarchies including the Ankole, Buganda and Bunyoro of Uganda, the Monomotapa of Zimbabwe, the Nyanga of Zaire, the Shilluk and Zande of Sudan, and the Fon of Dahomey in West Africa (Middleton, 1962; van den Berghe & Mesher, 1980; Shaw, 1992; Christensen, 1998). In African tribes, the king also had access to a royal harem and the children of unions with these non-incestuous wives could inherit the throne. This was a necessary pre-requisite for the continuation of the Ugandan and Sudanese monarchies because the incestuous unions of these kings were expected to remain childless (van den Berghe & Mesher, 1980).

Pharaonic Egypt

In Pharaonic Egypt, it was believed that the royal blood line would be maintained and strengthened through brother–sister unions. A further important factor was that the dowry of the Royal Heiress (i.e. the eldest daughter of the Pharaoh by his principal queen) literally included the throne, which itself was an object of considerable sanctity (Aldred, 1968, pp. 27–8). Given the high rates of infant mortality in Ancient Egypt, brother–sister unions were not always possible, even though marriage could take place in childhood. Therefore, to ensure continuity of the royal lineage, the Pharaoh had other unrelated brides and concubines.

It is difficult to comment with total conviction on the biological outcomes of dynastic incest because there is no contemporary evidence of the precise biological parentage of the Pharaoh's offspring. Nonetheless, there is quite compelling evidence that a number of 18th Dynasty Pharaohs (1570–1397 BC), including Ahmose I and Amenhotep I, married their sisters whereas others, such as Thutmose I, Thutmose II and Thutmose III, married half-sibs (Ruffer, 1921). Similarly, during the 19th Dynasty, Rameses II (1290–1223 BC) and Merneptah (1223–1211 BC) married several females who were sibs or half-sibs (Middleton, 1962).

The lives and reproductive records of some individuals from this period have attracted special attention, in particular Akhenaten (Amenophis IV) and his spouse Nefertiti (Aldred, 1968, pp. 88–99, 214; Tyldesley, 1998). Although said not to be brother and sister, it is claimed that the parents of Akhenaten, Amenophis III (1388–1351 BC) and Tiye were close relatives (Aldred, 1988, p. 146). In part, the special interest in Akenhaten stems from his devotion to a new form of monotheistic religion in which the Aten, the disc of the Sun, displaced the previous Theban deities of Amon-Re. As Pharaoh from 1351 to 1334 BC, Akhenaten also encouraged a more naturalistic style of artistic representation, illustrated by the sculpture of the Pharaoh reproduced in Plate 12.1.

The new artistic norms adopted in the portrayal of Akhenaten, however, have given rise to much speculation centred on his apparently unusual personal physical appearance (Plate 12.2), which also is shown in a family grouping of the Pharaoh, his wife Nefertiti and their three eldest daughters (Plate 12.3). From such evidence, Akhenaten has been described as having '. . . a typical feminine distribution of fat in the region of the breasts, abdomen, pubis, thighs and buttocks', and as platycephalic, with excessive growth of the jaw (Aldred, 1968, pp. 133–9).

On the basis of these claimed physical features, it was suggested that Akhenaten had Fröhlich's syndrome, a form of hypopituitary hypogonadism often associated with a pituitary tumour (Aldred, 1968, pp. 133–9). The same source, however, indicated that Akhenaten and Nefertiti as his chief wife had six daughters, which seems highly unlikely if indeed he had Fröhlich's syndrome. The improbability of this situation is further increased by the claims that Akhenaten had an extensive harem; that he had fathered children with two of his daughters, Meritaten and Ankhesenpaaten; and was the active partner in a homosexual relationship with his younger co-regent Smenkh-ka-Re (Aldred, 1968, 1988).

These multiple improbabilities notwithstanding, the Pharaoh's alleged androgynous nature was highlighted in the opera *Akhnaten* by the composer Philip Glass, with the role of the Pharaoh sung by a counter-tenor (Kobbé, 1987). As with the Spanish Habsburg King Charles II (Chapter 1), from a scientific perspective, it would be sensible to err on the side of caution in assessing the claims made with respect to the physiognomy of this long-dead individual, let alone attempt to judge his sexual practices and reproductive record from surviving effigies. Indeed, given the limited nature of the information on royal incest during the Pharaonic period, it is somewhat difficult to assess just how prevalent the practice may have been.

Besides Fröhlich syndrome, subsequent authors have suggested a wide range of other genetic disorders affecting Akhenaten and other members of the 18th Dynasty, including his son and successor Tutankhamun (1333–1324 BC), who is perhaps even better known because of the discovery of his intact tomb replete with burial goods in the Valley of the Kings in 1922. The inferred genetic disorders range from Marfan syndrome, to Wilson–Turner X-linked mental retardation, Klinefelter syndrome, androgen insensitivity syndrome, aromatase excess syndrome in conjunction with sagittal craniosynostosis syndrome, Antley–Bixler syndrome or some variant thereof, and homocystinuria, all based on examination of the reliefs, statuettes and other sculptures of the Amarna period (circa 1353–1323 BC) (Walshe, 1973; Paulshock, 1980; Farag & Iskandar, 1998; Braverman *et al.*, 2009; Cavka *et al.*, 2010).

Comprehensive disease studies have been undertaken on the mummy of Tutankhamun using computed tomographic examination. In addition, genomic analysis on DNA samples obtained by biopsy from the mummy of Tutankhamun and ten other mummies believed to be his biological relatives were conducted using autosomal and Y-chromosome markers. The studies confirmed that Tutankhamun was indeed the son of Akhenaten and that his mother and Akhenaten were brother and sister, which may require revision of the earlier assertion that Nefertiti was a cousin to Akhenaten rather than his sibling (Aldred, 1968). There was no genomic evidence of either Marfan syndrome or Antley–Bixler syndrome, but Akhenaten was identified as having a cleft palate and scoliosis, and on radiological examination, Tutankhamun had a cleft palate, mild curvature of the spine, mild clubfoot, and various foot abnormalities diagnosed as Köhler disease II (Hawass *et al.*, 2010).

Although his foot abnormalities may have caused Tutankhamun walking difficulties, the genomic identification and typing of plasmodial DNA in the Pharaoh's mummy strongly suggested that he died not because of an inherited defect but rather following a severe attack of malaria caused by *Plasmodium falciparum* (Hawass *et al.*, 2010). And the inherited disorders diagnosed via examination of reliefs and sculptures? Because macroscopic and radiological examination of the mummies of Akhenaten and Tutankhamun showed no specific signs of gynaecomastia (i.e. excessive development of the male mammary glands) or craniosynostoses (i.e. premature closure of the skull sutures), it was deemed improbable that either individual had a significantly bizarre or feminine physique. Therefore, the representations of the Pharaohs used as diagnostic aids by numerous authors more probably conformed to a royally decreed artistic style related to the sweeping religious reforms introduced by Akhenaten (Hawass *et al.*, 2010).

Ptolemaic Egypt

Few doubts apply to the occurrence of dynastic incest during the Ptolemaic period. Under the Greek tradition, cross- and parallel-cousin marriages were regarded as normal, and half-sib marriages also could be contracted. In Athens, these marriages were between a couple with the same father but different mothers, and in Sparta between couples with the same mother but different fathers (Hopkins, 1980; Shaw, 1992). After the conquest of Egypt by the Macedonians under Alexander the Great in 332 BC, a Graeco-Macedonian dynasty was founded by Ptolemy, who was one of Alexander's generals. The second ruler of the dynasty, Ptolemy II, divorced his wife who had already borne him several children and married his elder sister Arsinoe, who previously had been married to her half-brother Ptolemy Keraunos (Hopkins, 1980; Shaw, 1992). Following her marriage to Ptolemy II, the new Queen took on the official title of *Arsinoe Philadelphus* (i.e. Arsinoe the Brother-Lover).

Thus initiated, brother–sister and half-sib marriage became a near-tradition in Ptolemaic Egypt and, of the 15 succeeding royal marriages, 10 or perhaps 11 were brother–sister unions, although with other intra-familial marital options also actively pursued. Thus, Ptolemy VI married his sister Cleopatra II who bore him at least three children. On his death in 145 BC, Cleopatra II married her younger brother Ptolemy VII, who then also married his niece Cleopatra III, the daughter of his brother Ptolemy VI and his sister (and wife) Cleopatra II, and this royal ménage à trois ruled together until the death of Ptolemy VIII in 116 BC (Ager, 2005). Subsequently, the brothers Ptolemy XIII and Ptolemy XIV were successively married to their sister Cleopatra VII, prior to her non-incestuous but ultimately ill-fated liaisons with Julius Caesar and Mark Anthony (Middleton, 1962; Shaw, 1992; Ager, 2005).

Reputedly from information recorded by the Greek philosopher and historian Athenaeos (170–230 BC), it has been deduced that at least six generations of the Ptolemic rulers, Ptolemy II, IV, V, VI, VIII and X, suffered from obesity and sleep-disordered breathing (Michalopolous *et al.*, 2003). More recently, on the basis of coins and sculptures of Ptolemy II, his wife Arsinoe II, and other direct family members, a presumptive diagnosis of a multi-organ fibrotic condition, such as Erdheim–Chester disease or familial multifocal fibrosclerosis with thyroiditis, obesity and proptosis (i.e. bulging of the eyes) has been suggested (Ashrafian, 2005). As with the flawed post hoc diagnoses advanced for the alleged pathologies exhibited by the Pharaoh Akhenaten, an appropriate level of caution would appear merited in considering these various claims – especially since commencing from mid 270 BC, sibling, half-sib and uncle–niece marriages among the Ptolemaic dynasty spanned eight generations and some 240 years, with no indication of an accompanying lack of reproductive capacity (Shaw, 1992; Ager, 2005).

Historical examples of non-dynastic incestuous marriage

Roman Egypt

It has been suggested that incestuous marriages occurred among the families of non-royal officials and even artisans during the 12th Dynasty in Egypt, from approximately 1900 to 1800 BC (Murray, 1934), but the supporting information on its prevalence is sparse. By comparison, there is quite compelling evidence that almost 2000 years later, non-royal brother–sister marriages were contracted in Roman Egypt.

Roman attitudes towards consanguineous unions were significantly more proscriptive than those favoured by the Greeks, with parallel-cousin and closer-kin marriages regarded with disfavour, and incestuous (from the Latin *in castus*, 'not chaste') unions absolutely forbidden (Shaw, 1992). Papyrus records containing genealogical data from census returns, supplemented by marriage contracts, wedding invitations, personal letters and records of divorces, have however provided information on incestuous marriage outside the Egyptian ruling classes during the Roman period (Middleton, 1962; Hopkins, 1980; Bagnall & Frier, 1994; Scheidel, 1996, 2002, 2004). Brother–sister marriages occurred across socioeconomic strata in both urban and rural areas, and in several cases they had been contracted in multiple successive generations (Hopkins, 1980; Shaw, 1992). The most extreme example so far reported is a marriage between twins which appears to have produced an offspring (Gonis, 2000).

On average, incestuous couples married while in their mid 20s, which is much older than in the western half of the Roman Empire, with the bridegrooms some three years older than their spouses. However, the age difference between the spouses varied quite widely, from 1 to 17 years (Scheidel, 2004). Bridewealth was commonly paid, especially during the earlier Roman period, but bridewealth and dowry apparently coexisted for several centuries and could not be differentiated by region or social status (Hopkins, 1980).

From the records so far identified in Arsinoe, the district capital of the Fayum in Lower Egypt, and surrounding villages from AD 11 to 257, brother–sister unions ($n = 102$) accounted for 19.6% of marriages, with a further 3.9% of marriages between half-sibs and 1.0% between first cousins (Scheidel, 1997). This level of consanguineous marriage is equivalent to a mean coefficient of inbreeding in a single generation of $\alpha = 0.0545$. Furthermore, in the city of Arsinoe, all of the brother–sister marriages ($n = 17$) took place during the period AD 103–187. To assess the true impact of these data and hence the prevalence of incestuous marriages, it would be necessary to allow for the probability that many families would not have had a son and daughter or half-sibs of acceptable, marriageable age to betrothe. It has been suggested that 40% of families might have been in this situation (Hopkins, 1980), but in the absence of a thorough knowledge of the contemporary levels of fertility, and deaths in infancy and childhood, it is difficult to assess the accuracy of the estimate.

The precise reasons for the high rate of legally sanctioned incestuous marriage in Roman Egypt are unclear. But it has been claimed that a majority of the brother–sister unions occurred within the comparatively small and socially isolated Greek settler community, thus suggesting a restricted choice of marriage partners (Shaw, 1992). This option was not available to Roman citizens in Egypt for whom brother–sister marriage was banned as incestuous. If such a marriage did take place and it was discovered, the property of the offending couple was confiscated by the state (Middleton, 1962).

From a demographic perspective, it has been calculated that for early human groups with very restricted numbers of potential marriage partners, including the early Out-of-Africa migrants, an incest taboo could result in considerable costs in terms of reduced fertility through enforced delay of marriage (Hammel *et al.*, 1979). The small Greek communities in Roman Egypt may therefore have resorted to incestuous marriages as a means of overcoming this demographic difficulty. Some commentators have also claimed that the practice of incest in Roman Egypt was primarily economic in origin, through the avoidance of dowries. Although both of these ideas have been criticized as improbable (Scheidel, 1996), the economic benefits of brother–sister marriage, including the maintenance of land-holdings, were later favourably considered by the Cathars of Occitania in the late Medieval period (Le Roy Ladurie, 1980). Unfortunately for the Cathars, their sympathetic attitude towards brother–sister marriage may well have proved instrumental in provoking the Albigensian Crusade declared against them by Pope Innocent III in AD 1207 and their subsequent persecution as heretics by the newly created Inquisition (Burman, 1984).

Zoroastrian Iran

Zoroastrianism (Mazdaism) was the predominant religion in Iran from the time of the Archaemenid empire (550–331 BC) until the overthrow of the Sassanid dynasty (AD 224 to 651) by the invading Muslim Arabs (Scheidel, 1996). The rationale for incestuous unions within Zoroastrianism appears to have been based solely on religious principles and to have been associated with their attitude to marriage, which was regarded as divinely favoured and almost akin to a religious duty. The practice of mother–son and brother–sister marriage within early Persian society was apparently mentioned by contemporary Greek and Roman writers (Adam, 1865a, 1865b), and *xvétxvét* or *xwedodah* (usually translated as next-of-kin marriage) was discussed in the Pahlavi texts of the sixth to the ninth century AD with all three types of incestuous union – father–daughter, mother–son and brother–sister – advocated (Gray, 1915). *Xvétxvét* was described as being of especial religious merit, for example, being the ninth of a possible 33 ways of gaining entry to heaven, and its practice was accepted as a means of expiating mortal sin. Conversely, cessation of *xvétxvét* ranked fourth of the 30 worst sins that could be committed (Scheidel, 2002).

By comparison with Roman Egypt, although there are many extant and often critical references to *xvétxvét*, no data are available on either its prevalence or outcomes (Scheidel, 1996). Therefore, it has been suggested that although incestuous relationships may have been highly meritorious from a religious perspective, their occurrence in Zoroastrian Iran was probably very rare and thus inconsequential in terms of the numbers of offspring produced (Scheidel, 2002).

The incidence and biological outcomes of incest in modern societies

The incidence of incest

In most contemporary societies, incest is widely regarded not only as a breach of the civil law but a major cause of moral outrage, especially when the sexual union is cross-generational. Although understandable, a background of outraged public opinion is seldom a helpful starting point for the collection of unbiased data. This is certainly the case with incestuous relationships, where information on the incidence is often dependent on case studies conducted on individuals referred because of intellectual disability, consultations with patients in psychiatric clinics, or individual cases of young children diagnosed with a rare inherited disorder (Bittles, 2004).

Population-wide estimates of the incidence of incest have varied widely, and even unbelievably, from 1 case/million per year in the USA during the period 1910–1930, to 6.2–6.6 cases/million in Great Britain during 1974–1975, and 5000 cases/million per year in the USA during the decades 1945 to 1965 (Noble & Mason, 1978; Nakashima & Zakus, 1979). Even the latter USA figure may be an under-estimate because a survey of 796 college undergraduates in New England reported that 15% of females and 10% of males had some type of sexual experience involving a sibling; in 4.8% of cases, this had taken the form of intercourse or attempted intercourse (Finkelhor, 1980). Complementary data anonymously obtained in Finland indicated that 5% of 15-year-old girls had reported sexual experiences with a male parent, 2.0% with their biological father and 3.7% with a stepfather (Sariola & Uutela, 1996).

Psychiatric outcomes of incest

Information on psychiatric outcomes is equally varied with some studies identifying incest as the causative factor for low self-esteem in females, difficulty in intimate relationships and repeated victimization (Herman, 1981), female hysterical seizures/pseudoseizures (Goodwin *et al.*, 1979; Gross, 1979; Rosenfeld, 1979; Goodwin & Gross, 1979), and anorexia nervosa (Kumar & Agarwal, 1988). In most cases, the incestuous experience had occurred in the subject's preadolescent years (Husain & Chapel, 1983). The inevitability of a subsequent adverse psychological reaction to incest, however, has been questioned (Weitzel *et al.*, 1978), and a study of women who had experienced paternal incest reported variant outcomes, ranging from promiscuity, frigidity, personality disorders which rarely were neurotic and never psychiatric in nature, to no apparent ill-effects (Lukianowicz, 1972a).

In other forms of incest, mainly brother–sister, there appeared to be no major psychiatric sequelae, and the perpetrators were described as showing no marked deviations in terms of their personality, intelligence and social class from a random population sample (Lukianowicz, 1972b). However, in another UK study on women who had given birth within incestuous unions, a majority were of below-average intelligence (Jancar & Johnston, 1990). No long-term effects on adult sexual behaviour were observed in the USA following sibling and non-sibling sexual experiences during childhood (Greenwald & Leitenberg, 1989), but a second multi-centre study reported increased rates of childhood incestuous relationships among gay, lesbian and bisexual individuals (Cameron & Cameron, 1995).

In a situation such as father–daughter incest, the adults involved may themselves have been psychologically inadequate or showed psychopathological impairment prior to the event (Henderson, 1983). More generally, as with the extreme variation in the reported incidence of incest, the principal message conveyed by existing psychiatric studies would seem to be a need for access to unbiased representative data, if such material exists.

Clinical and genetic outcomes of incest

A similar conclusion applies in terms of the biological outcomes among progeny born to incestuous relationships for which five quite old and small-scale published studies are available (Adams & Neel, 1967; Carter, 1967; Seemanová, 1971; Knight, I.G., personal communication cited in Bundey, 1979; Baird & McGillivray 1982). The total sample size comprises 236 subjects with follow-up ranging from 0.5 year to 37 years (Table 12.1). The combined data suggest that 105 (44.5%) of the incestuous progeny were normal; 30 (12.7%) had died prematurely; 65 (27.5%) were diagnosed with a severe defect, including individuals with an autosomal recessive disorder, congenital malformation or non-specific severe intellectual disability; and a further 36 persons (15.3%) had mild intellectual disability.

Table 12.1 *Health status of the progeny of incestuous relationships (F = 0.25)*

Country of origin	Sample size	Follow-up (year)	Deaths	Severe defects/severe intellectual disability	Other defects/mild intellectual disability	Normal	Authors
USA	18	0.5	3	3	5	7	Adams & Neel (1967)
UK	13	4–6	3	1	4	5	Carter (1967)
UK	23	4–25	?	11	5	7	Knight, I.G., personal communication cited by Bundey (1979)
Czechoslovakia	161	1–37	24	41	18	78	Seemanová (1971)
Canada	21	0.5–19	0	9	4	8	Baird & McGillivray (1982)
Totals	236		30 (12.7%)	65 (27.5%)	36 (15.3%)	105 (44.5%)	

Table 12.2 *Non-genetic factors associated with incestuous relationships*

Very young maternal age
Advanced paternal and maternal ages
Physical or mental abnormality, or both, in one or both parents
Low socioeconomic status
Specific adverse social environmental effects, including unsuccessful attempted illegal abortion

With such a small combined sample size and very varied periods of follow-up, the results must be interpreted with caution. Interpretation of the data also is hampered by poor or in some cases no control for non-genetic variables, such as very young or advanced maternal and paternal ages, evidence of parental disease, and unsuccessful attempts to induce an abortion (Table 12.2). This problem is illustrated in the largest study, which was based on data retrospectively collected on births in Czechoslovakia between 1933 and 1970 (Seemanová, 1971). Of the 141 females who had given birth within incestuous unions, 20 had an intellectual disability (of whom 2 additionally were deaf-mutes, 2 had congenital syphilis and 2 were epileptics), a further 2 women were deaf-mute, and 3 were schizophrenic (Bittles, 1979). By comparison, of the 46 control mothers (i.e. women who also had pregnancies with unrelated males), 2 were intellectually impaired (one additionally being a deaf-mute) and 2 others were deaf-mute.

A similar picture was apparent among the 138 fathers of incestuous progeny, 8 of whom had an intellectual disability, 13 were chronic alcoholics, 2 had syphilis, and 4 had subsequently committed suicide. This compares with the male non-incestuous control group, none of whom had an intellectual disability, two were chronic alcoholics, and there was one person with polydactyly. Given the small total number of cases, the overall birth outcomes would have been significantly biased by the inclusion of a single individual who was alcoholic and had fathered five children with three of his daughters. Each of the children born of these unions had been diagnosed with varying types and degrees of abnormality, and three died within the first ten days of life.

The very young age of many of the mothers is a further complicating factor that could have adversely affected the viability and health of the incestuous progeny. Thus, in father–daughter unions, the mean and modal maternal ages at the time of birth were 18.9 and 16 years, by comparison with 19.9 and 14 years in brother–sister matings and 24.9 and 21 years in the married non-incestuous control group, respectively. Young maternal age is often associated with gynaecological immaturity and a poor pregnancy outcome. Given the almost 40-year time-span of the study and the marked negative secular trend with regard to menarche in Central Europe during the middle years of the twentieth century, it is probable that many girls in incestuous unions had just entered their reproductive phase when the pregnancy commenced and therefore they were at high risk of a problematic pregnancy outcome (Bittles, 2004).

In fact, if the data by Seemanová (1971) are adjusted to exclude physical and mental abnormalities among the male and female parents, and major disparities in terms of maternal and paternal ages, few differences remain in the poor overall health outcomes recorded for the progeny of both the incestuous and non-incestuous unions. It therefore is very difficult to know how best to interpret the data revealed by the five studies collated in Table 12.1.

An alternative method of assessing the probable levels of defect in incestuous unions is using information derived from legal consanguineous unions, in particular first-cousin marriages for which plentiful data are available and ascertainment bias is less pronounced. If the excess pre-reproductive mortality rate of 3.7% in first-cousin progeny ($F = 0.0625$) derived from Figure 9.3 is interpolated to children born of incestuous unions ($F = 0.25$), then an excess mortality rate of approximately 15% would be expected. Similarly, from Table 10.1, the median and mean excess morbidity due to congenital defects among first-cousin offspring were 3.3% and 4.1%, respectively, which would be equivalent to 13% to 16% excess morbidity at $F = 0.25$. Combining these mortality and morbidity estimates would therefore suggest a total excess rate for death and major disability in incestuous offspring of between 28% and 31%.

Commentary

A thorough examination of the genetic risks posed to the offspring of incestuous relationships is hampered by the currently inadequate empirical data. The genetic ill-effects of incest may be exaggerated by non-recognition of children born of an incestuous relationship with no obvious abnormality who are raised within the family as a younger 'brother' or 'sister' or are offered for adoption (Bittles, 2004; Port *et al.*, 2005). But deducing risk estimates from first-cousin data also needs to be undertaken with appropriate care, particularly because they depend on the assumption that there is no disproportionate increase in the prevalence of deaths and defects at closer levels of consanguinity due to the expression of conditional lethal genes, which may not be the case in the uniquely stressful circumstances of an incestuous pregnancy. Whatever the excess risks of physical and psychological ill-health associated with incest, it is the prevailing legal and the perceived moral status of such unions that ultimately decide how couples in incestuous unions are judged and treated. But, as will be discussed in Chapter 14, the categories of sexual unions that fall within the definition of incest vary widely among different countries, and within the USA among states, which can give rise to understandable confusion and misunderstanding in the minds of the general public.

13 Genetic screening, education and counselling in consanguineous marriage

Introduction

Different human societies exhibit quite varied attitudes towards illness and its causes, e.g. within Islam, illness may be regarded as the will of Allah, whereas many Christians believe that ill-health is a sign of God's judgement, possibly associated with sins committed in a previous life – or, alternatively, as an indication of God's blessing for an individual, by choosing them to suffer on behalf of others while secure in the knowledge of reward in the after-life. Given these core beliefs, in broaching the topic of genetic disease, it is important that attention is paid to the religious, cultural and social values of the individuals and communities for whom genetics services are planned and offered. How this aim can best be approached and achieved, however, may prove problematic and at least initially an empirical approach may be needed.

Genetic screening and diagnostic testing

Genetic knowledge in different societies

Both knowledge and an understanding of genetics remains low in most societies, which can undermine the provision of programmes to reduce the burden of genetic disease. Besides outright scepticism regarding inherited disease and modes of genetic inheritance, in the Middle East and the Indian subcontinent, there are prevalent folk beliefs in the power of the 'evil eye' – the casting of a spell by a female – which can turn a child who is normal at birth and apparently developing satisfactorily into one who is seriously ill (Panter-Brick 1991; Hussain, 2002).

Parents, especially those lacking a formal education, can find the concept of autosomal recessive disease difficult to understand and, for example, believe that a one in four chance of having an affected child implies that the birth of one abnormal child will naturally be followed by three normal infants (Mahdi, 1991). In many Islamic societies, there also is a belief that a child receives more genetic material from the father than the mother (Shaw & Hurst, 2008).

Therefore, it is quite commonly accepted that inherited disorders only arise through cousin marriages contracted on the paternal side of the family (Bittles & Hamamy, 2010), which as noted in Chapter 5 is the preferred form of consanguineous marriage in Arab societies. Parents also can find the concept that an individual could be a carrier of a disease gene and yet remain healthy difficult to accept, with members of the father's family particularly resistant to this idea (Mahdi, 1991), and so mothers may be held to be exclusively responsible for an inherited defect in a son or daughter (Gomaa, 2007).

Against this background, it is not surprising that attempted explanations of the recurrence risk of a genetic disorder in a consanguineous family are frequently misunderstood, especially because in many developing countries there is an acute shortage both of clinical staff trained in medical genetics and of genetic counsellors (Hamamy & Bittles, 2009). A key recommendation of the World Health Organization Report on Genomics and World Health was therefore to adapt medical education to place an appropriate emphasis on genomics (WHO, 2002). This initiative is being followed up by the World Health Organization with the identification of priority areas in genomic research and the formulation of associated strategies to improve public health.

Problems with staff whose knowledge of medical genetics is at best incomplete are not restricted to developing countries, e.g. a survey conducted among clinicians in the USA revealed a substantial level of misunderstanding of quite basic factual aspects of inheritance (Bennett *et al.*, 1999). The findings prompted the introduction of a genetics education programme for general practitioners (GPs) based on lectures, training in genetics clinics, seminars and journal clubs (Clyman *et al.*, 2007), with specialist certification in specific aspects of medical genetics provided via the American Board of Medical Genetics and the National Society of Genetic Counselors.

Consanguinity and screening for genetic carrier status in schoolchildren

The *Dor Yeshorim* programme of anonymized premarital screening for Tay-Sachs disease started in the Hasidic and Orthodox Jewish communities of metropolitan New York in the early 1980s and has since spread to other similar communities across the USA and in Jewish communities in many other countries, including Australia, Canada, Israel and the UK (Ekstein & Katzenstein, 2001). Females in Hasidic communities typically marry around 18 years of age and males in their early 20s, with the services of marriage match-makers frequently used (Greenstein & Bernstein, 1996). Thus, if information on the Tay-Sachs carrier status of both potential partners is to be included in the *shidduch* (i.e. marriage match), it is necessary to test boys and girls at an appropriate premarital age, which effectively entails testing teenagers in their senior high school years.

Considerations of a similar nature apply in many Middle Eastern countries where marriage also is undertaken at a young age, especially by females, and there is the added issue of high rates of intra-familial marriage. With these factors in mind, screening for inherited haemoglobin disorders has been undertaken among schoolchildren in a number of countries, including β-thalassaemia trait in Gaza (Sirdah *et al.*, 1998) and sickle cell disease and trait in Bahrain (Al Arrayed, 2005), with basic information on inherited diseases in general and haemoglobin disorders in particular provided before blood sampling.

From a general Western perspective, the sampling of children for their genetic status prior to the age at which they themselves can legally decide whether or not they agree to testing raises significant ethical issues. However, in the case of the *Dor Yeshorim* programme, sampling is conducted with the strong support of the community that wishes to minimize the risks of children being born with Tay-Sachs disease. For communities in which early marriage is traditional and where the community is known to have a high carrier frequency of a severe inherited disorder, such as transferase-deficient galactosaemia in the Irish Travellers (Murphy *et al.*, 1999), these general ethical concerns have to be balanced against the specific needs and declared wishes of the community.

Consanguinity and premarital population screening programmes

A basic lack of knowledge of genetics also has been apparent in many individuals and couples enrolled in premarital screening programmes (e.g. for β-thalassaemia and other haemoglobinopathies), which have become mandatory or quasi-mandatory in a number of Mediterranean and Middle Eastern countries (Angastiniotis & Hadjiminas, 1981; Samavat & Modell, 2004; Al Arrayed, 2005; Meyer, 2005; Canatan *et al.*, 2006; El-Hazmi, 2006; Al Hamdan *et al.*, 2007; Hamamy *et al.*, 2007; Tarazi *et al.*, 2007; Cowan, 2009) and in The Maldives (Firdous, 2005; Firdous *et al.*, 2011). Premarital screening for lysosomal storage disorders also is offered to the general population of Israel, with most high-risk couples additionally opting for prenatal diagnosis (Bach *et al.*, 2007).

Although premarital carrier-screening programmes for haemoglobinopathies have been implemented in Bahrain, Jordan, Tunisia and the UAE (Al-Gazali *et al.*, 2006), information on their effectiveness in the prevention of these disorders remains incomplete. A possible reason is that, to date, pretest genetic counselling to fully explain the possible sequelae of screening programmes frequently has not been provided (Hassan *et al.*, 2010). This omission represents an important ethical and practical shortcoming because genetic counselling and the options open to a couple can be significantly influenced by the stage in life at which they learn of their potential risks of conceiving a child with a genetic disorder (Bittles & Hamamy, 2010). It also is contrary to World Health Organization guidelines, which have recommended that counselling should be offered before any genetic testing is initiated and should continue thereafter if the results entail reproductive choices for the person and family tested (WHO, 1997, 2006).

In Saudi Arabia, a study of community perceptions of a premarital carrier-screening programme for thalassaemia and sickle cell anaemia reported that 94.3% of participants believed premarital examination was an effective means of preventing blood genetic diseases (El-Hazmi, 2006). But a survey conducted among male and female university students in the city of Jeddah revealed widespread misunderstanding on premarital screening, with 53.9% of all interviewees and 49.1% of students who had already undergone screening believing that the tests could detect all hereditary diseases, and a majority of students not knowing what the test results indicated (Al-Aama *et al.*, 2008).

Although most cases of genetic screening for haemoglobin disorders could not be initially followed-up to evaluate the outcome of the premarital screening programme (Al Aqeel, 2007), the situation has become significantly more positive according to a retrospective review of more than 1.57 million premarital couples tested for sickle cell disease and β-thalassaemia from 2004 to 2009, with 4.5% and 1.8%, respectively, identified as carriers or cases of sickle cell disease and β-thalassaemia (Memish & Saeedi, 2011). As part of the programme, the test results were shared with the couples examined, with genetic counselling offered to at-risk couples.

Although marriage certificates were issued irrespective of the result obtained, during the six-year screening period, the voluntary cancellation of marriage proposals between at-risk couples for either disorder increased from 9.2% in 2004 to 51.9% in 2009 (Memish & Saeedi, 2011). However, given the results of the interview-based survey conducted among highly educated students in the industrialized city of Jeddah (Al-Aama *et al.*, 2008), it is unclear whether the couples who opted either to proceed or to cancel a planned marriage fully understood the information provided.

When premarital counselling was offered to Bedouin couples in Israel judged to be at high risk of an autosomal recessive disorder, the information was reported to be useful in deciding whether or not to proceed with marriage to a biological relative whatever the actual decision taken (Shiloh *et al.*, 1995; Raz & Atar, 2004). An additional, beneficial side-effect of premarital haemoglobinopathy carrier-screening programmes has been to raise public awareness of genetic disease, with requests for premarital genetic counselling for other inherited conditions increasing in number and as a proportion of total referrals (Hamamy & Bittles, 2009).

Reproductive options for carrier couples identified by premarital screening

The current options for couples who both are carriers of a detrimental recessive mutation would include not marrying, to marry but not to have children, to marry with access to prenatal diagnosis and selective termination of an affected pregnancy, or to adopt a child (Hamamy & Bittles, 2009). Because reproduction is considered an essential requirement of marriage in many more traditional Arab communities, with sons particularly desired, a couple would rarely opt to marry and not have children. But because, as discussed herein, medical termination of pregnancy is often subject to strong personal religious and social convictions, it also may not be an acceptable option or solution for many couples (Bittles & Hamamy, 2010).

Contrary to the experience in Saudi Arabia (Memish & Saeedi, 2011), and with family pressures to continue with the planned wedding, it is not surprising that in many of the mandatory haemoglobinopathy testing programmes introduced in Middle Eastern countries, couples who have been identified as carriers for an inherited disorder have elected to proceed with their wedding. This outcome has often occurred when both partners are members of the same community, and it is even more probable if they are close biological relatives because failure to fulfil a contracted marriage arrangement could significantly disrupt close-knit family relationships. The decision to proceed with a marriage even if both partners have been identified as carriers of β-thalassaemia also can arise in regions where the disorder is not prevalent, with the potential costs perceived to be outweighed by the social and economic benefits offered by the agreed match (Acemoglu *et al.*, 2008; El-Tayeb *et al.*, 2008).

Prenatal diagnosis of genetic disorders

Even if couples, whether consanguineous or non-consanguineous, opt to ignore their carrier status for a detrimental recessive disorder, they have effectively been forewarned of the attendant risk of diseases faced by any child they might conceive. The timely provision of prenatal diagnosis is, however, dependent on appropriate access to specialist expertise and the requisite test facilities.

As previously discussed, in many developing countries where premarital and prenatal diagnosis is undertaken, they often occur without pre-test counselling, which means that couples may first be given advice as to the genetic status of their fetus at an early stage in the pregnancy even though medical termination would not be regarded as an acceptable option. Furthermore, even in high-income countries, women who opt for prenatal examinations are frequently unaware of the possibility of false positive or false negative results, or the risk of miscarriage following amniocentesis (Dahl *et al.*, 2006).

Despite these issues, prenatal diagnosis for genetic disorders has been successfully introduced in a number of Middle Eastern countries with a high prevalence of consanguineous marriage, including maternal serum screening for neural tube effects and Down syndrome in the West Bank (Husseini & Akkawi, 2005). Prenatal cytogenetic testing via amniocentesis has been adopted in Egypt, Tunisia and Lebanon for aneuploidies in at-risk pregnancies (Abdel-Meguid *et al.*, 2000; Chaabouni *et al.*, 2001; Eldahdah *et al.*, 2007), and referrals for maternal serum α-fetoprotein testing and amniocentesis were accepted by some Negev Bedouin women in Israel (Lewando-Hundt *et al.*, 2001). With support from the relevant Islamic religious authorities, prenatal screening programmes for the diagnosis of β-thalassaemia were established in Iran (Najmabadi *et al.*, 2006), and they also have been conducted in Pakistan (Baig *et al.*, 2006).

A study of attitudes towards the provision of genetic diagnosis in Pakistan revealed major differences in opinion between the participating clinicians, medical students, parents of thalassaemic children, lawyers and members of parliament. Although a majority of participants (77%) favoured premarital testing, by far the highest level of support for prenatal screening was indicated by the parents of thalassaemic children (94%), with a majority of these parents also in favour of medical termination of pregnancy in the case of an affected fetus (Gilani *et al.*, 2007).

Remarkable success in reducing the burden imposed by thalassaemia and other haemoglobinopathies has been achieved in The Maldives, an Islamic Republic located in the Indian Ocean comprising 200 inhabited islands and with a population of just 280 000. In 1992, it was estimated that the β-thalassaemia carrier rate across the country was 18.1%, with three mutations accounting for 98.7% of the β-thalassaemia chromosomes studied (Furuumi *et al.*, 1998), and an α-thalassaemia rate of 28.0% (Firdous, 2005). To reduce the associated burden of disease, a Government National Thalassaemia Centre was created with the inclusion of teaching on thalassaemia in the school curriculum, the legal requirement of screening before marriage, legalization of prenatal diagnosis and medical termination of pregnancy, and the establishment of prenatal diagnostic services (Firdous, 2005; Firdous *et al.*, 2011).

In a multi-faith, multi-ethnic society such as Israel, where differing attitudes towards prenatal diagnosis and termination of pregnancy may largely reflect specific religious beliefs, a community-based carrier-screening programme for severe genetic disorders present at high frequency in certain communities was deemed to be successful, with stigmatization avoided. However, a small number of couples declined prenatal diagnosis and in two of these families, affected infants were subsequently born (Zlotogora *et al.*, 2009).

Medical termination of pregnancy

Although it has been widely believed that pregnancy termination is absolutely prohibited within Islam, abortion to save the life of the mother has historically been permitted (Hedayat *et al.*, 2006). A significant initiative was introduced in 1990 by the Islamic Jurisprudence Council of the Islamic World League, which issued a *fatwa* permitting medical termination of pregnancy within 120 days of conception (i.e. before ensoulment of the fetus). Preconditions for the *fatwa* were that the fetus was proven beyond doubt to be affected with a severe malformation not amenable to therapy; if born, the child would lead a life of misery both for him or herself and his/her family; both parents gave their assent (Alkuraya & Kilani, 2001; Albar, 2002).

Despite this religious ruling, selective medical termination of pregnancy remains largely unavailable in a majority of Arab countries (UN, 2001; Hamamy & Bittles, 2009), and members of many Islamic communities believe that there is a total religious ban on medical termination of pregnancy (Raz & Atar, 2003, 2004). For this reason, preimplantation genetic diagnosis has increasingly been adopted as an alternative strategy to avoid the birth of an affected child in countries where Islam is the sole recognized or majority religion (Hellani *et al.*, 2004; Ozand *et al.*, 2005; Alsulaiman & Hewison, 2006; Al-Sayed *et al.*, 2007).

The process of decision-making in Islam on termination of pregnancy and other major bioethical issues may be influenced by the prevailing School of Islamic Jurisprudence to which an individual or community belongs, with four major Sunni Schools: Hanafi, Maliki, Shaf'ei and Hanbali; three Shia Schools: Twelve Imami, Zaydiya and Ismaili; and several population-specific Schools, such as the Ibadi in Oman and parts of North Africa (Roudi-Rahimi, 2004; Zuhar, 2005; Al Aqeel, 2007). In effect, each School of Jurisprudence represents a unique path to the interpretation of Islam, with the diversity of views among Muslims derived in part from the concepts integral to the various Schools, besides being influenced by differences in their cultural backgrounds and community organization (El-Hazmi, 2007; Hamamy & Bittles, 2009).

Within the Shia branch of Islam, the Grand Ayatollah Khamenei issued a *fatwa* in 1997 giving permission for therapeutic abortion of the fetus under limited circumstances and initially only for β-thalassaemia, which then passed into law following approval by the Iranian Parliament in 2003 and by the Islamic Guardian Council in 2005 (Hedayat *et al.*, 2006; Najmabadi *et al.*, 2006; Yari *et al.*, 2011). The bill that was finally approved stipulated that 'Therapeutic abortions may be performed under the following conditions. First, the fetus must be less than four months of age, that is, before the spirit is breathed into it. Second, the fetus must be suffering from profound developmental delay or profound deformations or malformations. Third, these fetal problems must be causing *extreme* suffering or hardship for the mother or the fetus. Fourth, the life of the mother should be in danger. Fifth, both the mother and the father give their consent to the procedure. The physician performing the abortion shall not be penalized for the performance of these services' (Hedayat *et al.*, 2006).

Besides β-thalassaemia major, under the terms of the legislation, a detailed list of maternal medical and fetal conditions was provided for which termination of pregnancy was permitted prior to four months gestation (Tables 13.1 and 13.2). Although the rationale behind specific inclusion or exclusion from these quite highly defined listings has not been published, they clearly were devised with specialist clinical genetics advice. It has been proposed that additional genetic disorders for which therapeutic termination is permissible could be added in future (Bazmi *et al.*, 2008). Given the multi-ethnic nature of Iranian society, and the very high rates of consanguineous marriage in some communities (Saadat *et al.*, 2004; Abbasi-Shavazi *et al.*, 2008), this would seem to represent a logical progression.

Passage of the 2005 legislation has radically changed the balance of approval for medical termination of pregnancy within Iran. Thus, during 1999–2000 in Teheran, 17% of all permissions for abortion were for fetal disorders (all for β-thalassaemia major) and 83% for maternal disorders, but during 2006/2007, the indications given for pregnancy termination had altered to 85% for fetal disorders and 15% for cases involving maternal disease (Bazmi *et al.*, 2008). However, in surveys both of clinicians (Karimi *et al.*, 2008) and the parents of children with β-thalassaemia major (Karimi *et al.*, 2010), a small minority declared themselves not to be in favour of the early termination of pregnancy on personal religious grounds.

From a global religious perspective, it is important to note that the unacceptability of medical termination of pregnancy also applies to the followers of many other faiths. For example, there are extensive prohibitions to pregnancy termination within Buddhism, the Hasidic branch of Judaism, and for members of the Christian Roman Catholic and Evangelical Churches, who believe that

Table 13.1 *Iran: Maternal diseases in which medical termination of pregnancy is permitted*

Maternal diseases
Valvular heart disease, function class III or IV heart failure, which is not reversible to function class II
Non-coronary acute heart problem, function class III or IV failure, such as myocarditis and pericarditis
Any history of dilated cardiomyopathy in previous pregnancies
Marfan syndrome accompanied by ascending aorta diameter > 5 cm
Eisenmenger syndrome
Gestational fatty liver
Oesophageal varices (grade III)
History of bleeding from oesophageal varices after portal hypertension
Uncontrollable autoimmune hepatitis
Renal failure
Hypertension (uncontrollable with permitted drugs in gestation period)
Any pulmonary diseases such as emphysema, fibrosis, kyphoscoliosis, diffuse bronchiectasis with pulmonary hypertension even in the mild form
Hypercoagulability in which heparin administration may worsen the mother's existing life-threatening disease
HIV infection that has entered the AIDS disease phase
Uncontrollable active lupus with the involvement of a major organ
Vasculites with the involvement of major organs
All space-occupying lesions of the central nervous system in which beginning treatment causes danger to the fetus and not beginning treatment causes danger to the mother
Phemphigus vulgaris, severe and generalised psoriasis and end-stage melanoma, which leads to serious fatal danger for the mother
Epilepsy that is resistant to treatment, despite giving multiple drugs

Source: Hedayat *et al.* (2006)

human life begins at conception and so any attempt to interfere with its natural development is proscribed.

Besides their religious beliefs and level of religious commitment, parental attitudes to pregnancy termination may be influenced by the probable expression of the disease, with some possibility of flexibility when faced with the in utero diagnosis of a severe disorder that would be incompatible with postnatal life (Saleem *et al.*, 1998; Ahmed *et al.*, 2006). This situation was observed in Israel, with both Jewish and non-Jewish couples usually opting to terminate a pregnancy when Down syndrome or another clinically significant aneuploidy was diagnosed by chorionic villus sampling or amniocentesis. But Arab couples in particular were less likely to consider termination of pregnancy if a haemoglobinopathy had been diagnosed (Zlotogora, 2002b).

Table 13.2 *Iran: Fetal deformities and disorders incompatible with life for which medical termination of pregnancy is permitted*

Fetal conditions
Osteogenesis imperfecta
Osteochondrodysplasia
Infantile osteopetrosis (malignant form)
Bilateral renal agenesis
Polycystic kidney (recessive form)
Multicystic dysplastic kidneys
Potter's syndrome
Congenital nephritic syndrome (with hydrops)
Chromosomal disorders leading to degenerative lesions and brain and kidney involvement, such as vertebrae, anus, cardiovascular tree, trachea, oesophagus, renal system and limb buds (VACTERL) syndrome
Severe bilateral hydronephrosis
α-thalassaemia with hydrops fetalis
Homozygote thrombotic disorders (i.e. protein C or factor V Leiden deficiency)
Trisomies 3, 8, 13, 16, 18
Anencephaly
Fetal hydrops
Cri du chat syndrome
Holoprosencephaly
Syringomyelia
Cranioschisis
Meningoencephalocele or hydroencephalocele
Thanatophoric dysplasia
Ichthyosis congenital neonatum
Schizencephaly
Exencephalia

Source: Hedayat *et al.* (2006)

Newborn screening for genetic disease

To date, newborn screening for genetic disease has largely been restricted to wealthier countries, an early exception being a pilot programme in the state of Karnataka, South India, throughout the 1980s, which provided free testing for amino-acid disorders on more than 112 000 neonates born in public hospitals (Appaji Rao *et al.*, 1988), with a smaller associated study offering neonatal screening for glucose 6-phosphate dehydrogenase deficiency (G6PD). Probably because of the mixed religious and caste background of the study population, and despite approximately 30% uncle–niece or first-cousin parentage (Bittles *et al.*, 1991), no strong association was identified between consanguinity and the incidence of amino-acid disorders such as hyperphenylaninaemia, tyrosinaemia, homocystinaemia, glycinaemia and branched-chain amino acidaemia.

But the high prevalence of positive results obtained with female neonates was an unexpected aspect of the G6PD screening programme (Ramadevi *et al.*, 1994).

Neonatal screening for genetic disorders is now quite well established in a number of Middle Eastern and North African countries with high levels of consanguineous marriage, although the spectrum of tests offered vary markedly and just four countries have national screening programmes (Hamamy & Bittles, 2009). For example, in Oman, screening for hearing impairment (Khandekar *et al.*, 2006) and congenital hypothyroidism (Joshi & Venugopalan, 2007) are offered on a national basis, with selective screening for haemoglobinopathies and metabolic disorders, the latter via tandem mass spectrometry. Egypt and Saudi Arabia also offer national screening for congenital hypothyroidism (Saadallah & Rashed, 2007), Kuwait has a selective screening programme for inborn errors of metabolism (Abdel-Hamid *et al.*, 2007), and Qatar has established a collaborative neonatal screening programme with a centre in Germany to investigate metabolic and endocrine disorders (Lindner *et al.*, 2007).

It has been recognized that a number of immigrant communities in the UK have higher rates of haemoglobinopathies (Hickman *et al.*, 1999; Modell *et al.*, 2001), which in some cases may be associated with a preference for consanguineous marriage. Since 2007/2008, the National Health Service in England has operated a Sickle Cell and Thalassaemia Screening Programme that incorporates a Family Origin questionnaire (Ryan *et al.*, 2010), with prenatal diagnosis offered for both disorders and neonatal screening for sickle cell disease (SCD). By 2008/2009, some 670 000 neonates were screened annually for SCD with an average of 360 cases identified, which indicates that in England, sickle cell disease is now more prevalent than cystic fibrosis (NHS, 2010). In the UK as a whole, it has been estimated that approximately 1250 individuals have SCD and there are some 1000 cases of severe thalassaemia, a large majority of whom are of Afro-Caribbean and Asian origin, respectively (APPG, 2009).

Inductive screening in consanguineous families

Where consanguineous marriage is prevalent, it has been proposed that the screening of extended families for recessively inherited genetic disorders can be a highly efficient process, because once the causative mutation has been identified, all carrier and affected members of the family would be expected to exhibit the same mutation (Ahmed *et al.*, 2002; Modell & Darr, 2002). The conclusions of the original study involving ten Pakistani families, each identified through an index case of β-thalassaemia (Ahmed *et al.*, 2002), were subsequently verified in similar investigations also conducted in Pakistan, with

Table 13.3 *Definition of genetic counselling*

Genetic counseling is the process of helping people understand and adapt to the
medical, psychological and familial implications of genetic contributions to
disease. This process integrates the following:
Interpretation of family and medical histories to assess the chance of disease
occurrence or recurrence.
Education about inheritance, testing, management, prevention, resources and research.
Counseling to promote informed choices and adaptation to the risk or condition.

Source: Resta *et al.* (2006)

genetic counselling provided to family members to reduce the future burden of
disease (Baig *et al.*, 2008).

Some caution, however, is merited since an earlier study of β-thalassaemia
in Pakistan reported that, on average, 11.7% of cases born to first-cousin parents were compound heterozygotes rather than homozygotes, and in Punjab
province, 16.5% of cases were compound heterozygotes. Among patients
whose parents were related but not first cousins ($F < 0.0625$), 41.6% were
heterozygotes for the disorder, rising to 69.2% among Pathans (Ahmed *et al.*,
1996), which strongly suggests that inter-family and/or inter-community marriages had occurred in a prior generation(s). On the basis of these results, and
similar findings with respect to lipase H (*LIPH*) mutations causing autosomal recessive woolly hair/hypothichosis in two large consanguineous Pakistani
families (Petukhova *et al.*, 2009), the construction of a detailed and extended
family pedigree would be an advisable first step before embarking on induc-
tive screening – especially because in other populations, multiple different
mutations for specific single gene disorders have been identified within consanguineous families and close-knit communities (Table 10.2).

Family- and community-oriented counselling in consanguineous marriages

In Western societies, genetic counselling practice has traditionally been based
on principles of personal autonomy, non-directiveness, confidentiality, benefi-
cence and non-maleficence. In their revised guidelines for genetic counsellors,
the US National Society of Genetic Counselors (Resta *et al.*, 2006) developed
the comprehensive definition reproduced in Table 13.3 which, by its general
nature, should be applicable to developing as well as developed societies.

As described in Chapters 9–11, in societies with a high prevalence of consanguineous marriage, rare genetic disorders frequently cluster in specific families

and communities. Perhaps for this reason, premarital genetic counselling is in high demand in Iran and accounted for 80.0% of all referrals in Shiraz, southern Iran, with consanguinity given as the reason for referral by 89.1% of these clients (Fathzadeh *et al.*, 2008). Although it might be supposed that couples would decide to revise their marriage plans in light of the risk estimates provided, as previously discussed with respect to premarital diagnosis of carrier status for thalassaemia in several Middle Eastern countries, in reality the majority of couples who requested counselling proceeded with marriage in deference to family and social ties.

Under these circumstances, and contrary to counselling practice in Western countries, family- or community-oriented genetic counselling may offer a more productive approach to the prevention of genetic disorders rather than counselling provided to individuals or couples (Bittles, 2008). Within many families and communities, it has been found that there is a high information yield with respect to carriers and couples at-risk, and family members can better understand the nature of the condition if they have had direct contact with an affected child (Modell *et al.*, 1997; Modell & Darr, 2002; Al-Gazali *et al.*, 2006).

Problems can arise, however, in the provision of advice to relatives concerning their at-risk status. This situation emerged in the UK Pakistani community where, despite a high level of consanguinity, many couples regarded genetic information as private to themselves and were unwilling to share their knowledge with the wider family on the grounds that it could lead to their personal stigmatization, be damaging to their child and their marriage, and adversely affect the marriage prospects of other family members (Shaw & Hurst, 2009).

Following the birth of a child with a genetic disorder, a wife may express concern that her husband and his family will consider divorce and remarriage (Gomaa, 2007), especially if the child is male and the couple is advised that the mode of inheritance of the disorder is X-linked recessive and the disease gene has therefore been transmitted by the wife to their son (Hamamy & Bittles, 2009). As discussed in Chapter 5, divorce would be a less likely outcome when the couple are consanguineous, but in polygynous societies, the husband may consider an additional marriage to a distant relative or to a female to whom he is unrelated.

The right of family members to access genetic information may conflict with a proband's right of confidentiality (Rantanen *et al.*, 2008). Initial World Health Organization guidelines on ethical issues in medical genetics suggested that where appropriate, and as part of their general duty to educate, counsellors should inform clients of genetic information that could be of importance to their relatives and, in turn, these individuals might reasonably be invited to request their relatives to seek genetic counselling (WHO, 1997).

But a more recent report changed the balance of the advice recommended for highly consanguineous communities in which arranged marriage is commonly practised (WHO, 2006). Because a genetic disease potentially could impact on all members of the extended family, respect for patient confidentiality may need to be considered alongside the rights of other family members for whom information about genetic risk could influence decision making on their own health, marital arrangements and reproductive choices. The importance of community opinions was highlighted in an Israeli Arab Muslim community where parents were principally concerned about what people might say about them and their image in society with regard to parenthood, regardless of the disease that had been identified in the family (Saleem *et al.*, 1998).

Consanguinity and genetic disease registers

Genetic registers incorporating details of long-term follow-up and a proactive approach to at-risk subjects have been recommended as a means of improving access to genetic counselling for families with inherited disorders (Wright *et al.*, 2002). The primary aim is to provide information to families on the implications of carrier status for an inherited disorder. Where carrier status cannot be diagnosed, a secondary aim may be to inform them of the possibility that future affected pregnancies could be circumvented, either by avoidance of marriage with close biological relatives or, where acceptable, via prenatal diagnosis and therapeutic termination of pregnancy. Registers should optimally be designed and organized as part of an ongoing consultation process with the communities at highest risk because during the 1990s, a lack of understanding of the religious and cultural beliefs of the UK Pakistani community significantly hindered the delivery of appropriate genetic services (Modell *et al.*, 2000, 2001).

Genetic registers are especially useful in countries and migrant communities where community endogamy and consanguineous marriage have long been the norm because, as previously noted, individual inherited disorders may be restricted to individual families or sub-communities. Examples of genetic registers in the Middle East include the Arab Genetic Disease Database (Teebi *et al.*, 2002) and online websites such as the Catalogue of Transmission Genetics in Arabs (CTGA), which is maintained by the Centre for Arab Genomic Studies (CAGS) in the United Arab Emirates (Tadmouri *et al.*, 2006). In both cases, healthcare providers in Arab communities can access information on relevant genetic disorders freely and confidentially. The Israeli National Genetic database provides a similar comprehensive service, with comparative details of genetic disorders diagnosed in the different religious and ethnic communities within that country, along with information on the mutation frequencies of specific disorders in each religious community (Zlotogora *et al.*, 2007).

The potential wider application of disease registers to developing countries with large populations has recently been demonstrated with respect to β-thalassaemia in India, where patients' mutation profiles clearly demonstrated the effect of their background characteristics, including the influence of geography and of ethnic, religious and caste endogamy. In study populations from South India, the long-term preference for consanguineous marriage also was obvious in the high levels of observed mutation homozygosity (Sinha *et al.*, 2009; Black *et al.*, 2010; Sinha *et al.*, 2011).

Consanguineous marriage in a public health context

Considerable strains are placed on existing health services as a result of the recent rapid transition of developing countries from a largely communicable to a non-communicable pattern of disease, and the increasing life expectancy of affected individuals who in former generations would have died in infancy or early childhood. The level of genetic literacy within the general population is a very important factor, but health professionals and the capacity of national and local health infrastructures also play critical roles. In the latter respect, the beneficial role of preconception care and periconceptual clinics has been emphasized in developed countries (Czeizel *et al.*, 2005; Dunlop *et al.*, 2007). Although preconceptional ancestry-based carrier screening for disorders such as cystic fibrosis and haemoglobinopathies was perceived as desirable by couples in the Netherlands, the effort and time needed for participation were negative factors, as was the possibility of some form of genetic discrimination (Lakeman *et al.*, 2009).

A need for urgency in the introduction of genetics into the health services offered in developing countries has been recognized (Alwan & Modell, 2003). The communication of quite complex genetic concepts in a manner that is readily comprehensible to members of the general public, a significant proportion of whom may be illiterate, nevertheless can be problematic and, in the Middle East, it has largely been undertaken by local health workers and primary healthcare centres (Al Arrayed, 2005; Al-Gazali *et al.*, 2006). The picture is somewhat different in Western countries, where consanguineous marriage is largely restricted to immigrant communities and couples referred on grounds of consanguinity may at least initially encounter a lack of knowledge and/or an unsympathetic understanding of their marital relationship (Hoodfar & Teebi, 1996; Nelson *et al.*, 1997; Port *et al.*, 2005; Bishop *et al.*, 2008; Shaw, 2009).

The role of medical and ancillary staff

Irrespective of the preferred marital patterns, in most developing countries, the current shortage of staff trained in medical genetics means that genetic counselling responsibilities can be devolved to staff whose specialist training is in disciplines other than genetics, and who may have an incomplete grasp of the potential complexities of the subject. Clients and families may also request directive advice, and failure to provide the expected guidance can be misunderstood as indicating indifference or misinterpreted as a sign of self-doubt and a lack of medical knowledge on the part of the consultant (Raz & Atar, 2003a; Eldahdah *et al.*, 2007).

Additional problems that may be encountered include the legal, social and religious criteria that govern pregnancy termination, with an absence of up-to-date national guidelines on prenatal genetic diagnosis and selective termination of an affected fetus (Hessini, 2007; Hamamy & Bittles, 2009). As discussed in Chapter 6, in communities with a long tradition of endogamy and consanguineous marriage, the observed levels of genomic homozygosity in first-cousin progeny may be significantly higher than the coefficient of inbreeding $F = 0.0625$ usually assumed in Western populations. Therefore, in cases where direct diagnostic testing is not possible, care needs to be exercised in the cut-off risk levels calculated for genetic-counselling purposes.

Primary healthcare centres occupy a position of particular importance in developing countries because they are often the sole point of access to advice and treatment. In the absence of specialist medical geneticists, paediatricians provided with appropriate short training updates could quite effectively assume responsibility for counselling on common genetic disorders (Hamamy & Bittles, 2009). However, as has been consistently indicated, the adverse influence of consanguinity on health is more often associated with disorders that are uncommon and therefore may not have been previously encountered in the course of normal paediatric care. The very rapid development of e-medicine and mobile telephony may provide at least a partial answer to this problem in more remote regions of developing countries and for those whose access to care is limited by financial considerations.

Lessons in improving genetic services in primary care can be learned from developed countries, where the caseload occupied by patients with inherited disorders has been steadily increasing. Initial misgivings by clinicians regarding involvement in the care of patients with genetic disorders can be overcome through minimal training in the taking of family histories, in which most general practitioners (GPs) already have considerable prior experience, together with integrated GP educational programmes in genetics, the involvement of complementary genetic nurse specialist outreach clinics, and computer-based decision support systems (Emery *et al.*, 1999; Rose *et al.*, 1999; Emery, 2005; Qureshi *et al.*, 2005; Clyman *et al.*, 2007).

International knowledge transfer

Transfer of the collective medical genetics experience gained in developed countries to developing nations through exchange placement visits; DVD-based educational programmes; the narrow-casting of lectures, tutorials and clinics; online electronic communication; and mobile telephony should be readily achievable and cost-effective. Clearly, the involvement of major international institutions, more specifically the World Health Organization and the World Bank, will be of major importance because they can operate and advise at the highest governmental levels. However, knowledge transfer between developing countries and the intervention of local, national and international non-governmental organizations (NGOs) specializing in healthcare delivery can also play a major role, especially in topics such as intellectual and developmental disability which otherwise may receive limited attention from clinicians and major funding agencies.

To facilitate these endeavours in countries and communities where consanguineous marriage continues to be favoured, there is an urgent need for up-to-date comprehensive, multi-disciplinary research and for sympathetic and clear-minded decision making by national legislators. Currently, inadequate data on the magnitude and economic burden of genetic disorders; low genetic literacy among the public and health professionals; the mistaken belief that diagnosis and prevention of genetic diseases require sophisticated technology; and cultural, social and religious restrictions all contribute to a notable lack of progress (Hamamy & Bittles, 2009). This situation is especially important in the case of consanguineous marriage, where developing countries all too often inappropriately copy and adopt what they suppose to be enlightened Western attitudes and standards.

Commentary

A basic lack of understanding of genetics remains a problem in developed and developing countries alike but, to an important extent, the public lack of knowledge of genetic disease can be overcome through the placing of accurate and responsible articles in the mass media. The ability of health practitioners to convey risk in an unambiguous and readily understood manner is a critical facet of genetic screening, education and counselling programmes, and different tactics and approaches may be needed according to circumstances. Risk and recurrence estimates expressed as relative risks, odds ratios, or attributable risks (i.e. the fraction of cases in a population that can be attributed to a particular risk factor) are very useful in epidemiological studies. Their application may, however, be confusing in a genetic counselling setting where the probability of an adverse outcome needs to take into account factors such as the background population risk, degree of consanguinity and relevant family history (Bennett *et al.*, 2002; Hamamy *et al.*, 2011). Especially when dealing with a topic as potentially sensitive as consanguineous marriage, the avoidance of any potential misunderstanding or misinterpretation by clients and their families becomes paramount.

14 *Whither consanguineous marriage?*

Introduction

The topic of consanguineous marriage logically encompasses and contributes to virtually all aspects of human life, including our evolutionary and historical past, literature and culture, social and economic structures, civil and religious law and, very important, health and well-being. Consanguinity also impacts significantly on less obvious but important aspects of contemporary human life, such as gender (in)equality. As was discussed in Chapter 4, it is instructive that in the often-heated scientific and medical debates on consanguinity during the late nineteenth and early twentieth centuries, eminent medical personages such as Sir Arthur Mitchell and Sir Archibald Garrod counselled strongly against the assumption that if a child born to first-cousin parents was diagnosed with a specific illness, parental consanguinity necessarily was the causative factor (Mitchell, 1862, 1864–5; Garrod 1902, 1908).

An even more forthright opinion was expressed by Dr John Langdon Down, who first clinically identified and described trisomy 21, the eponymous Down syndrome. In an article on consanguinity originally published in the London Hospital Reports of 1866, Down stated: 'No-one, I think, with a previously unbiased mind, can read the numerous examples which are cited to prove as the result of such unions sterility, deaf-mutism, idiocy and other characteristics of degeneration, without coming to the conclusion that the cases from which they argue are selected ones, and that the cause they advocate is damaged by special pleading where there should have been judicial deliberation' (Down, 1887).

In recent years, a comparable reticence search for balance has seldom been displayed by political commentators and the popular press, with discussion on consanguineous marriage frequently reduced to 'Is consanguinity good or bad?', and 'Should consanguineous marriage be permitted or banned?' Simplistic dichotomies of this nature may be suitable for newspaper headlines, but they singularly fail to grasp the complexity of the topic as discussed in earlier chapters and summarized in two-dimensional form only in Figure 14.1 – a more appropriate and realistic representation of the different possible and actual interactions would require a multi-dimensional display beyond the present capacity of the printed page.

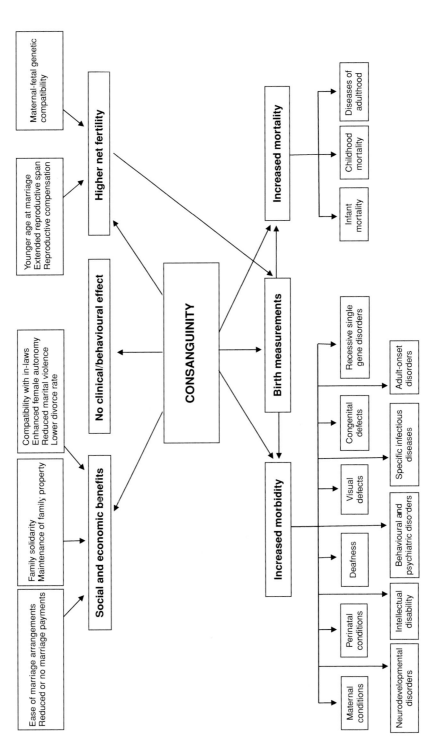

Figure 14.1 Consanguineous marriage: social, economic, behavioural and biological influences and clinical outcomes

Unfortunately, public suspicion of consanguinity has further been compounded by failure to distinguish between the concept of an arranged marriage and enforced marriage. Until the twentieth century, arranged marriage was very much a part of rural life in Western countries (Chapter 6) and among European nobility (Chapter 1), and it remains a valued tradition in many Asian, African and South American societies. By comparison, the enforced marriage of a woman or girl against her will is justifiably regarded as a serious crime which merits unambiguous condemnation. Therefore, in discussing the rights or wrongs of cousin unions, it is important to emphasize that where marriages have been arranged, whether consanguineous or non-consanguineous, they have proceeded with the approval or at least the acquiescence of both the bride and the groom.

Can consanguinity be beneficial to human health?

As indicated in Chapters 8–11, in overall terms, the data suggest that non-consanguineous progeny have a modest but statistically significant health advantage over their consanguineous counterparts, which is in agreement with the genetic concept of heterozygote advantage. However, individuals who are homozygous at specific loci, or even across large segments of their genome, may not necessarily be disadvantaged in all cases and under all circumstances, as demonstrated by the homozygous expression in nonagenarians and centenarians of specific histocompatibility alleles (Takata *et al.*, 1987) and the *YTHDF2* gene in Alu-rich genomic domains (Bonafè *et al.*, 2001; Cardelli *et al.*, 2006).

Consanguinity and major haemoglobinopathies

Although earlier predicted on a theoretical basis (Haldane, 1949), the first and still one of the few unambiguous examples of heterozygote advantage in humans was reported with respect to transmission of the protective HbS mutation by people with sickle-cell trait living in malarial regions (Allison, 1954). Subsequently, it was shown that individuals who possess the HbAS genotype have enhanced resistance to malaria due to the reduced ability of the *Plasmodium falciparum* parasite to grow in HbAS erythrocytes conferred by the combined action of innate and malaria-specific acquired immunity (Williams *et al.*, 2005).

Using data on the pre-intervention distribution of the HbS gene, a strong positive relationship between HbS allele frequencies and malaria endemicity has been shown for Africa, thus providing geographical confirmation of the malaria-resistance hypothesis (Piel *et al.*, 2010). At the population level, it could be assumed that consanguineous marriage would initially lead to increased numbers of offspring who are homozygous HbSS and so express sickle cell disease, with a resultant reduction in the frequency of the mutant allele. But according to the relative fitness benefit enjoyed by HbAS individuals and the negative impact of malarial parasitism on the survival and reproductive capacity of individuals who are HbAA, an equilibrium gene frequency for the HbS allele should result (i.e. the concept of balancing selection).

It has been widely assumed, although as yet without clear supporting evidence, that heterozygote advantage also explains the high frequencies of α- and β-globin chain variants which cause α- and β-thalassaemia, respectively, and are present in many populations living in regions with endemic malaria (Weatherall & Clegg, 2001). In an extended series of reports of computer-based simulations on the global distribution patterns and prevalence of α-thalassaemia, it has been proposed that consanguinity may actually exert an overall protective effect in such populations (Denic *et al.*, 2007a, 2007b, 2008b, 2011), and that this protective effect may extend to sickle cell disease (Denic *et al.*, 2008a).

There is some evidence that mild forms of α-thalassaemia (i.e. the α^+-thalassaemias) may be maintained by both homozygote and heterozygote protection (Weatherall & Clegg, 2001), and early childhood exposure to both *Plasmodium falciparum* and *Plasmodium vivax* also appears to offer a measure of protection against malaria and other non-malarial infections in later life (Allen *et al.*, 1997). More speculatively, it may be that the α-thalassaemias are an integral component of some as-yet-unidentified and beneficial coevolved gene complexes, especially because α-thalassaemia appears to be the most common monogenic disease in present-day or formerly malarial areas (Hedrick, 2011). Besides selection, the very large number and complexity of α- and β-globin chain variants present in different populations also reflect local founder effects, drift and migration (Sinha *et al.*, 2009; Black *et al.*, 2010).

As with all computer simulations, the veracity or otherwise of the end-result obtained in modelling α-thalassaemia is critically dependent on the assumptions and complexities that are incorporated into the programming. By their very nature, even the most sophisticated current models of human demographic and genetic structure are at best caricatures of the vicissitudes of everyday life faced by people living in the tropical regions, where malaria has long been most prevalent (Bittles, 2011b). Therefore, until such time as a causative relationship can be demonstrated between consanguinity and the incidence of α-thalassaemia or any other form of haemoglobinopathy, of necessity the conclusions of modelling exercises remain speculative.

Consanguinity and the role of epigenetics

The potential role of epigenesis in human gene expression and human disease has been a topic of increasing interest during the past decade, with the multi-part proposal that hereditary disorders of the cellular epigenetic apparatus lead to single gene developmental defects such as Beckwith–Wiedemann, Prader–Willi and Angelman syndromes; they are responsible for disruption of the stem-cell programme in cancer epigenetics; and are implicated in the interactions among the epigenome, the genome and the environment that result in common adult-onset disorders (Feinberg, 2007). If this is so, then epigenetic changes could play a significant role in many of the inherited disorders reviewed in Chapters 10 and 11. But, as yet, causal relationships between epigenetic expression and consanguinity have largely been hypothetical or inferential in nature, e.g. in the putative roles of noncoding microRNA (miRNA), DNA methylation and histone modifications in hereditary sensorineural nonsyndromic hearing loss (NSHL) (Friedman & Avraham, 2009) and how genomic autozygosity in consanguineous progeny could influence predisposition to cancer (Bacolod *et al.*, 2009).

It has been suggested that inbreeding depression may result from epigenetic mechanisms, including cell-specific DNA hyper- and hypomethylation, RNA interference, histone modifications, and chromatin remodelling and canalization/decanalization rather than chromosomal events leading to DNA sequence alterations (Biémont, 2010; Nebert *et al.*, 2010). For this reason, consanguineous families have been increasingly recruited for epigenetic studies to determine the nature of the mechanisms that underlie a variety of genetic disorders (Carr *et al.*, 2009; Hayward *et al.*, 2009; Meyer *et al.*, 2009; Turan *et al.*, 2010; Brun *et al.*, 2011).

The complexity of this situation in terms of hypomethylation alone, however, was illustrated by a report of two sisters with transient neonatal diabetes mellitus and Beckwith–Wiedemann syndrome born to highly consanguineous Turkish parents. Although differing in disease severity, both sibs exhibited a mosaic spectrum of hypomethylation at maternally methylated loci, which was interpreted as indicating a novel autosomal recessive defect of methylation (Boonen *et al.*, 2008).

What will be of major interest and potential application in terms of correcting at least some inherited disorders is the degree to which epigenetic memory occurs (i e. the ability to transfer epigenetic information between generations), and the extent to which epigenetic reprogramming is possible in different cell types and tissues (Migicovsky & Kovalchuk, 2011) – in effect, the extent to which information can be transferred to progeny on the basis of non-Mendelian inheritance rather than reliance on the transmission of DNA sequences.

Consanguinity and altruism

In its initial form, the theory of altruistic behaviour sought to reconcile natural selection with behaviour by an animal which was not to its own advantage, yet appeared to be advantageous to other members of its species that were not its direct descendants (i.e. the concept of inclusive fitness) (Hamilton, 1963, 1964). From a genetic perspective, a gene causing altruistic behaviour towards siblings, who on average have 50% of their genes in common with an altruist (coefficient of relationship, $r = 0.5$), will only be positively selected if the behaviour results in a gain that is more than twice the loss suffered by the altruist. For half-sibs and uncles, aunts, and nieces ($r = 0.25$), the necessary gain would have to be more than four-fold, whereas for first cousins ($r = 0.0125$), a greater than eight-fold gain would be needed.

Through time, the theory underlying the evolution of altruistic behaviour has been expanded to cover a significantly wider range of possible scenarios (Axelrod & Hamilton, 1981; Matessi & Karlin, 1984), with the roles of social selection, cultural evolution, group competition, religious affiliation and even morality variously canvassed as driving factors in encouraging and enabling large-scale human prosociality (Simon, 1990; Castro & Toro, 2004; Bowles, 2006; Boehm, 2008; Norenzayan & Shariff, 2008; Bell *et al.*, 2009; Ayala, 2010).

As discussed in Chapter 5 and illustrated in Figure 14.1, among humans the benefits of altruistic behaviour would seem to have their most obvious application in terms of the high levels of resource-sharing, resource-preservation and mutual protection typified by highly consanguineous communities. However, it appears that a measure of altruistic behaviour may be evoked even in distant relationships, as indicated in a study using names as cues for kinship. When assistance was requested from a large group of individuals via their email addresses, ostensibly by a same-sex person, the addressees were most likely to respond helpfully when they shared the same first name and family name as the person seeking assistance and least likely when they shared neither (Oates & Wilson, 2001).

After control for the effects of population stratification, a map of adolescent friendship networks established through the US National Longitudinal Study of Adolescent Health has shown homophily (i.e. a positive association between individuals who shared the same genotype), at one of the six genotypes tested, the dopamine D2 receptor locus of the gene *DRD2*. By comparison, the same network showed heterophily (i.e. a negative association indicating significant avoidance in friendship network terms), at the cytochrome P4502A6 locus of *CYP2A6* (Fowler *et al.*, 2011).

Intriguingly, the same results were obtained in an independent survey of individuals who were members of the Framingham Heart Study Social Network, which is based on individuals who are members of older generations. If reproducible across societies, it would therefore appear that besides the phenotypic influences that have been identified in assortative mating, the establishment of non-reproductive human friendships is subject to underlying genotypic selection, at least with respect to the genotypes tested, all of which are known to exert significant effects on behavioural and personality traits (Fowler *et al.*, 2011).

Consanguinity and donor matching in organ and cell-mediated transplantation

The choice of donors for organ and cell-mediated transplantation is an issue in which consanguinity quite clearly offers major advantages, and it also is a subject which impinges both on the concept of altruistic behaviour and on ethical and moral imperatives in different societies (Gohh *et al.*, 2001; Al-Khader, 2002; Kishore, 2005; Concejero & Chen, 2009). The potential choice of successful donors is extensive in countries with a high prevalence of consanguineous marriage and large family sizes and, following the long-term practice of intra-familial marriage, successful matching of the HLA repertoires of recipients and donors can be predicted with some confidence.

This concept was raised in a paper on the immunological reconstitution of a Lebanese child with severe combined immunodeficiency whose parents were first cousins and whose father acted as an HLA-identical donor (Geha *et al.*, 1976) and followed-up by a bone-marrow transplantation for a South Asian child in the UK whose donor was a first cousin (Taylor *et al.*, 1995). A subsequent study in Israel to identify potential allogeneic stem-cell transplant donors showed that in 62% of Arab cases and 40% of Jewish patients, the volunteer core family donors (i.e. first-degree relatives), were HLA-identical (Klein *et al.*, 2005). Further, in extended family searches among consanguineous families in both the Arab and Jewish communities (e.g. based on second-degree relatives such as grandparents, uncles and aunts), there was a 20% HLA match among the Jewish donors but a 64% match in the Arab families reflecting their greater cumulative level of consanguinity.

The increased availability of family donors for bone-marrow and stem-cell transplantation in populations with a high prevalence of consanguinity has been verified for a wide range of inherited disorders in Turkey (Balci *et al.*, 2010) and Iran (Hamidieh *et al.*, 2011). This option would be very feasible in many other Asian countries where allogeneic graft programmes are established, including the Eastern Mediterranean region of the World Health Organization (EMRO) (Aljurf *et al.*, 2010), especially as family sizes decline and matched-sibling donors reduce in number. For Muslim communities (e.g. in EMRO), it also is important that transplantation from living unrelated donors is ethically acceptable within Islam (Al-Khader, 2002).

Extended family searches for HLA-matched donors have been most effective in families and communities which are known to be consanguineous. The success of the strategy, however, has led to the recommendation that family searches also are adopted for the recruitment of volunteer donors in the populations of more developed countries, where consanguinity is generally believed to be rare or may be restricted to specific ethnic and/or religious communities (Schipper *et al.*, 1996; Hurley *et al.*, 1999; Heemskerk *et al.*, 2005; de Medeiros *et al.*, 2006; CCLG, 2010). In such cases, runs of homozygosity (ROH) principally focused within the major histocompatibility complex (MHC) could usefully be employed to screen potential donors.

Human health and the declining prevalence of consanguineous marriage

As discussed in Chapter 11, long-term studies conducted on the Dalmatian Islands of Croatia indicated a positive association between village and island endogamy and an unexpectedly wide range of common adulthood disorders, including hypertension, coronary heart disease, stroke, cancer, uni/bipolar depression, asthma, gout, peptic ulcer, renal stones and osteoporosis (Rudan *et al.*, 2002a, 2003a, 2003b, 2004, 2006).

Analysis of genealogical data for these populations covering four to five generations showed substantial levels of consanguinity in some communities, with mean coefficients of inbreeding ranging from $\alpha = 0.002$ to 0.049 calculated at village level, indicating major variations in local marriage patterns that reflected the history and geographical location of each settlement (Rudan *et al.*, 2003b). The genealogical data for Dalmatian residents also were supported by individual autozygosity (Froh) values from high-density genome scans of uninterrupted ROHs (McQuillan *et al.*, 2008).

It follows that a secular decline in consanguineous marriage would result in increasing genome-wide heterozygosity, with a consequent decline in the expression of rare recessive disease genes and, in turn, to a widespread reduction in the burden of common genetic diseases involving partially recessive genetic variants of smaller effect (Rudan *et al.*, 2004; Campbell *et al.*, 2007; Rudan *et al.*, 2008; Campbell *et al.*, 2009).

Significant changes in autozygosity during the course of the twentieth century have been reported in two US study populations, with steady decreases in the size and frequency of ROHs greater than 1 Mb long ascribed to sustained isolate break-up and thus an increase in reproductive relationships between people from different regional, ethnic and religious backgrounds (Nalls *et al.*, 2009). At least in theoretical terms, a reduction in endogamous and consanguineous unions will result in reduced selection against recessive disease genes with, ultimately, the establishment of new equilibrium gene frequencies (Haldane, 1939). The average timespan originally envisaged by Haldane for such a change was approximately 4000 years (i.e. some 160 generations). But, given the predicted ongoing increases in the global population (PRB, 2011) and the apparent acceleration in the rate of human adaptive evolution (Hawks *et al.*, 2007), this theoretical approximation could benefit from recalculation.

Consanguineous marriage and contemporary legal issues

The current legal status of consanguineous marriage

As was illustrated in Chapter 3, with a number of specific exceptions such as China, both Koreas and the Philippines, in all other countries there is no civil prohibition on consanguineous marriage up to and including first-cousin unions at national level. The USA represents a specific exception in that the laws which govern differing types and degrees of consanguineous unions vary markedly from state to state (Figure 3.1), with only a few examples where reason or scientific rationale is readily apparent. According to the National Conference of State Legislatures: 'States generally recognize marriages of first cousins married in a state where such marriages are legal' (NCSL, 2011), which echoes the 'place of celebration principle' that has been held to apply within the USA since the nineteenth century (i.e. 'A marriage valid where celebrated is valid everywhere'). But some exceptions did occur in the past, with a District Court Judge in Kansas reporting that in a case of first-cousin marriage, the couple were 'divorced on the grounds of consanguinity' (Arner, 1908).

Despite the commonsense approach of the 'place of celebration principle', some current state laws on consanguineous relationships appear illogical from legal and genetic perspectives, with transgressors facing punishments that appear to be gratuitously non-commensurate with the nature of the alleged offence. Thus, in the State of Wisconsin, which maintains a general ban on first-cousin marriage but allows such unions if the woman is over 55 years of age or where one partner is unable to reproduce (Table 3.3), Section 994.06 of the State Laws on Crimes Against Sexual Morality applies to: 'Nonmarital sexual intercourse with a blood relative related in a degree within which the marriage of the parties is prohibited (generally a kinship of first cousins or closer)' (LRB, 2001). Under the terms of this legislation, the act of consensual sexual intercourse between unmarried adult first cousins is categorized as Incest and is a Grade C Felony, with guilty parties potentially subject to a fine not to exceed US$10 000, or imprisonment not to exceed 15 years, or both (LRB, 2001).

By comparison, in another state of the Union, a man already polygamously married to two wives secretly 'married' both his 12-year-old daughter and his 13-year-old stepdaughter, with the apparent approval of both of his existing adult wives. The incestuous ceremonies allegedly were undertaken by the man in response to a directive from God that he should father two children with each of his child 'brides'. Although his parental rights were terminated in a family court, no criminal charges were pursued because his biological daughter, with whom he had consummated his 'marital' relationship, refused to testify against him (Myers & Brasington, 2002).

Current laws governing sexual relationships and marriage between biological relatives in Western Europe also exhibit a notable lack of coherence that strongly suggests pressures of a primarily social, political or religious nature, with a reasoned genetic perspective often trailing far behind. In the UK, a number of Members of Parliament and the House of Lords, branches of the media, and individual members of the medical profession have encouraged legislation to control or ban first-cousin marriage ($F = 0.0625$) (Dyer, 2005; Rastogi, 2010; Alidina & Walji, 2010; Deech, 2011), and in Norway in 2011, the Foreign Minister proposed the prohibition of first-cousin unions, notwithstanding the fact that the parents of the present Norwegian monarch King Harald V were first cousins (Goll, 2011).

In Germany, the 2010 publication of a book by a Board Member of the Bundesbank, Thilo Sarrazin, entitled *Deutschland schaft sich ab* (*Germany does itself in*), provoked a major public controversy. A major theme of the book was that Germany was in a phase of decline because of immigration from Muslim countries where: 'Entire clans have a long tradition of inbreeding and a correspondingly high rate of disability. It is known that the percentage of congenital disabilities among Turkish and Kurdish immigrants is well above average'. '... perish the thought that genetic factors could be partially responsible for the failure of parts of the Turkish population in the German school system' (Bode *et al.*, 2010). In discussing the heritability of intelligence, Sarrazin's further speculation that there was perhaps 'a specific gene' that 'all Jews share' was judged to be at best insensitive as well as scientific nonsense. The book nonetheless had a highly successful first print run of 250 000 copies (Bode *et al.*, 2010).

Very different attitudes towards different levels of consanguineous unions, including incest, prevail or are under consideration for revision in various Western European countries. Thus, in accordance with the Napoleonic Code of 1811, there are no criminal incest provisions in France; in the Netherlands, special permission may be granted by the Ministry of Justice for a male or female to marry their adopted sibling; and under the terms of the Swedish Marriage Law of 1987, half-sibs ($F = 0.125$) can marry, once again subject to special approval by the Government or an appropriate Authority designated by the Government. Having recognized the legality of uncle–niece and aunt–nephew marriage ($F = 0.125$) in Switzerland some 10 years ago, a change in the Swiss Criminal Code has been proposed to allow sexual relations between consensual adults who are first-degree relatives ($F = 0.25$) that would replace Article 213 of the existing Criminal Code under which sexual intercourse with a direct lineal blood relative or with a full or half-sib is prohibited (Federal Authorities of the Swiss Confederation, 2010; Berthoud, 2011).

In 2007, the German Federal Constitutional Court commissioned the Max Planck Institute for Foreign and International Criminal Law to conduct a detailed comparative analysis of limits on the protection of legal interest in the criminalization of incest. As part of the analysis, the Institute noted that in many countries besides France, including the Netherlands, Turkey, Russia, China, Spain and Israel and the US states of Rhode Island, New Jersey and Michigan, there is no criminal provision for the prohibition of a consensual incestuous relationship between adults (Max Planck Institute, 2011a). As part of its interim findings, the Institute therefore concluded that ' . . . decriminalization of incest is worth considering . . . ', in part because sexual activity between consensual first-degree adults is very rare, and protection against exploitative sexual relationships involving a minor is adequately provided under other criminal legislative measures (Max Planck Institute, 2011b).

It has been observed that what may principally be at stake in such legal considerations is whether the harm that could result from an incestuous relationship is primarily genetic and involve an increased risk of inherited disabilities among offspring, or is mainly considered an offence in social terms. If the former, bringing criminal charges against consenting adults for choosing consanguineous partners who could increase the risk of conceiving a child with a disability could establish an awkward legal precedent for all couples at risk of transmitting a genetic disorder to their progeny, including known carriers of disorders such as cystic fibrosis and sickle cell disease (Farrelly, 2008).

Although couched in public health terms, recent campaigns in Delhi and the northern Indian states of Harayana, Rajasthan and Uttar Pradesh for the Hindu Marriage Act to be amended to ban marriages between couples from the same *gotra* (sub-caste) appear to be mainly socio-religious in nature. As detailed in Chapter 2, any such change in the law would predictably be met with strong opposition from the Dravidian Hindu populations of South India, where uncle–niece and first-cousin marriages remain popular. To date, petitions by proponents of the ban have been rejected by the Delhi Court and the High Courts of Harayana and Punjab on the grounds that there is no such provision on their Statute Books (Bansal, 2010).

Many of these examples seem to reflect varying degrees of intolerance towards couples or communities perceived as minorities, with xenophobia and religious doctrinal differences as additional driving factors that can emerge under specific circumstances. Unfortunately, attempts to pursue balanced and reasoned debate on the topic of consanguineous marriage frequently are thwarted by the inadequacy of the alleged scientific data (Bittles, 2008, 2011c), with hearsay adopted as a poor and inappropriate but all too common substitute for credible information and unbiased analysis.

Consanguinity, international legislation on the right to marry, and genetic discrimination

First-cousin marriage has been legal in England and Wales since the sixteenth century (Chapter 2), apparently without imposing significant dysgenic effects on the population. During the past ten years, most discussion in the UK has centred on the perceived undesirability of cousin marriage in the Pakistani community, with similar sentiments expressed in continental European countries regarding Moroccan, Turkish, Kurdish and other minorities. Any attempt to introduce legislation to ban first-cousin unions would therefore principally affect these recent migrant communities in which consanguineous marriage continues to be practised.

As detailed in Chapter 3, legislation to prohibit first-cousin marriage predictably would lead to major legal challenges in the UK and other Western European countries which are signatories of the United Nations Declaration of Human Rights and the European Convention on Human Rights, both of which prohibit any form of limitation on the right to marry and found a family (UN, 1948; Council of Europe, 1953). It also would contravene the Universal Declaration on the Human Genome and Human Rights, which bans any form of genetic discrimination (UNESCO, 1997). Given these quite explicit internationally agreed protocols, calls for the introduction of bans to prevent couples who are first cousins from marrying would, at best, appear to be futile and have little chance of legislative success.

Assisted reproduction and inadvertent consanguinity

The possibility of consanguinity and even inadvertent sibling incest occurring via anonymous sperm donation has been a controversial topic for several decades, with recommendations on the number of permitted donations per male widely debated in genetic, medical and ethical terms (Curie-Cohen, 1980; de Boer *et al.*, 1995; Le Lannou *et al.*, 1998; Shenfield, 1998; Pennings, 2001; Janssens, 2003), along with the additional risk of miscarriage (Egbase *et al.*, 1996), stillbirths (Skeie *et al.*, 2003) and imprinting disorders (Halliday *et al.*, 2004) following assisted reproductive techniques. More recently, attention also has focused on the number of cycles or donations that an oocyte donor should or may undergo and, as with male sperm donors, the possibility of inadvertent consanguinity (ASRM, 2008).

These matters mostly have been discussed on a somewhat ad hoc basis, with cultural and social concerns predominating (Wang *et al.*, 2007), as in Jewish communities where oocyte donation may create difficulties in establishing the child's religious status within Jewish law (*halacha*) (Feuer, 2011). As has previously been shown, there are no substantial data to indicate that excess prenatal losses (Table 7.2) or stillbirths (Figure 9.1) occur at the level of first-cousin pregnancies, and imprinting disorders such as Beckwith–Wiedemann syndrome are epigenetic rather than genetic in origin (Meyer *et al.*, 2009).

Regulations governing the numbers of offspring a single sperm donor can father vary widely between countries (e.g. with a limit of just five sperm donations permitted in China) (Ping *et al.*, 2011). From 1992 in the Netherlands, an upper limit of 25 offspring per sperm donor was applied, based on the principle that children conceived in this manner would be at similar risk to members of the general population of unintentionally entering a consanguineous relationship in adulthood (Janssens, 2003). Following the introduction of legislation in 2004 which removed the right of donor anonymity, the numbers of donors and sperm banks declined but the permitted number of successful inseminations remained at 25 per donor (Janssens *et al.*, 2006, 2011). At the same time, the demographic composition of the Netherlands has altered quite dramatically, with many migrants of child-bearing age originating in regions where consanguineous marriage is strongly favoured. On the assumption that some of these migrants may request artificial insemination and specify a preference for a donor from their own community, should the permitted number of inseminations per donor be reduced to account for the greater likelihood of inadvertent consanguinity?

The overall situation with respect to possible donation-enabled consanguinity is made more complicated by the fact that according to the findings of the European Society of Human Reproduction and Embryology Task Force, some couples who require assisted reproduction do prefer to enlist another family member as a donor, whether of sperm or oocytes (ESHRE, 2011). Given the increasing use of male and female surrogacy, this issue of intra-familial medically assisted reproduction (IMAR) needs clear national and preferably international guidelines and regulation, both on the permitted genetic closeness of donors and recipients and whether IMAR could be inter- as well as intra-generational (ESHRE, 2011).

In the interim, the Practise Committee of the American Society for Reproductive Medicine has recommended that the number of stimulated cycles for a oocyte donor be limited to approximately six and that to prevent possible future consanguinity, successful donations from a single oocyte donor be restricted to 25 families per population of 800 000 persons (ASRM, 2008). This recommendation contrasts with sperm donation in the USA, where a recent press report indicated that a single donor had fathered 150 offspring. To avoid unknowingly entering into a close consanguineous relationship, children conceived by sperm donation are encouraged to seek information on the identity of half-sibs via the Donor Sibling Registry, www.donorsiblingregistry.com (Mroz, 2011).

Commentary

Current legislation governing consanguineous marriage is an international and, in some cases, a national hotch-potch. In the USA, much of the confusion can be ascribed to the questionable and confused data and motives on which the early legislation was based, but this can scarcely be an excuse for laws banning first-cousin unions that have been promulgated within the last decade. Data on the health outcomes of consanguineous marriage still require verification, with a balanced emphasis needed on positive socioeconomic aspects of consanguinity (Hamamy *et al.*, 2011), including altruistic perspectives. The role of epigenetics is an intriguing topic that will undoubtedly increase in significance, and initiatives such as the 1000 Genomes Project are revealing important detail regarding genetic variation at the individual level. Contemporary legal issues on the right to marry, and sperm and oocyte donation, also need to be addressed, with the UK Warnock Report (Warnock, 1984), which examined many aspects of human fertilization, a helpful and appropriate starting-point.

15 *Consanguinity in context*

Introduction

The principal aim of this book has been to present detailed, balanced and representative information on the history, prevalence and effects of consanguineous marriage, and by so doing to cast appropriate light on past factual errors and misunderstandings that have bedevilled the subject. It also was intended that a holistic approach be devoted to the study of consanguinity and, to this end, Figure 14.1 illustrates the many contributory topics that need to be considered, both those 'above the line' that largely indicate the perceived benefits accruing from consanguineous marriage, and those 'below the line' that summarize the adverse health issues. Considered from this overall perspective, it is clear that over-concentration on any single facet of consanguineous marriage can readily result in unsustainable conclusions being drawn.

Assessing the health outcomes of consanguineous marriage

Although remarkable progress in genomic analysis has been made during the past 20 years, our understanding and knowledge of the human genome and its expression is far from complete. The initial mapping of two human genomes provided evidence of the enormous heterogeneity that exists at the individual level (Levy *et al.*, 2007; Wheeler *et al.*, 2008), and preliminary results from the 1000 Genomes Project have greatly expanded that knowledge with, for example, the information that each individual carries some 250–300 loss-of-function variants in annotated genes and 50–100 variants previously implicated in inherited disorders (The 1000 Genomes Project Consortium, 2010). Important data also are accumulating on the diversity of human copy-number variants and their role in human and disease phenotypic variation (Sudmant *et al.*, 2010).

The application of next-generation sequencing to the detection of structural variations in the human genome will add greatly to our information base, especially when complemented by bioinformatic filtering of exome data to help identify causal mutations (Xi *et al.*, 2010; Rödelsperger *et al.*, 2011). At the same time, the identification and investigation of epigenetic mechanisms in genetic diseases is still at an early stage, especially with respect to consanguinity, and the 'missing heritability' of complex diseases that play major roles in the lives of many individuals has yet to be resolved (Maniolo *et al.*, 2009). The distribution of disease alleles largely along ethnic lines (Moore *et al.*, 2011) is another topic that has resonance in consanguinity studies, especially in the multi-ethnic and multi-community populations of South Asia.

As is readily apparent from the bibliography, since the late nineteenth century, the role of consanguinity in health and disease has been the primary focus of most investigations on consanguineous marriage worldwide. Figures 9.2 and 9.3 and the extensive discussions in Chapters 10 and 11 show that the progeny of consanguineous parents are at greater *average* risk of an adverse health outcome than their non-consanguineous peers. However, it is easier to demonstrate the adverse effects of recessive mutant gene expression than identify possible minor incremental beneficial outcomes, and a significant problem shared with other health topics is the ongoing issue of publication bias, with articles reporting a statistically significant positive outcome more likely to be submitted and accepted for publication than data yielding an equivocal or a negative finding (Easterbrook *et al.*, 1991; Dwan *et al.*, 2008; Hopewell *et al.*, 2009; Loder, 2011).

Consanguinity and population stratification

Major concerns remain that many published data on consanguinity were collected in an inappropriate manner, and they often may not be representative of the study populations from which they were drawn. Failure to differentiate between consanguinity and endogamy has been a very common shortcoming and it is a particularly fraught issue in countries and populations that are characterized by long-standing population subdivisions, whether of a geographical, ethnic, religious or a community basis (e.g. including caste, clan, tribal or *biraderi* membership based on hereditary male lineages). To simply compare health outcomes in the progeny of first-cousin parents with those of unrelated couples without controlling for population stratification implicitly assumes genetic homogeneity and in light of genomic studies is unsustainable (Chapter 6).

Socioeconomic, demographic and environmental variables and consanguinity outcomes

In a majority of consanguinity studies there continues to be no credible control for non-genetic variables, even in the investigation of complex disorders in which social and 'environmental' factors are known to operate. At best, the reader is advised that the cases and controls are of 'similar', 'equal' or 'equivalent' socioeconomic status which does not, and should not, suffice. The eerie absence of appropriate control for factors such as parental age, education and occupation, residence and socioeconomic status that are known to affect early health and survival is especially puzzling in assessments of the influence of consanguinity on infant and childhood morbidity and mortality (Hamamy *et al.*, 2011).

The similar widespread failure to investigate the causative role of common infectious agents, whether bacteria, viruses or intra-cellular parasites, in studies of congenital anomalies has been a serious flaw and may invalidate the claimed results to a significant extent. Likewise, failure to control for the effects of maternal illness (e.g. maternal insulin-dependent diabetes mellitus or maternal epilepsy treated with the anti-convulsant drugs sodium valproate or phenytoin during early pregnancy), can lead to fetal abnormalities mistakenly being ascribed to consanguinity. Although readily acknowledging and personally all too well aware how difficult and time-consuming multidisciplinary investigations can be, in their absence the observed outcome is uni-dimensional rather than multi-dimensional in nature and the results need to be interpreted as such.

Towards a more balanced assessment of consanguineous marriage

Because a minority of first-cousin progeny have greater health risks than the average child of non-consanguineous parents, it could be argued that on these grounds alone, prohibition of marriage between first cousins would be justified. Quite apart from the legal difficulties that almost certainly would arise (see Chapters 3 and 14), it is helpful to consider the derived estimates for adverse consanguinity-associated health effects both from a family viewpoint and on a comparative basis.

Current data indicate that on average, first-cousin offspring experience an additional 3.7% mortality from approximately 28 weeks gestation to 10–12 years of age (see Figure 9.3). In terms of birth defects, they have an additional median risk of 3.3% (Table 10.4) (i.e. a total additional risk of 7% prereproductive morbidity or mortality), which in light of the preceding criticisms of study design and data collection is best regarded as an upper-bound estimate.

Whereas an estimate of this nature has meaning in population terms, its significance is considerably more difficult to convey at the individual or family level. For example, the mean family size in the UK and USA is 2.0 children, and in India and Pakistan family sizes average 2.6 and 3.6 children, respectively (PRB, 2011). In terms of the health or survival of their offspring, what would an estimated additional health risk of 7% actually mean to first-cousin families with 2.0, 2.6 or 3.6 children? And how would an additional health risk of 7% be regarded and assessed in communities where ~20% of children die in the first five years of life?

The perspective is obviously different in high-income countries with low levels of infant and childhood mortality or morbidity. From a Western viewpoint, it is more helpful to contrast the average health effects of consanguinity at first-cousin level with the well-described sequelae of disorders such as Down syndrome (trisomy 21) and fetal alcohol syndrome.

Down syndrome

The additional risk of conceiving and bearing a child with Down syndrome is significantly correlated with maternal age, and composite data from the USA indicate that although the incidence of a 20-year-old woman bearing a child with Down syndrome is 1 in 2000 (0.05%), this increases to 1 in 350 (0.3%) at age 35 years, 1 in 100 (1.0%) at age 40, 1 in 30 (3.3%) at age 45, and 1 in 10 (10.0%) by age 49 years (NDSS, 2011).

People with Down syndrome have a significantly increased incidence of many serious health disorders, including intellectual disability, congenital heart disease, acute leukaemia, recurrent infections, visual and hearing impairments, hypothyroidism and epilepsy (Bittles *et al.*, 2007; Rasmussen *et al.*, 2008). This extensive disease profile causes a major health burden for the individual and it also represents a significant impost in terms of community healthcare during childhood and adolescence (Dye *et al.*, 2011a). Because the median life expectancy of a person with Down syndrome is now approximately 60 years of age, despite accelerated biological ageing and the early onset of Alzheimer disease, ensuring the care of their offspring in later life also is a cause of major practical concern for elderly parents or other carers (Bittles & Glasson, 2004; Bittles & Glasson, 2010).

Because many women in industrialized societies are delaying marriage and child-bearing, and despite the increasing proportion of Down syndrome pregnancies that are medically terminated, there has been no corresponding decline in the number of Down syndrome livebirths in Australia (Bittles *et al.*, 2007), whereas in the UK it has been estimated that without medical terminations, a 48% increase in Down syndrome pregnancies would have occurred between 1989–1991 and 2005–2007 because of increases in maternal age (Morris & Alberman, 2009).

Fetal alcohol syndrome

Fetal alcohol syndrome (FAS) caused by maternal alcohol ingestion during pregnancy was first described in 1973 and despite a wide array of associated defects, including intellectual disability; birth defects; abnormal facial features; growth and neurodevelopmental problems; vision, hearing and behaviour problems; and learning difficulties, its diagnosis has often been missed in affected children (Bertrand *et al.*, 2005). In the USA, > 50% of women of child-bearing age reported alcohol consumption during the previous month with ≥13% classifiable as moderate or heavy drinkers (Bertrand *et al.*, 2005), and 12% of pregnant women admitted drinking alcohol in the previous month, 2% of whom reported binge drinking during this time (NIH, 2011). Because of incomplete diagnosis, incidence data for FAS are not robust, but estimates of 0.5 to 2.0 per 1000 pregnancies have been calculated for the USA, whereas in countries with significantly higher levels of alcohol consumption, such as South Africa, the incidence of FAS is thought to exceed 60 per 1000 pregnancies (NIH, 2011).

Given the fact that the risk of bearing a child with Down syndrome is 12 times greater for a 40-year-old mother than for a 25-year-old mother in the USA (NDSS, 2011) and 16 times greater in the UK (Morris & Alberman, 2009), would these statistics warrant legal prohibitions being enacted in either country to restrict pregnancies by older women? Equally, should legislation be enacted to prevent the > 50% of women of child-bearing age in the USA from drinking alcohol, which for most people represents an avoidable risk? If the answer to either or both of these examples is in the negative, would it be logical or ethical to sponsor equivalent legislation to prohibit first-cousin unions?

The changing community perceptions of consanguineous marriage

For many of the world's impoverished rural and urban populations, consanguineous marriage continues to be perceived and employed as a means of maintaining the often scarce financial resources available within family networks – the more so in societies in which dowry payments are required, with dowry expected not only at marriage but often after one year of marriage, the birth of a son, or other significant event. For a family with several daughters, arranging their marriages to unrelated bridegrooms will almost inevitably be detrimental to the well-being of all other family members, and it could reasonably be claimed that consanguineous unions, in which dowry is excused or minimal, are a beneficial, more equitable, and readily understandable marriage strategy benefiting the entire family (Figure 15.1).

Biological disadvantages of   **vs.**   *Social benefits of*
            consanguinity   *consanguinity*

Pre-industrial

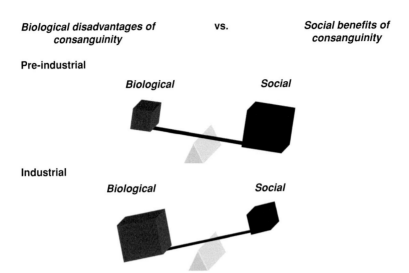

Industrial

Figure 15.1 Biological and social outcomes of consanguineous marriage in pre-industrial urban and rural communities and modern industrialized societies. Reproduced from Bittles & Black (2010a) with permission.

The picture changes quite radically with the transition in the global pattern of disease from a profile that predominantly encompasses communicable and nutritional disorders to disorders of an increasingly non-communicable nature. Diseases which in previous generations may have caused ill-health and death but remained unrecognized and undiagnosed are now given a name, an identity and a prognosis (Bittles, 2008). Particularly in families where consanguineous unions have been a long-standing tradition or in small, highly endogamous communities, the diagnosis of a recessive genetic disorder in one individual probably is indicative of the presence of the causative mutation in other family and community members, with major implications both for their health and for future marriage arrangements. Under these challenging circumstances for all family members, consanguinity could be viewed as a significant problem when in previous generations it was regarded as a boon.

What to do? Seeking a beneficial compromise

With smaller family sizes, which restrict the numbers of family members available for marriage, and an increasing emphasis on female education that implies a delay in marriage until later ages, a gradual global decline in consanguineous marriage seems inevitable. Additionally, in at least some sections of immigrant communities in Western countries, cross-generational tensions can be expected between a wish to retain customary family practices and a desire to blend more closely with the majority population, which in some cases will result in refusal to proceed with proposed consanguineous unions.

Meanwhile, across Asia and Africa, genetic disorders are becoming increasingly obvious as deaths due to infectious diseases decline, and early childhood mortality is being replaced by multi-level morbidity that extends into adulthood. With an excess of 1100 million people world-wide either in consanguineous unions or the progeny of consanguineous relationships, there is an urgent need for multi-disciplinary studies to complement and extend the classic work of Schull, Neel and their colleagues in Japan some 50 years ago – studies which will benefit greatly from the remarkable advances in genomic analysis during the last decade.

Prospective investigative programmes that can encompass the many aspects of consanguinity demonstrated to influence human health and social well-being (Figure 14.1), or for which at least preliminary evidence has been forthcoming, need to be planned and instituted. To achieve an optimal outcome in these programmes, it is essential that community members and their advisors are involved as fully as possible in all aspects of the planning and implementation (e.g. as in the UK Born in Bradford programme, www.borninbradford.nhs.uk). Due acknowledgement of individual and community social and religious beliefs is vital in all bioethical matters, and it is critical that community leaders and religious advisors ensure that the views and opinions of the communities they serve are appropriately incorporated into all deliberations.

Ultimately, families and communities need to be prepared to take responsibility for their own health and well-being and, while respecting their social and religious obligations, to take full advantage of the professional services and health facilities available to them. It also is important that, within reason, individuals and couples should be allowed to make autonomous decisions that may impinge on their own health and that of their children. Enforced legislation, however well meaning, could in time lead us all back along unintended eugenic pathways that are best left in the past and, with respect to consanguineous marriage, the advice offered by the WHO (2006), 'Preference for consanguineous marriage is a feature of the socio-cultural context within which medical genetic services must work', is apposite.

References

Abbasi-Shavazi, M.J., McDonald, P. & Hosseini-Chavoshi, M. (2008). Modernization or cultural maintenance: the practice of consanguineous marriage in Iran. *Journal of Biosocial Science*, **40**, 911–33.

Abdel-Hamid, M., Tisocki, K., Sharaf, L. & Ramadan, D. (2007). Development, validation and application of tandem mass spectrometry for screening of inborn metabolic disorders in Kuwaiti infants. *Medical Principles and Practice*, **16**, 215–21.

Abdel-Meguid, N., Zaki, M.S.A. & Hammad, S.A. (2000). Premarital genetic investigations: effect of genetic counseling. *Eastern Mediterranean Health Journal*, **6**, 652–60.

Abdulrazzaq, Y.M., Bener, A., al-Gazali, L.I. *et al.* (1997). A study of possible deleterious effects of consanguinity. *Clinical Genetics*, **51**, 167–73.

Abel, L., Plancoulaine, S., Jouanguy, E. *et al.* (2010). Age-dependent Mendelian predisposition to herpes simplex virus type 1 encephalitis in childhood. *Journal of Pediatrics*, **157**, 623–9.

Abelson, A. (1978). Population structure in the Western Pyrenees: social class, migration and the frequency of consanguineous marriage, 1850 to 1921. *Annals of Human Biology*, **5**, 165–78.

Abolfotouh, M.A., Nofal, L.M. & Safwat, H. (1990). Growth and nutritional status of preschool children attending the well-baby clinics. *Journal of the Egyptian Public Health Association*, **65**, 167–73.

Abouda, H., Hizem, Y., Gargouri, A. *et al.* (2010). Familial form of typical childhood absence epilepsy in a consanguineous context. *Epilepsia*, **51**, 1889–93.

Abul-Einem, M. & Toppozada, H. K. (1966). Aspects of births in the Shatby Hospital, Alexandria. *British Journal of Preventive and Social Medicine*, **20**, 176–80.

Abu-Rabia, S. & Maroun, L. (2005). The effect of consanguineous marriage on reading disability in the Arab community. *Dyslexia*, **11**, 1–21.

Accetturo, M., Creanza, T.M., Santoro, C. *et al.* (2010). Finding new genes for non-syndromic hearing loss through an *in silico* prioritization study. *PLoSOne*, **25**, e12742.

Acemoglu, H., Beyhoun, N.E., Vancelik, S., Polat, H. & Guraksin, A. (2008). Thalassaemia screening in a non-prevalent region of a prevalent country (Turkey): is it necessary? *Public Health*, **122**, 620–4.

Adaimy, L., Chouery, E., Mégarbané, H. *et al.* (2007). Mutation in WNT10A is associated with an autosomal recessive ectodermal dysplasia: the odonto-onycho-dermal dysplasia. *American Journal of Human Genetics*, **81**, 821–8.

Adam, W. (1865a). Consanguinity in marriage, Part I. *The Fortnightly Review*, **2**, 710–30.

Adam, W. (1865b). Consanguinity in marriage, Part II. *The Fortnightly Review*, **3**, 74–88.

Adams, M.S. & Neel, J.V. (1967). Children of incest. *Pediatrics*, **40**, 55–62.

Adinolfi, M. (1986). Recurrent habitual abortion, HLA sharing and deliberate immunization with partner's cells: a controversial topic. *Human Reproduction*, **1**, 45–8.

Afzal, M. (1988). Consequences of consanguinity on cognitive behavior. *Behavior Genetics*, **18**, 583–94.

Agarwal, N., Sinha, S.N. & Jensen, A.R. (1984). Effects of inbreeding on Raven matrices. *Behavior Genetics*, **14**, 579–85.

Agarwal, S.S., Singh, U., Singh, P.S. *et al.* (1991). Prevalence and spectrum of congenital malformations in a prospective study at a teaching hospital. *Indian Journal of Medical Research*, **94**, 413–9.

Ager, S.L. (2005). Familiarity breeds: incest and the Ptolemaic dynasty. *Journal of Hellenistic Studies*, **125**, 1–34.

Ahmed, A.H. (1979). Consanguinity and schizophrenia in Sudan. *British Journal of Psychiatry*, **134**, 635–6.

Ahmed, S., Atkin, K., Hewison, J. & Green, J. (2006). The influence of faith and religion and the role of religious and community leaders in prenatal decsisions for sickle cell disorders and thalassaemia major. *Prenatal Diagnosis*, **26**, 801–9.

Ahmed, S., Petrou, M. & Saleem, M. (1996). Molecular genetics of beta-thalassaemia in Pakistan: a basis for prenatal diagnosis. *British Journal of Haematology*, **94**, 476–82.

Ahmed, S., Saleem, M., Modell, B. & Petrou, M. (2002). Screening extended families for genetic hemoglobin disorders in Pakistan. *New England Journal of Medicine*, **347**, 1162–8.

Ahmed, T., Ali, S.M., Aliaga, A. *et al.* (1992). *Pakistan Demographic and Health Survey 1990/91*. Islamabad and Columbia, MD: Pakistan National Institute of Population Studies and Macro International.

Ai, Q., Haligiamu, Ke, Q. *et al.* (1985). A survey of five minority nationalities' consanguineous marriage in Yili, Xinjiang. *Acta Anthropologica Sinica*, **4**, 242–9. [In Chinese.]

Aiyappan, A. (1934). Cross-cousin and uncle–niece marriage in South India. *Proceedings of the International Congress of Anthropological and Ethnological Sciences*. London: Royal Institute of Anthropology, pp. 281–2.

Akbayram, S., Sari, N., Akgün, C. *et al.* (2009). The frequency of consanguineous marriage in eastern Turkey. *Genetic Counseling*, **20**, 207–14.

Akhavan Karbasi, S., Fallh, R. & Golestan, M. (2011). The prevalence of speech disorder in primary school students in Yazd – Iran. *Acta Medica Iranica*, **49**, 33–7.

Akramı, S.M., Montazeri, V., Shomali, S.R., Heshmat, R. & Larijani, B. (2009). Is there a significant trend of consanguineous marriage in Tehran? A review of three generations. *Journal of Genetic Counseling*, **18**, 82–6.

Akramı, S.M. & Osati, Z. (2006). Is consanguineous marriage religiously encouraged? Islamic and Iranian considerations. *Journal of Biosocial Science*, **39**, 313–6.

al-Abdulkareem, A.A. & Ballal, M.D. (1998). Consanguineous marriage in an urban area of Saudi Arabia: Rates and adverse health effects on the offspring. *Journal of Community Health*, **23**, 75–83.

Alam, N., Saha, S.K., Razzaque, A. & van Ginneken, J.K. (2001). The effect of divorce on infant mortality in a remote area of Bangladesh. *Journal of Biosocial Science*, **33**, 271–8.

Al-Aama, J.Y., Al-Nabulsi, B.K., Alyousef, M.A. *et al.* (2008). Knowledge regarding the national premarital screening program among university students in western Saudi Arabia. *Saudi Medical Journal*, **29**, 1649–53.

Al-Ani, Z.R., Al-Hiali, S.J. & Al-Mehimdi, S.M. (2010). Neural tube defects among neonates delivered in Al-Ramadi Maternity and Children's Hospital, western Iraq. *Saudi Medical Journal*, **31**, 163–9.

al-Ansari, A. (1993). Etiology of mild mental retardation among Bahraini children: a community-based case control study. *Mental Retardation*, **31**, 140–3.

Al Aqeel, A. (2007). Islamic ethical framework for research into and prevention of genetic diseases. *Nature Genetics*, **39**, 1293–8.

Al Arrayed, S. (2005). Campaign to control genetic blood diseases in Bahrain. *Community Genetics*, **8**, 52–5.

Al-Awadi, S.A., Naguib, K.K., Moussa, M.A. *et al.* (1986). The effects of consanguineous marriages on reproductive wastage. *Clinical Genetics*, **29**, 384–8.

Albar, M.A. (1999). Counselling about genetic disease: an Islamic perspective. *Eastern Mediterranean Health Journal*, **5**, 1129–33.

Albar, M.A. (2002). Ethical considerations in the prevention and management of genetic disorders with special emphasis on religious considerations. *Saudi Medical Journal*, **23**, 627–32.

al-Bustan, S.A., El-Zawahri, M.M., Al-Adsani, A.M. *et al.* (2002). Epidemiological and genetic study of 121 cases of oral clefts in Kuwait. *Orthodontic and Craniofacial Research*, **5**, 154–60.

Aldahmesh, M.A., Abu-Safieh, L., Khan, A.O. *et al.* (2009a). Allelic heterogeneity in inbred populations: the Saudi experience with Alström syndrome as an illustrative example. *American Journal of Medical Genetics Part A*, **149A**, 662–5.

Aldamesh, M.A., Safieh, L.A., Alkuraya, H. *et al.* (2009b). Molecular characterization of retinitis pigmentosa in Saudi Arabia. *Molecular Vision*, **15**, 2464–9.

Aldred, C. (1968). *Akhenaten, Pharaoh of Egypt, a New Study*. London: Thames and Hudson.

Aldred, C. (1988). *Akhenaten, King of Egypt*, pp. 279–90. London: Thames and Hudson.

Aldrich, C.L., Stephenson, M.D., Karrison, T. *et al.* (2001). HLA-G genotypes and pregnancy outcomes in couples with unexplained recurrent miscarriage. *Molecular Human Reproduction*, **7**, 1167–72.

Al-Eissa, Y.A. (1995). Febrile seizures: rate and risk factors of recurrence. *Journal of Childhood Neurology*, **10**, 315–9.

Al-Eissa, Y.A. & Ba'Aqeel, H.S. (1994). Risk factors for spontaneous preterm birth in a Saudi population. *European Journal of Obstetrics & Gynecology and Reproductive Biology*, **57**, 19–24.

Al-Eissa, Y.A., Ba'Aqeel, H.S. & Haque, K.N. (1991). Low birthweight in Riyadh, Saudi Arabia: incidence and risk factors. *Annals of Tropical Paediatrics*, **11**, 75–82.

Alfi, O.S., Ghang, R. & Azen, S.P. (1980). Evidence for genetic control of non-disjunction in man. *American Journal of Human Genetics*, **32**, 477–83.

Alford, R.L. (2011). Nonsyndromic hereditary hearing loss. *Advances in Otorhinolaryngology*, **70**, 37–42.

Al-Gazali, L.I. (1998). A genetic aetiological survey of severe childhood deafness in the United Arab Emirates. *Journal of Tropical Pediatrics*, **44**, 157–60.

Al-Gazali, L.I., Bakir, M., Hamid, Z. *et al.* (2003). Birth prevalence and pattern of ostoechondrodysplasias in an inbred high risk population. *Birth Defects Research (Part A)*, **67**, 125–32.

Al-Gazali, L.I., Bener, A., Abdulrazzaq, Y.M. *et al.* (1997). Consanguineous marriages in the United Arab Emirates. *Journal of Biosocial Science*, **29**, 491–7.

Al-Gazali, L., Hamamy, H. & Al-Arrayed, S. (2006). Genetic disorders in the Arab world. *British Medical Journal*, **333**, 831–4.

Al-Gazali, L.I., Sztriha, L., Dawodu, A. *et al.* (1999). Pattern of central nervous system anomalies in a population with a high rate of consanguineous marriages. *Clinical Genetics*, **55**, 95–102.

Al Hamdan, N.A., Al Mazrou, Y.Y., Al Swaidi, F.M. & Choudry, A.J. (2007). Premarital screening for thalassemia and sickle cell disease in Saudi Arabia. *Genetics in Medicine*, **9**, 372–7.

Al-Herz, W., Naguib, K.K., Noatangelo, L.D., Geha, R.S. & Alwadaani, A. (2011). Parental consanguinity and the risk of primary immunodeficiency disorders: report from the Kuwait National Primary Immunodeficiency Disorders Registry. *International Archives of Allergy and Immunology*, **154**, 76–80.

al Husain, M. & al Bunyan, M. (1997). Consanguineous marriages in a Saudi population and the effect of inbreeding on prenatal and postnatal mortality. *Annals of Tropical Paediatrics*, **17**, 155–60.

Ali, A., Feroze, A.H., Rizvi, Z.H. & Rehman, T.U. (2003). Consanguineous marriage resulting in homozygous occurrence of X-linked retinoschisis in girls. *American Journal of Ophthalmology*, **136**, 767–9.

Alidina, R. & Walji, M. (2010). The challenge of genetics and consanguinity in General Practice. *BioNews*, no. 578. London: Progress Educational Trust.

Al-Idrissi, I., Al-Kaff, A.S. & Senft, S.H. (1992). Cumulative incidence of retinoblastoma in Riyadh, Saudi Arabia. *Ophthalmic Paediatrics*, **13**, 9–12.

Al-Issa, I. (1989). Psychiatry in Algeria. *Psychiatric Bulletin*, **13**, 240–5.

Aljohar, A., Ravichandran, K. & Subhani, S. (2008). Pattern of cleft lip and palate in hospital-based population in Saudi Arabia: retrospective study. *The Cleft Palate – Craniofacial Journal*, **45**, 592–6.

Aljurf, M., Zaidi, S.Z., Hussian, F. *et al.* (2010). Status of hematopoietic stem cell transplantation in the WHO Eastern Mediterranean Region (EMRO). *Transfusion and Apheresis Science*, **42**, 169–75.

Al-Kandari, Y.Y. & Crews, D.E. (2011). The effect of consanguinity on congenital disabilities in the Kuwaiti population. *Journal of Biosocial Science*, **43**, 65–73.

Al-Khader, A.A. (2002). The Iranian transplant programme: comment from an Islamic perspective. *Nephrology, Dialysis and Transplantation*, **17**, 213–5.

Alkuraya, F.S. & Kilani, R.A. (2001). Attitude of Saudi families affected with hemoglobinopathies towards prenatal screening and abortion and the influence of religious ruling (Fatwa). *Prenatal Diagnosis*, **21**, 448–51.

Al-Kuraya, K., Schraml, P., Sheikh, S. *et al.* (2005). Predominance of high-grade pathway in breast cancer development of Middle East women. *Modern Pathology*, **18**, 891–7.

Allen, G. & Redekop, C.W. (1987). Old Colony Mennonites in Mexico: migration and inbreeding. *Social Biology*, **34**, 166–79.

Allen, S.J., O'Donnell, A., Alexander, N.D.E. *et al.* (1997). α+-thalassemia protects children against disease caused by other infections as well as malaria. *Proceedings of the National Academy of Sciences of the United States of America*, **94**, 14736–41.

Allison, A.C. (1954). Protection afforded by sickle-cell trait against subtertian malareal infection. *British Medical Journal*, **1**, 290–4.

Al-Mazrou, Y.Y., Fraid, S.M. & Khan, M.U. (1995). Changing marriage age and consanguineous marriage in Saudi females. *Annals of Saudi Medicine*, **15**, 481–5.

Al-Nassar, K.E., Kelley, C.L. & El-Kazimi, A.A. (1989). Patterns of consanguinity in Kuwait. *American Journal of Medical Genetics*, **45** Suppl. 4, A0915.

Alper, Ö.M., Erengin, H., Manguoğlu, A.E. *et al.* (2004). Consanguineous marriages in the province of Antalya, Turkey. *Annales de Génétique*, **47**, 129–38.

al Rajeh, S., Bademosi, O., Ismail, H. *et al.* (1993). A community survey of neurological disorders in Saudi Arabia: the Thugbah study. *Neuroepidemiology*, **12**, 164–78.

Al-Rifai, M. & Woody, R.C. (2007). Marriage patterns and pediatric neurologic disease in Damascus, Syria. *Pakistan Journal of Neurological Sciences*, **2**, 136–40.

Al-Salehi, S.M., Al-Hifthy, E.H. & Ghaziuddin, M. (2009). Autism in Saudi Arabia: presentation, clinical correlates and comorbidity. *Transcultural Psychiatry*, **46**, 340–7.

Al-Salem, M. & Rawashdeh, N. (1992). Pattern of childhood blindness and partial sight among Jordanians in two generations. *Journal of Pediatric Ophthalmology and Stabismus*, **29**, 3661–5.

Al-Salem, M. & Rawashdeh, N. (1993). Consanguinity in North Jordan: prevalence and pattern. *Journal of Biosocial Science*, **25**, 553–6.

Al-Sayed, M., Al-Hassan, S., Rashed, M., Qeba, M. & Coskun, S. (2007). Preimplantation diagnosis for Zellweger syndrome. *Fertility and Sterility*, **87**, 1468.e1–3.

Al-Sharbati, M.M., Al-Hussaini, A.A. & Antony, S.X. (2003). Profile of child and adolescent psychiatry in Oman. *Saudi Medical Journal*, **24**, 391–5.

Alston, C.L., He, L., Morris, A.A. *et al.* (2011). Maternally inherited mitochondrial DNA disease in consanguineous families. *European Journal of Human Genetics*, doi: 01.1038/ejhg.2011.124.

Alsulaiman, A. & Hewison, J. (2006). Attitudes to prenatal and preimplantation diagnosis in Saudi patients at genetic risk. *Prenatal Diagnosis*, **26**, 1010–4.

Alter, A., Grant, A., Abel, L., Alcaïs, A. & Schurr, E. (2011). Leprosy as a genetic disease. *Mammalian Genome*, **22**, 19–31.

Al-Thakeb, F.T. (1985). The Arab family and modernity: evidence from Kuwait. *Current Anthropology*, **26**, 575–80.

Altukhov, Y.P., Sheremet'eva, V.A. & Rychov, Y.G. (2000). Heterosis as the cause of the secular trend in humans. *Doklady Biological Sciences*, **370**, 130–3.

Alvarez, G., Ceballos, F.C. & Quintero, C. (2009). The role of inbreeding in the extinction of a European royal dynasty. *PLoS One*, **4**, e5174.

Alwan, A. & Modell, B. (2003). Recommendations for introducing genetics services in developing countries. *Nature Reviews. Genetics*, **4**, 61–8.

Al-Zayir, A.A. & Al-Amro Alakloby, O.M. (2004). Primary hereditary ichthyoses in the Eastern Province of Saudi Arabia. *International Journal of Dermatology* **43**, 415–9.

Amos, W., Wilmer, J.W., Fullard, K. *et al.* (2001). The influence of parental relatedness on reproductive success. *Proceedings of the Royal Society of London Series B*, **268**, 2021–7.

Anderson, N.F. (1982). The "Marriage with a Deceased Wife's Sister Bill" controversy: incest anxiety and the defense of family purity in Victorian England. *Journal of British Studies*, **21**, 67–86.

Anderson, N.F. (1986). Cousin marriage in Victorian England. *Journal of Family History*, **11**, 285–301.

Anderson, R.C. (1951). Familial leukemia: a report of leukemia in five siblings, with a brief review of the genetic aspects of this disease. *American Journal of Diseases of Children*, **81**, 313–22.

Anderson, R.R. (2009). Religious traditions and prenatal genetic counseling. *American Journal of Medical Genetics C Seminars in Medical Genetics*, **15**, 52–61.

Angastiniotis, M.A. & Hadjiminas, M.G. (1981). Prevention of thalassaemia in Cyprus. *Lancet*, **1**, 369–71.

Antonisamy, B., Raghupathy, P., Christopher, S. *et al.* (2009). Cohort profile: The 1969–73 Vellore Birth Cohort Study in South India. *International Journal of Epidemiology*, **38**, 663–9.

Anvar, Z., Namavar-Jahromi, B. & Saddat, M. (2011). Association between consanguineous marriages and risk of pre-eclampsia. *Archives of Gynecology and Obstetrics*, **283** Suppl. 1, 5–7.

Aoki, K. & Feldman, M.W. (1997). A gene-culture coevolutionary model for brother–sister mating. *Proceedings of the National Academy of Sciences of the United Sates of America*, **94**, 13046–50.

Appaji Rao, N., Radha Rama Devi, A., Savithri, H.S., Venkat Rao, S. & Bittles, A.H. (1988). Neonatal screening for amino acidaemias in Karnataka, South India. *Clinical Genetics*, **34**, 60–3.

APPG (2009). *Sickle Cell Disease and Thalassaemia; a Health Check*. London: All-Party Parliamentary Group on Sickle Cell Disease and Thalassaemia.

Aquaron, R. (1980). L'albinisme oculo-cutané au Cameroun. *Revue d'épidémiologie et de santé publique*, **28**, 81–8. [In French.]

Aquaron, R. (1990). Oculocutaneous albinism in Cameroon: a 15-year follow-up study. *Ophthalmic Paediatrics and Genetics*, **11**, 255–63.

Aribi, M., Moulessehoul, S., Benabadji, A.B. & Kendoucitani, M. (2004). HLA DR phenotypic frequencies and genetic risk of type 1 diabetes in west region of Algeria, Tlemcen. *BMC Genetics*, **5**, 24.

Armstrong, K. (1991). *Muhammad: A Western Attempt to Understand Islam*. London: Gollancz.

Arner, G.B.L. (1908). *Cousin Marriages in the American Population*. New York: Columbia University Faculty of Political Science.

Arnold, F., Choe, M.K., & Roy, T.K. (1998). Son preference, the family-building process and child mortality in India. *Population Studies*, **52**, 310–15.

Arnold, F., Kishor, S., & Roy, T.K. (2002). Sex-selective abortions in India. *Population and Development Review*, **28**, 759–85.

Arnold, F. & Zhaoxiang, L. (1986). Sex preference, fertility, and family planning in China. *Population and Development Review*, **12**, 221–46.

Arnos, K.S., Welch, K.O., Tekin, M. *et al.* (2008). A comparative analysis of the genetic epidemiology of deafness in the United States in two sets of pedigrees collected more than a century apart. *American Journal of Human Genetics*, **83**, 200–7.

Arokiasamy, P. (2004). Regional patterns of sex bias and excess female child mortality in India. *Population*, **59**, 833–64.

Asadi-Pooya, A.A. (2005). Epilepsy and consanguinity in Shiraz, Iran. *European Journal of Paediatric Neurology*, **9**, 383–6.

Asha Bai, P.V. & John, T.J. (1982). The effect of consanguinity on the gestation period and anthropometric traits of the new-born in Southern India. *Tropical and Geographical Medicine*, **34**, 225–9.

Asha Bai, P.V., John, T.J. & Subramaniam, V.R. (1981). Reproductive wastage and developmental disorders in relation to consanguinity in south India. *Tropical and Geographical Medicine*, **33**, 275–80.

ASHG. (1998). *Eugenics and the Misuse of Genetic Information to Restrict Reproductive Freedom*. Bethesda, MD: American Society for Human Genetics.

Ashrafian, H. (2005). Familial proptosis and obesity in the Ptolemies. *Journal of the Royal Society of Medicine*, **98**, 85–6.

Asindi, A. & Shehri, A. (2001). Neural tube defects in the Asir region of Saudi Arabia. *Annals of Saudi Medicine*, **21**, 26–9.

Assaf, S. & Khawaja, M. (2009). Consanguinity trends and correlates in the Palestinian Territories. *Journal of Biosocial Science*, **41**, 107–24.

Assaf, S., Khawaja, M., DeJong, J., Mahfoud, Z. & Yunis, K. (2009). Consanguinity and reproductive wastage in the Palestinian Territories. *Paediatric and Perinatal Epidemiology*, **23**, 107–15.

ASRM (2008). Repetitive oocyte donation: conclusions of the Practice Committee of the American Society for Reproductive Medicine. *Fertility and Sterility*, **90**, S194–5.

Atmaca, L.S., Şayli, B.S., Akarsu, N. & Gündüz, K. (1995). Genetic features of retinitis pigmentosa in Turkey. *Documenta Ophthalmologica*, **89**, 387–92.

Attias, J., Al-Masri, M., Abukader, L. *et al.* (2006). The prevalence of congenital and early-onset hearing loss in Jordanian and Israeli infants. *International Journal of Audiology*, **45**, 528–36.

Audinarayana, N. & Krishnamoorthy, S. (2000). Contribution of social and cultural factors to the decline of consanguinity in South India. *Social Biology*, **47**, 189–200.

Avenarius, M.R., Hildebrand, M.S., Zhang, Y. *et al.* (2009). Human male infertility caused by mutations in the CATSPER1 channel protein. *American Journal of Human Genetics*, **84**, 505–10.

Avent, N.D., Madgett, T.E., Lee, Z.E. *et al.* (2006). Molecular biology of Rh proteins and relevance to molecular medicine. *Expert Reviews in Molecular Medicine*, **8**, 1–20.

Axelrod, R. & Hamilton, W.D. (1981). The evolution of cooperation. *Science*, **211**, 1390–6.

Ayala, F. (2010). The difference of being human: Morality. *Proceedings of the National Academy of Sciences of the United States of America*, **107**, 9015–22.

Ayoub, M.R. (1959). Parallel cousin marriage and endogamy: a study in sociometry. *Southwestern Journal of Anthropology*, **15**, 266–75.

Azarsiz, E., Gulez, N., Edeer Karaca, N., Aksu, G. & Kutukculer, N. (2011). Consanguinity rate and delay in diagnosis in Turkish patients with combined immunodeficiencies: a single-center study. *Journal of Clinical Immunology*, **31**, 106–11.

Baccetti, B., Capitani, S., Collodel, G. *et al.* (2001). Genetic sperm defects and consanguinity. *Human Reproduction*, **16**, 1365–71.

Bach, G., Zeigler, M. & Zlotogora, J. (2007). Prevention of lysosomal storage disorders in Israel. *Molecular and Genetic Metabolism*, **90**, 353–7.

Bacolod, M.D., Schemmann, G.S., Giardina, S.F. *et al.* (2009). Emerging paradigms in cancer genetics: some important findings from high-density single nucleotide polymorphism array studies. *Cancer Research*, **69**, 723–7.

Badaruddoza (2004a). Inbreeding effects on metrical phenotypes among North Indian children. *Collegium Antropologicum*, **28** Suppl. 2, 311–8.

Badaruddoza (2004b). Effect of inbreeding on Wechsler intelligence scores among North Indian children. *Asia Pacific Journal of Public Health*, **16**, 99–103.

Badaruddoza & Afzal, M. (1993). Inbreeding depression and intelligence quotient among North Indian children. *Behavior Genetics*, **23**, 343–7.

Badaruddoza, Afzal, M. & Akhtaruzzaman (1994). Inbreeding and congenital heart diseases in a north Indian population. *Clinical Genetics*, **45**, 288–91.

Badaruddoza, Afzal, M. & Ali, M. (1998). Inbreeding effects on the incidence of congenital disorders and fetal growth and development at birth in North India. *Indian Pediatrics*, **35**, 1110–3.

Badii, R., Bener, A., Zirie, M. *et al.* (2008). Lack of association between the Pro12Ala polymorphism of the PPAR-gamma 2 gene and type 2 diabetes mellitus in the Qatari consanguineous population. *Acta Diabetologica*, **45**, 15–21.

Badria, L.F., Abu-Heija, A., Zayed, F., Ziadeh, S.M. & Alchalabi, H. (2001). Has consanguinity any impact on occurrence of pre-eclampsia and eclampsia? *Journal of Obstetrics & Gynaecology*, **21**, 358–60.

Badria, L.F. & Amarin, Z.O. (2003). Does consanguinity affect the severity of pre-eclampsia? *Archives of Gynecology & Obstetrics*, **268**, 117–20.

Badshah, S., Mason, L., McKelvie, K., Payne, R. & Lisboa, P.J.G. (2008). Risk factors for low birthweight in the public-hospitals at Peshawar, NWFP Pakistan. *BMC Public Health*, **8**, 197.

Baghdassarian, S.A. & Tabbara, K.H. (1975). Childhood blindness in Lebanon. *American Journal of Ophthalmology*, **79**, 827–30.

Bagnall, R.S. & Frier, B.W. (1994). *The Demography of Roman Egypt*. Cambridge: Cambridge University Press.

Bahakim, H.M. & El-Idrissy, I.M. (1989). Epidemiological observations of consanguinity and retinoblastoma in Arabia. A retrospective study. *Tropical and Geographical Medicine*, **41**, 361–4.

Baig, S.A., Azhar, A., Hassan, H. *et al.* (2006). Prenatal diagnosis of β-thalassaemia in Southern Punjab, Pakistan. *Prenatal Diagnosis*, **26**, 903–5.

Baig, S.M., Din, M.A., Hassan, H. *et al.* (2008). Prevention of beta-thalassemia in a large Pakistani family through cascade testing. *Community Genetics*, **11**, 68–70.

Bailit, H.L., Damon, S.T. & Damon, A. (1966). Consanguinity on Tristan da Cunha in 1938. *Eugenics Quarterly*, **13**, 30–3.

Baird, P.A. & McGillivray, B. (1982). Children of incest. *Journal of Pediatrics*, **101**, 854–7.

Bajaj, Y., Sirimanna, T., Albert, D.M. *et al.* (2009). Causes of deafness in British Bangladeshi children: a prevalence twice that of the UK population cannot be accounted for by consanguinity alone. *Clinical Otolaryngology*, **34**, 113–9.

Balci, Y.I., Tavil, B., Tan, C.S. *et al.* (2010). Increased availability of family donors for hematopoietic stem cell transplantation in a population with increased incidence of consanguinity. *Clinical Transplantation*, **25**, 475–80.

Bandipalliam, P. (2005). Syndrome of early onset colon cancers, hematologic malignancies & features of neurofibromatosis in HNPCC families with homozygous mismatch repair gene mutations. *Familial Cancer*, **4**, 323–33.

Bandyopadhyay, A.R., Chatterjee, D., Chatterjee, M. & Ghosh, J.R. (2011). Maternal fetal interaction in the ABO system: a comparative analysis of healthy mother and couples with spontaneous abortion in Bengalee population. *American Journal of Human Biology*, **23**, 76–9.

Bansal, P.K. (2010). *Same Gotra Marriages Lead to Genetic Disorders*. Available at www.merinews.com/. [5 February 2011.]

Barbari, A., Stephan, A., Masri, M. *et al.* (2003). Consanguinity-associated kidney diseases in Lebanon: an epidemiology study. *Molecular Immunology*, **39**, 1109–14.

Barbour, B. & Salameh, P. (2009). Consanguinity in Lebanon: prevalence, distribution and determinants. *Journal of Biosocial Science*, **41**, 505–17.

Baronciani, L., Cozzi, G., Canciani, M.T. *et al.* (2003). Molecular defects in type 3 von Willebrand disease: updated results from 40 multiethnic patients. *Blood Cells, Molecules and Diseases*, **30**, 264–70.

Barooha, V., Do, Q-T., Iyer, S. & Joshi, S. (2009). Missing women and India's religious demography. *Working Paper Series 5096*. Washington, DC: The World Bank.

Barrai, I., Mi, M.P., Morton N.E., Yasuda, N. (1965). Estimation of prevalence under incomplete selection. *American Journal of Human Genetics*, **17**, 221–36.

Barth, F. (1954). Father's brother's daughter marriage in Kurdistan. *Southwestern Journal of Anthropology*, **10**, 164–71.

Başaran, N., Artan, S., Yazicioglu, S. & Şayli, B.S. (1994). Effects of consanguinity on anthropometric measurements of newborn infants. *Clinical Genetics*, **45**, 208–11.

Başaran, N., Cenani, A., Şayli, B.S. *et al.* (1992). Consanguineous marriages among parents of Down patients. *Clinical Genetics*, **42**, 13–5.

Başaran, N., Hassa, H., Başaran, A. *et al.* (1989). The effect of consanguinity on the reproductive wastage in the Turkish population. *Clinical Genetics*, **36**, 168–73.

Basel-Vanagaite, L., Alkelai, A., Struassberg, R. *et al.* (2003). Mapping of a new locus for autosomal recessive non-syndromic mental retardation in the chromosomal region 19p13.12-p13.2: further genetic heterogeneity. *Journal of Medical Genetics*, **40**, 729–32.

Basel-Vanagaite, L., Attia, R., Yahav, M. *et al.* (2006). The CC2D1A, a member of a new gene family with C2 domains, is involved in autosomal recessive non-syndromic mental retardation. *Journal of Medical Genetics*, **43**, 203–10.

Basel-Vanagaite, L., Rainshtein, L., Inbar, D. *et al.* (2007a). Autosomal recessive mental retardation syndrome with anterior maxillary protrusion and strabismus: MRAMS syndrome. *American Journal of Medical Genetics A*, **143A**, 1687–91.

Basel-Vanagaite, L., Taub, E., Halpern, G.J. *et al.* (2007b). Genetic screening for autosomal recessive nonsyndromic mental retardation in an isolated population in Israel. *European Journal of Human Genetics*, **15**, 250–3.

Bashi, J. (1977). Effects of inbreeding on cognitive performance. *Nature*, **266**, 440–2.

Bashyam, M.D., Bashyam, L., Gorinabele, R., Sangal, M.G.V. & Rama Devi, A.R. (2004). Molecular genetic analysis of β-thalassaemia in South India reveals rare mutations in the β-globin gene. *Journal of Human Genetics*, **49**, 408–13.

Basu, A.M. (1989). Is discrimination in food really necessary for explaining sex differentials in childhood mortality? *Population Studies*, **43**, 193–210.

Basu, A.M. (1993). Cultural influences on the timing of first births in India: large differences that add up to little difference. *Population Studies*, **47**, 85–95.

Basu, S.K. (1975). Effect of consanguinity among North Indian Muslims. *Journal of Population Research*, **2**, 57–68.

Basu, S.K. (1978). The effects of consanguinity among Muslim groups of India. In *Medical Genetics in India*, vol. 2, ed. I.C. Verma. Pondicherry: Auroma, pp. 173–87.

Basu, S.K. & Roy, S. (1972). Change in the frequency of consanguineous marriages among the Delhi Muslims after Partition. *The Eastern Anthropologist*, **25**, 21–8.

Bateson, W. (1902). *Mendel's Principles of Heredity*. Cambridge, Cambridge University Press.

Bateson, W. (1909). Heredity and disease. *Proceedings of the Royal Society of Medicine*, **2**, 22–30.

Bayoumi, R.A., Al-Yahyaee, S.A., Albarwani, S.A. *et al.* (2007). Heritability of determinants of the metabolic syndrome among healthy Arabs of the Oman family study. *Obesity (Silver Spring)*, **15**, 551–6.

Bazmi, S., Behnoush, B., Kiani, M. & Bazmi, E. (2008). Comparative study of therapeutic abortion permissions in Central Department of Tehran Legal

Medicine Organization before and after approval of law on abortion in Iran. *Iran Journal of Pediatrics*, **18**, 315–22.

Bear, J.C., Nemec, T.F., Kennedy, J.C. *et al.* (1988). Inbreeding in Outport Newfoundland. *American Journal of Medical Genetics*, **29**, 649–60.

Beaty, T.H. (2007). Invited Commentary: Two studies of genetic control of birth weight where large data sets were available. *American Journal of Epidemiology*, **165**, 753–5.

Beck, J.A., Lloyd, S., Hafezparast, M. *et al.* (2000). Genealogies of mouse inbred strains. *Nature Genetics*, **24**, 23–5.

Becker, S.M., Al Halees, Z., Molina, C. & Paterson, R.M. (2001). Consanguinity and congenital heart disease in Saudi Arabia. *American Journal of Medical Genetics*, **99**, 8–13.

Beckman, L. & Elston, R. (1962). Assortative mating and fertility. *Acta Genetica et Statistica Medica, Basel*, **12**, 117–22.

Bede c731. *The Ecclesiastical History of the English People*. London: Penguin Books [Revised edition, 1990], pp. 79–81.

Bejjani, B.A., Lewis, R.A., Tomey, K.F. *et al.* (1998). Mutations in *CYP1B1*, the gene for cytochrome P4501B1, are the predominant cause of primary congenital glaucoma in Saudi Arabia. *American Journal of Human Genetics*, **62**, 325–33.

Bell, A.G. (1883). *Memoir: Upon the Formation of a Deaf Variety of the Human Race*. Washington, DC: National Academy of Sciences, pp. 1–88.

Bell, A.G. (1889). Census of the defective classes. *Science*, **13**, 38–41.

Bell, A.V., Richerson, P.J. & McElreath, R. (2009). Culture rather than genes provide greater scope for the evolution of large-scale human prosociality. *Proceedings of the National Academy of Sciences of the United States of America*, **106**, 17671–4.

Bell, J. (1940). A determination of the consanguinity rate in the general hospital population of England and Wales. *Annals of Eugenics*, **10**, 370–91.

Bemiss, S.M. (1858). Report on influence of marriages of consanguinity upon offspring. *Transactions of the American Medical Association*, **11**, 319–425.

Benallègue, A. & Kedji, F. (1984). Consanguinity and public health. *Archives Francaises de Pédiatrie*, **412**, 435–40. [In French.]

Bener, A., Abdulrazzaq, Y.M., al-Gazali, L.I. *et al.* (1996). Consanguinity and associated socio-demographic factors in the United Arab Emirates. *Human Heredity*, **46**, 256–64.

Bener, A. & Alali, K.A. (2006). Consanguineous marriage in a newly developed country: the Qatari population. *Journal of Biosocial Science*, **38**, 239–46.

Bener, A., Al Qahtani, R., Teebi, A.S. & Bessisso, M. (2008). The prevalence of attention deficit hyperactivity symptoms in schoolchildren in a highly consanguineous community. *Medical Principles and Practice*, **17**, 440–6.

Bener, A., Ayoubi, H.R., Ali, A.I., Al-Kubaisi, A. & Al-Sulaiti, H. (2010a). Does consanguinity lead to decreased incidence of breast cancer? *Cancer Epidemiology*, **34**, 413–8.

Bener, A., Denic, S. & Al-Mazrouel, M. (2001). Consanguinity and family history of cancer in children with leukemia and lymphomas. *Cancer*, **92**, 106.

Bener, A., El Ayoubi, H.R., Chouchane, L. *et al.* (2009). Impact of consanguinity on cancer in a highly endogamous population. *Asian Pacific Journal of Cancer Prevention*, **10**, 35–40.

Bener, A. & Hussain, R. (2006). Consanguineous unions and child health in the State of Qatar. *Paediatric and Perinatal Epidemiology*, **20**, 372–8.

Bener, A., Hussain, R. & Teebi, A.S. (2007). Consanguineous marriages and their effects on common adult diseases: studies from an endogamous population. *Medical Principles and Practice*, **16**, 262–7.

Bener, A., Moore, M.A., Ali, R. & El Ayoubi, H.R. (2010b). Impacts of family history and lifestyle habits on colorectal cancer risk: a case control study in Qatar. *Asian Pacific Journal of Cancer Prevention*, **11**, 963–8.

Bener, A., Rizk, D.E., Ezimokhai, M. *et al.* (1998). Consanguinity and the age of menopause in the United Arab Emirates. *International Journal of Gynecology & Obstetrics*, **60**, 155–60.

Bener, A., Zirie, M. & Al-Rikabi, A. (2005). Genetics, obesity, and environmental risk factors associated with type 2 diabetes. *Croatian Medical Journal*, **46**, 302–7.

Bennett, R.L., Hudgins, L., Smith, C.O. & Motulsky, A.G. (1999). Inconsistencies in genetic counseling and screening for consanguineous couples and their offspring: the need for practice guidelines. *Genetics in Medicine*, **1**, 286–92.

Bennett, R.L., Motulsky, A.G., Bittles, A.H. *et al.* (2002). Genetic counseling and screening of consanguineous couples and their offspring: recommendations of the National Society of Genetic Counselors. *Journal of Genetic Counseling*, **11**, 97–119.

Bennett, S., Lienhardt, C., Bah-Sow, O. *et al.* (2002). Investigation of environmental and host-related risk factors for tuberculosis in Africa. II. Investigation of host genetic factors. *American Journal of Epidemiology*, **155**, 1074–9.

Berends, A.L., Steegers, E.A., Isaacs, A. *et al.* (2008). Familial aggregation of preeclampsia and intrauterine growth restriction in a genetically isolated population in the Netherlands. *European Journal of Human Genetics*, **16**, 1437–42.

Berkovic, S.F., Howell, R.A., Hay, D.A. & Hopper, J.L. (1998). Epilepsies in twins: genetics of the major epilepsy syndromes. *Annals of Neurology*, **43**, 435–45.

Berliner, S.A., Seligsohn, U., Zivelin, A., Zwang, E. & Sofferman, G. (1986). A relatively high frequency of severe (type III) von Willebrand's disease in Israel. *British Journal of Haematology*, **62**, 535–43.

Berra, T.M., Alvarez, G. & Ceballos, F.C. (2010). Was the Darwin/Wedgwood dynasty adversely affected by consanguinity? *BioScience*, **60**, 376–83.

Berthoud, J-M. (2011). *Should Incest Still be Considered a Crime?* Available at http://www.swissinfo.ch/eng/swiss_news [28 June 2011].

Bertram, L., Lange, C., Mullin, K. *et al.* (2008). Genome-wide association analysis reveals putative Alzheimer's disease susceptibility loci in addition to APOE. *American Journal of Human Genetics*, **83**, 623–32.

Bertrand, J., Floyd, R.L. & Weber, M.K. (2005). Guidelines for identifying and referring persons with fetal alcohol syndrome. *MMWR Recommendations and Reports*, **54** (RR11), 1–10.

Bevc, I. & Silverman, I. (1993). Early proximity and intimacy between siblings and incestuous behavior: a test of Westermarck theory. *Ethology and Sociobiology*, **14**, 171–81.

Bevc, I. & Silverman, I. (2000). Early separation and sibling incest: a test of the revised Westermarck theory. *Evolution and Human Behavior*, **21**, 151–61.

Beydoun, H. & Saftlas, A.F. (2005). Association of human leucocyte antigen sharing with recurrent spontaneous abortions. *Tissue Antigens*, **65**, 123–35.

Bhasin, M.K. & Nag, S. (2002). Consanguinity and its effects on fertility and child survival among Muslims of Ladakh in Jammu and Kashmir. In *Anthropology: Trends and Applications*, eds. M.K. Bhasin & S.L. Malik. New Delhi: Kamla-Raj, pp. 131–40.

Biémont, C. (2010). Inbreeding effects in the epigenetic era. *Nature Reviews. Genetics*, **11**, 234.

Bishop, M. Metcalfe, S. & Gaff, C. (2008). The missing element: consanguinity as a component of genetic risk assessment. *Genetics in Medicine*, **10**, 612–20.

Bittles, A.H. (1979). Incest re-assessed. *Nature*, **280**, 107.

Bittles, A.H. (1994). The role and significance of consanguinity as a demographic variable. *Population and Development Review*, **20**, 561–84.

Bittles, A.H. (2001). Consanguinity and its relevance to clinical genetics. *Clinical Genetics*, **60**, 89–98.

Bittles, A.H. (2002). Endogamy, consanguinity and community genetics. *Journal of Genetics*, **81**, 91–8.

Bittles, A.H. (2003). The bases of Western attitudes to consanguineous marriage. *Developmental Medicine and Child Neurology*, **45**, 135–8.

Bittles, A.H. (2004). Genetic aspects of inbreeding and incest. In *Inbreeding, Incest, and the Incest Taboo*, A. Wolf and W. Durham, eds., Stanford, CA: Stanford University Press, pp. 38–60.

Bittles, A.H. (2005). Endogamy, consanguinity and community disease profiles. *Community Genetics*, **8**, 17–20.

Bittles, A.H. (2008). A Community Genetics perspective on consanguineous marriage. *Community Genetics*, **11**, 324–30.

Bittles, A.H. (2009a). Commentary: The background and outcomes of the first-cousin marriage controversy in Great Britain. *International Journal of Epidemiology*, **38**, 1453–8.

Bittles, A.H. (2009b). Consanguinity, genetic drift and genetic diseases in populations with reduced numbers of founders. In *Human Genetics – Principles and Approaches*, 4th edn., M. Speicher, S.E. Antonarakis & A.G. Motulsky, eds. Heidelberg: Springer, pp. 507–28.

Bittles, A.H. (2011a). Assessing the influence of consanguinity on congenital heart disease. *Annals of Pediatric Cardiology*, **4**, 111–6.

Bittles, A.H. (2011b). Time to get real: investigating potential beneficial genetic aspects of consanguinity. *Public Health Genomics*, **14**, 169–71.

Bittles, A.H. (2011c). Here we go again: misinformation and confusion on consanguineous marriage. *BioNews*, No. 610. London: Progress Educational Trust.

Bittles, A.H. & Black, M.L. (2010a). Consanguinity, human evolution and complex diseases. *Proceedings of the National Academy of Sciences of the United States of America*, **107**, 1779–86.

Bittles, A.H. & Black, M.L. (2010b). The impact of consanguinity on neonatal and infant health. *Early Human Development*, **86**, 737–41.

Bittles, A.H. & Black, M.L. (2010c). Consanguineous marriage and evolution. *Annual Review of Anthropology*, **39**, 193–207.

Bittles, A.H., Bower, C., Hussain, R. & Glasson, E.J. (2007). The four ages of Down syndrome. *European Journal of Public Health*, **17**, 221–5.

Bittles, A.H. & Chew, Y-Y. (1998). Eugenics and population policies. In *Human Biology and Social Inequality*, S. Strickland & P. Shetty, eds. Cambridge: Cambridge University Press, pp. 272–87.

Bittles, A.H., Coble, J.M. & Appaji Rao, N. (1993). Trends in consanguineous marriage in Karnataka, South India, 1980–1989. *Journal of Biosocial Science*, **25**, 111–6.

Bittles, A.H. & Egerbladh, I. (2005). The influence of past endogamy and consanguinity on genetic disorders in northern Sweden. *Annals of Human Genetics*, **69**, 1–10.

Bittles, A.H. & Glasson, E.J. (2004). Clinical, social and ethical aspects of the changing life expectancy of people with Down syndrome. *Developmental Medicine and Child Neurology*, **46**, 282–6.

Bittles, A.H. & Glasson, E.J. (2010). Increased longevity and the comorbidities associated with intellectual and developmental disability. In *Clinics in Developmental Medicine No. 187: Comorbidities in Developmental Disorders*, M. Bax & C. Gillberg, eds., London: Mac Keith Press, pp. 125–41.

Bittles, A.H., Grant, J.C. & Shami, S.A. (1993). Consanguinity as a determinant of reproductive behaviour and mortality in Pakistan. *International Journal of Epidemiology*, **22**, 463–7.

Bittles, A.H., Grant, J.C., Sullivan, S.G. & Hussain, R. (2002). Does inbreeding lead to decreased human infertility? *Annals of Human Biology*, **29**, 111–31.

Bittles, A.H. & Hamamy, H. (2010). Endogamy and consanguineous marriage in Arab populations. In *Genetic Disorders among Arab Populations*, 2nd edn., A. Teebi, ed., Heidelberg: Springer, pp. 85–108.

Bittles, A.H. & Hussain, R. (2000). An analysis of consanguineous marriage in the Muslim population of India at regional and state levels. *Annals of Human Biology* **27**, 163–71.

Bittles, A.H. & Makov, U.E. (1985). The use of linear regressions in the calculation of lethal-equivalent genes in man. *Annals of Human Biology*, **12**, 287–90.

Bittles, A.H., Mason, W.M., Greene, J. & Appaji Rao, N. (1991). Reproductive behavior and health in consanguineous marriages. *Science*, **252**, 789–94.

Bittles, A.H., Mason, W.M., Singarayer, D.N., Shreeniwas, S. & Spinar, M. (1993). Sex ratio determinants in Indian populations: studies at national, state and district levels. In *Urban Ecology and Health in the Third World*, L.M. Schell, M.T. Smith & A. Bilsborough, eds., Cambridge: Cambridge University Press, pp. 244–59.

Bittles, A.H. & Matson, P. (2000). Genetic influences on human fertility. In *Infertility in the Modern World: Present and Future Prospects*, G. Bentley & C.G.N. Mascie-Taylor, eds., Cambridge: Cambridge University Press, pp. 46–81.

Bittles, A.H. & Neel, J.V. (1994). The costs of human inbreeding and their implications for variations at the DNA level. *Nature Genetics*, **8**, 117–21.

Bittles, A.H., Petterson, B.A., Sullivan, S.G. *et al.* (2002). The influence of intellectual disability on life expectancy. *Journal of Gerontology: Medical Sciences*, **57A**, M470–2.

Bittles, A.H., Radha Rama Devi, A. & Appaji Rao, N. (1988). Consanguinity, twinning and secondary sex ratio in the population of Karnataka, South India. *Annals of Human Biology*, **15**, 455–60.

Bittles, A.H., Shami, S.A. & Appaji Rao, N. (1992). Consanguineous marriage in South Asia: incidence, causes and effects. In *Minority Populations: Genetics, Demography and Health*, A.H. Bittles and D.F. Roberts, eds., London: Macmillan, pp. 102–17.

Bittles, A.H. & Smith, M.T. (1994). Religious differentials in post-Famine marriage patterns, Northern Ireland, 1840–1915. I. Demographic and isonymy analysis. *Human Biology*, **66**, 59–76.

Black, M.L., Sinha, S., Agarwal, S. *et al.* (2010). A descriptive profile of β-thalassaemia mutations in India, Pakistan and Sri Lanka. *Journal of Community Genetics*, **1**, 149–57.

Black, M.L., Wise, C.A., Wang, W. & Bittles, A.H. (2006). Combining genetics and population history in the study of ethnic diversity in the People's Republic of China. *Human Biology*, **78**, 277–93.

Blanco Villegas, M.J., Boattini, A., Rodriguez Otero, H. & Pettener, D. (2004). Inbreeding patterns in La Cabrera, Spain: dispensations, multiple consanguinity analysis, and isonymy. *Human Biology*, **76**, 191–210.

Bluglass, R. (1979). Incest. *British Journal of Hospital Medicine*, **22**, 152–7.

Boakes, E.H., Wang, J. & Amos, W. (2007). An investigation of inbreeding depression and purging in captive pedigreed populations. *Heredity*, **98**, 172–82.

Bode, K., Blech, J., Elger, K. *et al.* (2010). Why Sarrazin's integration demagoguery has many followers. Available at www.spiegel.de.international/ [9 June 2010].

Bodmer, W.F. & Cavalli-Sforza, L.L. (1976). *Genetics, Evolution, and Man*. San Francisco, CA: W.H. Freeman, pp. 375–7.

Boehm, C. (2008). Purposive social selection and the evolution of human altruism. *Cross-Cultural Research*, **42**, 319–52.

Boëtsch, G., Prost, M., & Rabino-Massa, E. (2002). Evolution of consanguinity in a French Alpine Valley: the Vallouise in the Briançon region (17th–19th centuries). *Human Biology*, **74**, 285–300.

Bonafè, M., Cardelli, M., Marchegiani, F. *et al.* (2001). Increase of homozygosity in centenarians revealed by a new inter-Alu PCR technique. *Experimental Gerontology*, **36**, 1063–73.

Bond, J., Scott, S., Hampshire, D.J. *et al.* (2003). Protein-truncating mutations in ASPM cause variable reduction in brain size. *American Journal of Human Genetics*, **73**, 1170–7.

Bongaarts, J., Frank, O. & Lesthaeghe, R. (1984). The proximate determinants of fertility in sub-Saharan Africa. *Population and Development Review*, **10**, 511–37.

Bonneau, D., Kaplan, J., Girard, G. & Dufier, J-L. (1992). Autosomal inheritance of "senile" retinitis pigmentosa. A report of a family with consanguinity. *Clinical Genetics*, **42**, 199–200.

Boonen, S.E., Pörksen, S., Mackay, D.J.G. *et al.* (2008). Clinical characterisation of the multiple maternal hypomethylation syndrome in siblings. *European Journal of Human Genetics*, **16**, 453–61.

Borck, G., Rainshtein, L., Helman-Aharony, S. *et al.* (2011). High frequency of autosomal-recessive DFNB59 hearing loss in an isolated Arab population in Israel. *Clinical Genetics*, doi: 10.1111/j.1399–0004.2011.01741.x.

Borhany, M., Pahore, Z., ul Qadr, Z. *et al.* (2010). Bleeding disorders in the tribe: results of consanguineous breeding. *Orphanet Journal of Rare Diseases*, **5**, 23.

Borhany, M., Shamsi, T., Naz, A. *et al.* (2011). Clinical features and types of von Willebrand disease in Karachi. *Clinical and Applied Thrombosis/Hemostasis*.

Boström, I., Callander, M., Kurtzke, J.F. & Landtblom, A.M. (2009). High prevalence of multiple sclerosis in the Swedish county of Värmland. *Multiple Sclerosis*, **15**, 1253–62.

Boudin, M., (1862). Dangers des unions consanguines et nécessité des croisements dans les l'espèce humaine et parmi les animaux. *Annales d'Hygiène publique et de Médecine légale*, 2nd series, **18**, 5–82. [In French.]

Bouton, M.J., Phillips, H.J., Smithells, R.W. & Walker, S. (1961). Congenital leukaemia with parental consanguinity: case report with chromosome studies. *British Medical Journal*, **2**, 866–9.

Bowles, S. (2006). Group competition, reproductive leveling, and the evolution of human altruism. *Science*, **314**, 1569–72.

Bradfield, J.P., Qu, H-Q., Wang, K. *et al.* (2011). A genome-wide meta-analysis of six type 1 diabetes cohorts identifies multiple associated loci. *PloS Genetics*, **7**, e1002293.

Brahmajothi, V., Pitchappan, R.M., Kakkanaiah, V. *et al.* (1991). Association of pulmonary tuberculosis and HLA in South India. *Tubercle*, **72**, 123–32.

Bratt, C.S. (1984). Incest statutes and the fundamental right of marriage: Is Oedipus free to marry? *Family Law Quarterly*, **18**, 257–309.

Braverman, I.M., Redford, D.B. & Mackowiak, P.A. (2009). Akhenaten and the strange physiques of Egypt's 18th dynasty. *Annals of Internal Medicine*, **150**, 556–60.

Brito, L.A., Cruz, L.A., Rocha, K.M. *et al.* (2011). Genetic contribution for non-syndromic cleft lip with or without cleft palate (NS CL/P) in different regions of Brazil and implications for association studies. *American Journal of Medical Genetics A*, **155**, 1581–7.

Britvić, D., Aleksić-Shihabi, A., Titlić, M. & Dolić, K. (2010). Schizophrenia spectrum psychosis on a Croatian genetic isolate: genealogical reconstructions. *Psychiatrica Danubina*, **22**, 51–6.

Bromiker, R., Glam-Baruch, M., Gofin, R., Hammerman, C. & Amitai, Y. (2004). Association of parental consanguinity with congenital malformations among Arab newborns in Jerusalem. *Clinical Genetics*, **66**, 63–6.

Brooks, C. (1856). Laws of reproduction, considered with particular reference to the inter-marriage of first cousins. *Proceedings of the American Association for the Advancement of Science*, Cambridge: Lovering, pp. 236–46.

Brooks, W.D.W., Heasman, M.A. & Lovell, R.R.H. (1949). Retinitis pigmentosa associated with cystinuria. Two uncommon inherited conditions occurring in a family. *Lancet*, **1**, 1096–8.

Brown, J.S. (1951). Social class, intermarriage, and church membership in a Kentucky community. *American Journal of Sociology*, **57**, 232–42.

Brownstein, Z., Friedman, L.M., Shahin, H. *et al.* (2011). Targeted genomic capture and massively parallel sequencing to identify genes for hereditary hearing loss in Middle Eastern families. *Genome Biology*, **12**, R89.

Brun, M-E., Lana, E., Rivals, I. *et al.* (2011). Heterochromatic genes undergo epigenetic changes and escape silencing in immunodeficiency, centromeric instability, facial anomalies (ICF) syndrome. *PLoS One*, **6**, e19464.

Buchan, J.C., Bradbury, J.A. & Sheridan, E. (2003). Consanguinity and disease coincidence. *Eye*, **17**, 280–2.

Bulayeva, K.B. (2006). Overview of genetic-epidemiological studies in ethnically and demographically diverse isolates of Dagestan, Northern Caucasus, Russia. *Croatian Medical Journal*, **47**, 641–8.

Bulayeva, K.B., Leal, S.M., Pavlova, T.A. *et al.* (2000). The ascertainment of multiplex schizophrenia pedigrees from Daghestan genetic isolates (Northern Caucasus, Russia). *Psychiatric Genetics*, **10**, 67–72.

Bundey, S. (1979). The child of an incestuous union. In *Medical Aspects of Adoption and Foster Care*, S. Wolkind, ed., London, Spastics International Medical Publications, pp. 36–41.

Bundey, S., Alam, H., Kaur, A., Mir, S. & Lancashire, R.J. (1990). Race, consanguinity and social features in Birmingham babies: a basis for prospective study. *Journal of Epidemiology and Community Health*, **44**, 130–5.

Bundey, S. & Crews, S.J. (1984). A study of retinitis pigmentosa in the City of Birmingham. I. Prevalence. *Journal of Medical Genetics*, **21**, 417–20.

Burman, E. (1984). *The Inquisition, Hammer of Heresy*. New York: Dorset Press.

Burzynski, N.J. & Escobar, V.H. (1983). Classification and genetics of numeric anomalies of dentition. *Birth Defects Original Article Series*, **19**, 95–106.

Byatt, A.S. (1992). *Angels and Insects*. London: Chatto and Windus.

Calderón, R. (1989). Consanguinity in the Archbishopric of Toledo, Spain, 1900–1979. I. Types of consanguineous mating in relation to premarital migration and its effects on inbreeding levels. *Journal of Biosocial Science*, **21**, 253–66.

Calderón, R., Peña, J.A., Delgado, J. & Guevara, J.I. (1993). Inbreeding patterns in the Basque Country (Alava Province, 1831–1980). *Human Biology*, **65**, 743 70.

Calderón, R., Peña, J.A., Delgado, J. & Morales, B. (1998). Multiple kinship in two Spanish regions: new model relating multiple and simple consanguinity. *Human Biology*, **70**, 535–61.

Caldwell, J.C., Reddy, P.H. & Caldwell, P. (1983). The causes of marriage change in South India. *Population Studies*, **37**, 343–61.

Çalişkan, M., Chong, J.X., Uricchio, L. *et al.* (2011). Exome sequencing reveals a novel mutation for autosomal recessive non-syndromic mental retardation in the *TECR* gene on chromosome 19q13. *Human Molecular Genetics*, **20**, 1285–9.

Callander, M. & Landtblom, A.M. (2004). A cluster of multiple sclerosis cases in Lysvik in the Swedish county of Värmland. *Acta Neurologica Scandinavica*, **110**, 14–22.

Cameron, P. & Cameron, K. (1995). Does incest cause homosexuality? *Psychological Reports*, **76**, 611–21.

Campbell, H., Carothers, A.D., Rudan, I. *et al.* (2007). Effects of genome-wide heterozygosity on a range of biomedically relevant human quantitative traits. *Human Molecular Genetics*, **16**, 233–41.

Campbell, H., Rudan, I., Bittles, A.H. & Wright, A.F. (2009). Human population structure, outbreeding and human health. *Genome Medicine*, **1**, 91.

Canatan, D., Kose, M.R., Ustundag, M., Haznedaroglu, D. & Ozbas, S. (2006). Hemoglobinopathy control program in Turkey. *Community Genetics*, **9**, 124–6.

Cardelli, M., Marchegiani, F., Cavallone, L. *et al.* (2006). A polymorphism of the YTHDF2 gene (1p35) located in an Alu-rich genomic domain is associated with human longevity. *Journal of Gerontology A Biological Sciences and Medical Sciences*, **61**, 547–56.

Carothers, A.D., Rudan, I., Kolcic, I. *et al.* (2006). Estimating human inbreeding coefficients: comparison of genealogical and marker heterozygosity approaches. *Annals of Human Genetics*, **70**, 666–76.

Carr, I.M., Sheridan, E., Hayward, B.E., Markham, A.F. & Bonthron, D.T. (2009). *IBDfinder* and *SNPsetter*: tools for pedigree-independent identification of autozygous regions in individuals with recessive inherited disease. *Human Mutation*, **30**, 960–7.

Carr-Hill, R., Campbell, D.M., Hall, M.H. & Meredith, A. (1987). Is birth weight determined genetically? *British Medical Journal*, **295**, 687–9.

Carroll, L.S. & Owen, M.J. (2009). Genetic overlap between autism, schizophrenia and bipolar disease. *Genome Medicine*, **1**, 102.

Carroll, L.S., Penn, D.J. & Potts, W.K. (2002). Discrimination of MHC-derived odors by untrained mice is consistent with divergence in peptide-binding region residues. *Proceedings of the National Academy of Sciences of the United States of America*, **99**, 2187–92.

Carter, C.O. (1967). Risk to offspring of incest. *Lancet*, **289**, 436.

Casanova, J-L., Jouanguy, E., Lamhamedi, S., Blanche, S. & Fischer, A. (1995). Immunological conditions of children with BCG disseminated infection. *Lancet*, **346**, 581.

Castilla, E.E., Gomez, M.A., Lopez-Camelo, J.S. & Paz, J.E. (1991). Frequency of first-cousin marriages from civil marriage certificates in Argentina. *Human Biology*, **63**, 203–10.

Castilla, E.E., López-Camelo, J.S. & Campaña, H. (1999). Altitude as a risk factor for congenital anomalies. *American Journal of Medical Genetics*, **86**, 9–14.

Castro, L. & Toro, M.A. (2004). The evolution of culture: from primate social learning to human culture. *Proceedings of the National Academy of Sciences of the United States of America*, **101**, 10235–40.

Cavalli Sforza, L.L., Moroni, A. & Zei, G. (2004). *Consanguinity, Inbreeding, and Genetic Drift in Italy*. Princeton, NJ: Princeton University Press.

Cavka, M., Kelava, T., Cavka, V. *et al.* (2010). Homocystinuria, a possible solution of the Akhenaten's mystery. *Collegium Anthropologicum*, **34**, 255–8.

CCLG (2010). *Guidelines for Selection and HLA Matching of Related Donors, Adult Unrelated, Donors, Umbilical Cord Units for Haematopoietic Stem Cell Transplantation.* Leicester: Children's Cancer and Leukaemia Group.

CCP (1993). *The Civil Code of the Philippines* (Republic Act No. 386 as amended) with related laws: Revised Edition. Available at www.unescap.org/esid/psis/population/database/poplaws/law_phi/phi_062.htm [28 February 2011].

Cereijo, A.I. & Martínez-Frías, M.L. (1993). Consanguineous marriages among parents of patients with Down syndrome. *Clinical Genetics*, **44**, 221–2.

Chaabouni, H., Chaabouni, M., Maazoul, F. *et al.* (2001). Prenatal diagnosis of chromosome disorders in Tunisian population. *Annales de Génétique*, **44**, 99–104.

Chaix, R., Cao, C. & Donnelly, P. (2008). Is mate choice in humans MHC-dependent? *PLoS Genetics*, **4**, e1000184.

Chakraborty, T. & Kim, S. (2010). Kinship institutions and sex ratios in India. *Demography*, **47**, 989–1012.

Chaleby, K. (1988). Traditional Arabian marriages and mental health in a group of outpatient Saudis. *Acta Psychiatrica Scandinavica*, **77**, 139–42.

Chaleby, K. & Tuma, T.A. (1987). Cousin marriages and schizophrenia in Saudi Arabia. *British Journal of Psychiatry*, **150**, 547–9.

Chantrelle, P. & Dupire, M. (1964). L'endogamie des peuples du Fouta-Djallon. *Population*, **19**, 529–58. [In French.]

Charfeddine, C., Mokni, M., Ben Mousli, R. *et al.* (2003). A novel missense mutation in the gene encoding SLURP-1 in patients with Mal de Meleda from northern Tunisia. *British Journal of Dermatology*, **149**, 1108–15.

Chautard-Freire-Maia, E.A., Freire-Maia, N., Krieger, H., Barbosa, C.A.A. & Müller, V.S. (1983). Inbreeding studies in Brasilian schoolchildren: V. Inbreeding effects on metrical traits. *American Journal of Medical Genetics*, **16**, 331–55.

Chéhab, G., Chedid, P., Saliba, Z. & Bouvagnet, P. (2007). Congenital cardiac disease and inbreeding: specific defects escape higher risk due to parental consanguinity. *Cardiology of the Young*, **17**, 414–22.

Chengal Reddy, P. (1983). Consanguinity and inbreeding effects on fertility, mortality and morbidity in the Malas of Chittoor district. *Zeitscrift für Morphologie und Anthropologie*, **74**, 45–51.

Child, G.W. (1863). Marriages of consanguinity. *Westminster Review*, **24**, 88–104.

Chirambo, M.C. & Benezra, D. (1976). Causes of blindness among students in blind school institutions in a developing country. *British Journal of Ophthalmology*, **60**, 665–8.

Chishti, M.S., Muhammad, D., Haider, M. & Ahmad, W. (2006). A novel missense mutation in MSX1 underlies autosomal recessive oligodontia with associated dental anomalies in Pakistani families. *Journal of Human Genetics*, **51**, 872–8.

Chowdhury, A., Ala'uddin, K.M., Khan, A.R. & Chen, L. (1976). The effect of child mortality experience on subsequent fertility in Pakistan and Bangladesh. *Population Studies*, **30**, 249–61.

Christensen, A.F. (1998). Ethnohistorical evidence for inbreeding among the Pre-Hispanic Mixtec royal caste. *Human Biology*, **70**, 563–77.

Christianson, A., Howson, C.P. & Modell, B. (2006). *March of Dimes Global Report on Birth Defects*. White Plains, NY: March of Dimes Birth Defects Foundation.

Christianson, A.L., Venter, P.A., Du Toit, J.L., Shipalana, N. & Gericke, G.S. (1994). Acrocallosal syndrome in two African brothers born to consanguineous parents. *American Journal of Medical Genetics*, **51**, 98–101.

CIA. (2011). *CIA World Factbook*. Available at https://www.cia.gov/library/publications/the-world-factbook/ [25 July 2011].

Clarimón, J., Djaldetti, R., Lleó, A. *et al.* (2009). Whole genome analysis in a consanguineous family with early onset Alzheimer's diseases. *Neurobiology of Aging*, **30**, 1986–91.

Clark, C.J., Hill, A., Jabbar, K. & Silverman, J.G. (2009). Violence during pregnancy in Jordan: its prevalence and associated risk and protective factors. *Violence Against Women*, **15**, 720–35.

Clark, D.A. & Daya, S. (1991). Trials and tribulations in the treatment of recurrent spontaneous abortion. *American Journal of Reproductive Immunology*, **25**, 81–4.

Clarke, B. & Kirby, D.R.S. (1966). Maintenance of histocompatibility polymorphisms. *Nature*, **211**, 999–1000.

Clyman, J.C., Nazir, F., Tarolli, S. *et al.* (2007). The impact of a genetics education program on physicians' knowledge and genetic counseling referral patterns. *Medical Teaching*, **29**, e143–50.

Coltman, D.W., Bowen, W.D. & Wright, J.M. (1998). Birth weight and neonatal survival of harbour seal pups are positively correlated with genetic variation measured by microsatellites. *Proceedings of the Royal Society of London Series B*, **265**, 803–9.

Coltman, D.W., Pilkington, J.G., Smith, J.A. & Pemberton, J.M. (1999). Parasite-mediated selection against inbred Soay sheep in a free-living, island population. *Evolution*, **53**, 1259–67.

Concejero, A.M. & Chen, C-L. (2009). Ethical perspectives on living donor organ transplantation in Asia. *Liver Transplantation*, **15**, 1658–61.

Constitutional Court of Korea (2001). *The First Ten Years of the Korean Constitutional Court* (1988–1998). Seoul: The Constitutional Court of Korea, pp. 242–5.

Coon, K.D., Myers, A.J., Craig, D.W. *et al.* (2007). A high-density whole-genome association study reveals that APOE is the major susceptibility gene for sporadic late-onset Alzheimer's disease. *Journal of Clinical Psychiatry*, **68**, 613–8.

Cooper, E. (1993). Cousin marriage in rural China: more and less than generalized exchange. *American Ethnologist*, **20**, 758–80.

Cooper, E. & Zhang, M. (1993). Patterns of cousin marriage in rural Zhejiang and in "Dream of the Red Chamber." *Journal of Asian Studies*, **52**, 90–110.

COSIT. (2005). *Iraq Living Conditions Survey 2004*, vol. II. *Analytical Report*. Baghdad: Ministry of Planning and Development Cooperation.

Costeff, H., Cohen, B.E., Weller, L.E. & Rahman, D. (1977). Consanguinity analysis in Israeli mental retardates. *American Journal of Human Genetics*, **29**, 339–49.

Costeff, H. & Dar, H. (1980). Consanguinity analysis of congenital deafness in northern Israel. *American Journal of Human Genetics*, **32**, 64–8.

Coulam, C.B., Moore, S.B. & O'Fallon, W.M. (1987). Association between major histocompatibility antigen and reproductive performance. *American Journal of Reproductive Immunology and Microbiology*, **14**, 54–8.

Council of Europe (1953). *The European Convention on Human Rights*. Strasbourg: Council of Europe.

Cowan, R.S. (2009). Moving up the slippery slope: mandated genetic screening in Cyprus. *American Journal of Medical Genetics C Seminars in Medical Genetics*, **151C**, 95–103.

Croll, E. (2010). *The Politics of Marriage in Contemporary China*. Cambridge: Cambridge University Press.

Crossman, E. (1861). Intermarriage of relations as a cause of degeneracy of the offspring. *British Medical Journal*, **2**, 401–2.

Crow, J.F. (1958). Some possibilities for measuring selection intensities in man. *Human Biology*, **30**, 1–13.

Crow, J.F. (1963). Genetic load: three views. 2. The concept of genetic load: a reply. *American Journal of Human Genetics*, **115**, 310–5.

Curie-Cohen, M. (1980). The frequency of consanguineous matings due to multiple use of donors in artificial insemination. *American Journal of Human Genetics*, **32**, 589–600.

Czeizel, A.E., Gasztonyi, Z. & Kuliev, A. (2005). Periconceptional clinics: a medical health care infrastructure of new genetics. *Fetal Diagnosis & Therapy*, **20**, 515–8.

da Fonseca, L.G. & Freire-Maia, N. (1970). Further data on inbreeding levels in Brazilian populations. *Social Biology*, **17**, 324–8.

Dahl, K., Kesmodel, U., Hvidman, L. & Olesan, F. (2006). Informed consent: attitudes, knowledge and information concerning prenatal examinations. *Acta Obstetrica et Gynecologica Scandinavica*, **85**, 1414–9.

Dahlberg, G. (1929). Inbreeding in man. *Genetics*, **14**, 421–54.

Dahlberg, G. (1947). *Mathematical Methods for Population Genetics*. Basel: Karger, pp. 61–7.

Dally, E. (1864). An inquiry into consanguineous marriages and pure races. *Anthropological Review*, **2**, 65–108.

Danubio, M.E., Piro, A. & Tagarelli, A. (1999). Endogamy and inbreeding since the 17th century in past malarial communities in the Province of Cosenza (Calabria, Southern Italy). *Annals of Human Biology*, **26**, 473–88.

Daoud, A.S., Batieha, A., Bashtawi, M. & El-Shanti, H. (2003). Risk factors for childhood epilepsy: a case control study from Irbid, Jordan. *Seizure*, **12**, 171–4.

Darlington, C.D. (1960). Cousin marriage and the evolution of the breeding system in man. *Heredity*, **14**, 297–332.

Darr, A. & Modell, B. (1988). The frequency of consanguineous marriage among British Pakistanis. *Journal of Medical Genetics*, **25**, 186–90.

Darwin, C. (1862). *On the Various Contrivances by which British and Foreign Orchids are Fertilized by Insects, and on the Good Effects of Intercrossing*, 1st edn., London: John Murray, p. 360.

Darwin, C. (1871). *The Descent of Man and Selection in Relation to Sex*, vol. 2, London: John Murray, p. 403.

Darwin, C. (1876). *The Effects of Cross and Self Fertilisation in the Vegetable Kingdom*. London: John Murray, pp. 460–1.

Darwin, C. (1877). *On the Various Contrivances by which Orchids are Fertilized by Insects*, 2nd edn. London: John Murray.

Darwin, C.R. (1874). *Letter to G.H. Darwin, 6th December 1874*. Darwin Correspondence Project, University of Cambridge, Letter no. 9746, Provenance CUL.DAR.210.1:42. Available at www.darwinproject.ac.uk/advanced-search? [4 February 2011].

Darwin, G.H. (1875a). Marriages between first cousins in England and Wales and their effects. *Journal of the Statistical Society*, **38**, 153–84.

Darwin, G.H. (1875b). Note on the marriages of first cousins. *Journal of the Statistical Society*, **38**, 344–8.

Darwin, L. (1926). *The Need for Eugenic Reform*, New York: Appleton, pp. 360–1.

Das, V. (1973). The structure of marriage preferences: an account from Pakistani fiction. *Man*, **8**, 30–45.

Das Gupta, M. (1987). Selective discrimination against female children in rural Punjab. *Population and Development Review*, **13**, 77–100.

Das Gupta, M. (1990). Death clustering, mothers' education and the determinants of child mortality in rural Punjab, India. *Population Studies*, **44**, 489–505.

Davidson, A.B. & Ekelund, R.B. (1997). The medieval church and rents from marriage market regulations. *Journal of Economic Behavior and Organization*, **32**, 215–45.

Dawson, D.V., Ober, C. & Kostyu, D.D. (1995). Extended HLA profile of an inbred isolate: the Schmiedeleut Hutterites of South Dakota. *Genetic Epidemiology*, **12**, 47–62.

de Aquino, S.N., Paranaiba, L.M.R., Martelli, D.R.B. *et al.* (2011). Study of patients with cleft lip and palate with consanguineous parents. *Brazilian Journal of Otorhinolaryngology*, **77**, 19–23.

de Boer, A., Oosterwijk, J.C. & Rigters-Aris, C.A.E. (1995). Determination of a maximum number of artificial inseminations by donor children per sperm donor. *Fertility and Sterility*, **63**, 419–21.

de Braekeleer, M., Landry, T. & Cholette, A. (1994). Consanguinity and kinship in Down syndrome in Saguenay Lac-Saint-Jean (Québec). *Annales de Génétique*, **37**, 86–8.

De Braekeleer, M. & Ross, M. (1991). Inbreeding in Saguenay-Lac-St-Jean (Quebec, Canada): a study of Catholic Church dispensations. *Human Heredity*, **41**, 379–84.

DeBruine, L.M. (2005). Trustworthy but not lust-worthy: context-specific effects of facial resemblance. *Proceedings of the Royal Society of London Series B*, **272**, 919–22.

Deech, R. (2011). *Cousin Marriage*. Available at http://www.gresham.ac/uk/lectures-and-events/cousin-marriage [28 April 2011].

Degos, L., Colombani, A., Chaventre, A., Bengtson, B. & Jacquard, A. (1974). Selective pressure on HL-A polymorphism. *Nature*, **249**, 62–3.

de Medeiros, C.R., Bitencourt, M.A., Zanis-Neto, J. *et al.* (2006). Allogeneic hematopoietic stem cell transplantation from an alternative stem cell source in Fanconi anemia patients: analysis of 47 patients from a single institution. *Brazilian Journal of Medical and Biological Research*, **39**, 1297–304.

Demir, E., Prud'homme, J.F. & Topçu, M. (2004). Infantile convulsions and paroxysmal choreoathetosis in a consanguineous family. *Pediatric Neurology*, **30**, 349–53.

Demirel, S., Kaplanoğlu, N., Acar, A., Bodur, S. & Paydak, F. (1997). The frequency of consanguinity in Konya, Turkey, and its medical effects. *Genetic Counseling*, **8**, 295–301.

Denic, S. & Bener, A. (2001). Consanguinity decreases risk of breast cancer – cervical cancer unaffected. *British Journal of Cancer*, **85**, 1675–9.

Denic, S., Bener, A., Sabri, S., Khatib, F. & Milenkovic, J. (2005). Parental consanguinity and risk of breast cancer: a population-based case-control study. *Medical Science Monitor*, **11**, CR415–9.

Denic, S., Frampton, C., Nagelwerke, N. & Nicholls, M.G. (2007b). Consanguinity affects selection of alpha-thalassaemia phenotypes and the size of population under selective pressure from malaria. *Annals of Human Biology*, **34**, 1–12.

Denic, S., Frampton, C. & Nicholls, M.G. (2007a). Risk of cancer in an inbred population. *Cancer Detection and Prevention*, **31**, 263–9.

Denic, S., Nagelwerke, N. & Agarwal, M.M. (2008a). Consanguineous marriages and endemic malaria: can inbreeding increase population fitness? *Malaria Journal*, **7**, 150.

Denic, S., Nagelwerke, N. & Agarwal, M.M. (2008b). Consanguineous marriages: do genetic benefits outweigh its costs in populations with alpha+-thalassemia, hemoglobin S, and malaria? *Evolution and Human Behavior*, **29**, 364–9.

Denic, S., Nagelwerke, N. & Agarwal, M.M. (2011). On some novel aspects of consanguineous marriages. *Public Health Genomics*, **14**, 162–8.

Denic, S. & Nicholls, M.G. (2006). Incestuous gene in consanguinophilia and incest: toward a consilience theory of incest taboo. *Medical Hypotheses*, **66**, 52–8.

Denic, S. & Nicholls, G. (2007). Genetic benefits of consanguinity through selection of genotypes protective against malaria. *Human Biology*, **29**, 145–58.

Dereköy, F.S. (2000). Etiology of deafness in Afyon School for the Deaf in Turkey. *International Journal of Pediatric Otorhinolaryngology*, **55**, 125–31.

de Roux, N., Genin, E., Carel, J-C. *et al.* (2003). Hypogonadotropic hypogonadism due to loss of function of the KiSS-derived peptide receptor GPR54. *Proceedings of the National Academy of Sciences of the United States of America*, **100**, 10972–6.

Derti, A., Cenik, C., Kraft, P. & Roth, F.P. (2010). Absence of evidence for MHC-dependent mate selection within HapMap populations. *PLoS Genetics*, **6**, e1000925.

De Vaal, O.M. (1955). Genetic intersexuality in three brothers, connected with consanguineous marriages in the previous generations. *Acta Paediatrica*, **44**, 35–9.

Devereux, G., Stellitano, L., Verity, C.M. *et al.* (2004). Variations in neurodegenerative disease across the UK: findings from the national study of Progressive Intellectual and Neurological Deterioration (PIND). *Archives of Disease in Childhood*, **89**, 8–12.

Devoto, M., Prosperi, L., Bricarelli, F.D. *et al.* (1985). Frequency of consanguineous marriages among parents and grandparents of Down patients. *Human Genetics*, **70**, 256–8.

de Wert, G., Dondorp, G., Dondorp, W. *et al.* (2011). Intrafamilial medically assisted reproduction. *Human Reproduction*, **26**, 504–9.

Dewey, W.J., Barrai, I., Morton, N.E. & Mi, M.P. (1965). Recessive genes in severe mental defect. *American Journal of Human Genetics*, **17**, 237–56.

Dhalla, M.N. (1938). *History of Zoroastrianism*. New York: Oxford University Press.

Do, Q-T., Iyer, S. & Joshi, S. (2011). The economics of consanguineous marriage. *Working Paper Series 4085*, updated 11 April 2011. Washington, DC: The World Bank.

Dobrusin, M., Weitzman, D., Levine, J. *et al.* (2009). The rate of consanguineous marriages among parents of schizophrenic patients in the Arab Bedouin population in Southern Israel. *World Journal of Biological Psychiatry*, **10**, 334–6.

Doherty, J.P., Norton, E.C. & Veney, J.E. (2001). China's one-child policy: the economic choices and consequences faced by pregnant women. *Social Science and Medicine*, **52**, 745–61.

Doherty, P.C. & Zinkernagel, R.M. (1975). A biological role for the major histocompatibility antigens. *Lancet*, **1**, 1406–9.

Dommaraju, P. & Agadjanian, V. (2009). India's North–South divide and theories of fertility change. *Journal of Population Research*, **26**, 249–72.

Donbak, L. (2004). Consanguinity in Kahramanmaras city, Turkey, and its medical impact. *Saudi Medical Journal*, **25**, 1991–4.

Dorsten, L.E., Hotchkiss, L. & King, T.M. (1999). The effect of inbreeding on early childhood mortality: twelve generations of an Amish settlement. *Demography*, **36**, 263–71.

dos Santos Silva, I., De Stavola, B. & McCormack, V. (2008). Birth size and breast cancer risk: re-analysis of individual participant data from 32 studies. *PLoS Medicine*, **5**, e193.

Down, J.L. (1887). Marriages of consanguinity in relation to degeneration of race. In *Mental Affections of Childhood and Youth*. London: J & A Churchill, pp. 185–209.

Downey, L.M., Keen, T.J., Jalili, I.K. *et al.* (2002). Identification of a locus on chromosome 2q11 at which recessive amelogenesis imperfecta and cone-rod dystrophy cosegregate. *European Journal of Human Genetics*, **10**, 865–9.

Du, R-B., Zhao, Z.-L., Xu, L.-J. *et al.* (1981). Percentage and types of consanguineous marriages of different nationalities and regions in China. *National Medical Journal of China*, **61**, 723–8.

Du, Z.D., Roguin, N., Barak, M., Bihari, S.G. & Ben-Elisha, M. (1996). High prevalence of muscular venticular septal defect in preterm neonates. *American Journal of Cardiology*, **78**, 1183–5.

Ducroq, D., Shalev, S., Habib, A. *et al.* (2006). Three different ABCA4 mutations in the same large family with several consanguineous loops affected with autosomal recessive cone-rod dystrophy. *European Journal of Human Genetics*, **14**, 1269–73.

Dull, J.K. (1978). Marriage and divorce in Han China: a glimpse at Pre-Confucian society. In *Chinese Family Law and Social Change in Historical and Comparative Perspective*, D.C. Buxbaum, ed., Seattle: University of Washington Press, pp. 23–74.

Dunlop, A.L., Jack, B. & Frey, K. (2007). National recommendations for preconception care: the essential role of the family physician. *Journal of the American Board of Family Medicine*, **20**, 81–4.

Durey, J. (2008). The Church, consanguinity and Trollope. *Churchman*, **122**, 125–46.

Durham, W.H. (1991). *Coevolution: Genes, Culture and Human Diversity*. Stanford, CA: Stanford University Press.

Durkin, M. (2002). The epidemiology of developmental disabilities in low-income countries. *Mental Retardation and Developmental Disability Research Reviews*, **8**, 206–11.

Durkin, M.S., Hasan, Z.M. & Hasan, K.Z. (1998). Prevalence and correlates of mental retardation among children in Karachi, Pakistan. *American Journal of Epidemiology*, **147**, 281–8.

Durkin, M.S., Khan, N.Z., Davidson, L.L. *et al.* (2000). Prenatal and postnatal risk factors for mental retardation among children in Bangladesh. *American Journal of Epidemiology*, **152**, 1024–33.

Duru, N., Iseri, S.A., Selçuk, N. & Tolun, A. (2010). Early-onset progressive myoclonic epilepsy with dystonia mapping to 16pter-p13.3. *Journal of Neurogenetics*, **24**, 207–15.

Dwan, K., Altman, D.G., Arnaiz, J.A. *et al.* (2008). Systematic review of the empirical evidence of study publication bias and outcome reporting bias. *PloS One*, **3**, e3081.

Dye, D., Brameld, K.J., Maxwell, S. *et al.* (2011a). The impact of single gene and chromosomal disorders on hospital admissions of children and adolescents: a population-based study. *Public Health Genomics*, **14**, 153–61.

Dye, D., Brameld, K.J., Maxwell, S. *et al.* (2011b). The impact of single gene and chromosomal disorders on hospital admissions in an adult population. *Journal of Community Genetics*, **2**, 81–90.

Dyer, O. (2005). MP is criticized for saying that marriage of first cousins is a health risk. *British Medical Journal*, **331**, 1292.

Dyson, T. & Moore, M. (1983). On kinship structure, female autonomy, and demographic behaviour in India. *Population and Development Review*, **9**, 35–59.

Eapen, V., Al-Gazali, L., Bin-Othman, S. & Abou-Saleh, M. (1998). Mental health problems among schoolchildren in United Arab Emirates: prevalence and risk factors. *Journal of the American Academy of Childhood and Adolescent Psychiatry*, **37**, 880–6.

Easterbrook, P.J., Gopalan, R., Berlin, J.A. & Matthews, D.R. (1991). Publication bias in clinical research. *Lancet*, **337**, 867–72.

Eckl, K.M., Stevens, H.P., Lestringant, G.G. *et al.* (2003). Mal de Meleda (MDM) caused by mutations in the gene for SLURP-1 in patients from Germany, Turkey, Palestine, and the United Arab Emirates. *Human Genetics*, **112**, 50–6.

Edmond, M. & De Braekeleer, M. (1993). Inbreeding effects on fertility and sterility: A case-control study in Saguenay-Lac-Saint-Jean (Quebéc, Canada) based on a population registry 1838–1971. *Annals of Human Biology*, **20**, 545–55.

Egbase, P.E., Al-Sharhan, M., Al-Othman, S., Al-Mutawa, M. & Grudzinkas, J.G. (1996). Outcome of assisted reproduction technology in infertile couples of consanguineous marriage. *Journal of Assisted Reproduction and Genetics*, **13**, 279–81.

Egeli, E., Çiçekci, G., Silan, F. *et al.* (2003). Etiology of deafness at the Yditepe School for the Deaf in Istanbul. *International Journal of Pediatric Otorhinolaryngology*, **67**, 467–71.

Egerbladh, I. & Bittles, A.H. (2008). The influence of consanguineous marriage on reproductive behaviour and early mortality in Skellefteå, Sweden, 1780–1899. In *Kinship and Demographic Behavior in the Past*, T. Bengtsson & G. Mineau, eds., Heidelberg: Springer, pp. 220–44.

Egerbladh, I. & Bittles, A.H. (2011). Socioeconomic, demographic and legal influences on consanguinity and kinship in northern coastal Sweden 1780–1899. *Journal of Biosocial Science*, **43**, 413–35.

Ekstein, J. & Katzenstein, H. (2001). The Dor Yeshorim story: community-based carrier screening for Tay-Sachs disease. *Advances in Genetics*, **44**, 297–310.

El-Alfi, O.S., Bahig, A.H., Abdul Salam, T. & Shaath, R. (1969). Birth weights in Kuwait, and their relation to consanguinity and to birth order. *Journal of the Kuwait Medical Association*, **3**, 227–31.

Elahi, M.M., Jackson, I.T., Elahi, O. *et al.* (2004). Epidemiology of cleft lip and cleft palate in Pakistan. *Plastic and Reconstructive Surgery*, **113**, 1548–55.

El-Badry, M.A. (1969). Higher female than male mortality in some countries of South Asia: a digest. *Journal of the American Statistical Association*, **64**, 1234–44.

Eldahdah, L., Ormond, K., Nassar, A., Hkalil, T. & Zahed, L. (2007). Outcome of chromosomally abnormal pregnancies in Lebanon: obstetricians' roles during and after prenatal diagnosis. *Prenatal Diagnosis*, **27**, 525–34.

Elder, M.J. & De Cock, R. (1993). Childhood blindness in the West Bank and the Gaza Strip: prevalence, aetiology and hereditary factors. *Eye*, **7**, 580–3.

Elderton, E.M. (1911). *On the Marriage of First Cousins*. London: Dulau.

Elderton, E.M. & Pearson, K. (1907). Eugenics Laboratory Memoirs IV: *On the Measure of the Resemblance of First Cousins*, London: Dulau.

Eldon, B.J., Axelsson, J., Sigurdsson, S.B. & Arnason, E. (2001). Cardiovascular risk factors and relatedness in an Icelandic subpopulation. *International Journal of Circumpolar Health*, **60**, 499–502.

El-Hazmi, M.A. (2006). Pre-marital examination as a method of prevention from blood genetic disorders. Community views. *Saudi Medical Journal*, **27**, 1291–5.

El-Hazmi, M.A. (2007). Islamic teachings of bioethics in relation to the practice of medical genetics. *Saudi Medical Journal*, **28**, 1781–7.

El-Hazmi, M.A.F., Jabbar, F.A., Al-Faleh, F.Z., Al-Swailem, A.R. & Warsy, A.S. (1991). Patterns of sickle cell, thalassaemia and glucose-6-phosphate dehydrogenase deficiency genes in north-western Saudi Arabia. *Human Heredity*, **41**, 26–34.

El-Hazmi, M.A.F. & Warsy, A.S. (1989). Frequency of glucose-6-phosphate dehydrogenase deficiency genes in Al-Baha. *Human Heredity*, **39**, 313–7.

El-Islam, M.F. (1976). Intergenerational conflict and the young Qatari neurotic. *Ethos*, **4**, 45–56.

El-Mouzan, M.I., Al-Salloum, A.A., Al-Herbish, A.S., Qurachi, M.M. & Al-Omar, A.A. (2007). Regional variations in the prevalence of consanguinity in Saudi Arabia. *Saudi Medical Journal*, **28**, 1881–4.

El-Mouzan, M.I., Al-Salloum, A.A., Al-Herbish, A.S., Qurachi, M.M. & Al-Omar, A.A. (2008). Consanguinity and major genetic disorders in Saudi children: a community-based cross-sectional study. *Annals of Saudi Medicine*, **28**, 169–73.

El-Orfi, A.H.A., Singh, M. & Gaisuddin, A.S.M. (1998). Conjugal leprosy among Libyan patients. *Dermatology*, **196**, 271–2.

El-Tayeb, E-N.H., Yaqoob, M., Abdur-Rahim, K. & Gustavson, K-K. (2008). Prevalence of β-thalassaemia and sickle cell trait in premarital screening in Al-Qassim, Saudi Arabia. *Genetic Counseling*, **19**, 211–8.

Ember, M. (1975). On the origin and extension of the incest taboo. *Behavior Science Research*, **10**, 249–81.

Emery, J. (2005). The GRAIDS Trial: the development and evaluation of computer decision support for cancer genetic risk assessment in primary care. *Annals of Human Biology*, **32**, 218–27.

Emery, J., Watson, E., Rose, P. & Andermann, A. (1999). A systematic review of the literature exploring the role of primary care in genetic services. *Family Practice*, **16**, 426–45.

Eppig, C., Fincher, C.L. & Thornhill, R. (2010). Parasite prevalence and the worldwide distribution of cognitive ability. *Proceedings of the Royal Society Series B*, **277**, 3801–8.

Eraksoy, M., Kurtuncu, M., Akman-Demir, G. *et al.* (2003). A whole genome screen for linkage in Turkish multiple sclerosis. *Journal of Immunology*, **143**, 17–24.

Erickson, M.T. (1993). Rethinking Oedipus: an evolutionary perspective of incest avoidance. *American Journal of Psychiatry*, **150**, 411–6.

ESHRE (2011). ESHRE Task Force on Ethics and Law: de Wert, G., Dondorp, G., Pennings, G. *et al.* Intrafamilial medically assisted reproduction. *Human Reproduction*, **26**, 504–9.

Eugenides, J. (2003) *Middlesex*. London: Bloomsbury.

Fallahian, M. (2003). Familial gestational trophoblastic disease. *Placenta*, **24**, 797–9.

Farag, T.I., al-Awadi, S.A., el-Badramary, M.H. *et al.* (1993). Disease profile of 400 institutionalized mentally retarded patients in Kuwait. *Clinical Genetics*, **44**, 329–34.

Farag, T.I. & Iskandar, A. (1998). Tutankhuman's paternity. *Journal of the Royal Society of Medicine*, **91**, 291–2.

Farag, T.I. & Teebi, A.S. (1988). Possible evidence for genetic predisposition to non-disjunction in man. *Journal of Medical Genetics*, **25**, 136–7.

Farhud, D.D., Kamali, M.S., Walizadeh, G.H. & Kamali Rusta, M. (1991). Some biological data on the Iranian infants. *South Asian Anthropologist*, **12**, 33–7.

Farrelly, C. (2008). The case for re-thinking incest laws. *Journal of Medical Ethics*, **34**, e11.

Farrer, L.A., Bowirrat, A., Friedland, R.P. *et al.* (2003a). Identification of multiple loci for Alzheimer disease in a consanguineous Israeli-Arab community. *Human Molecular Genetics*, **12**, 415–22.

Farrer, L.A., Friedland, R.P., Bowirrat, A. *et al.* (2003b). Genetic and environmental epidemiology of Alzheimer's disease in Arabs residing in Israel. *Journal of Molecular Neuroscience*, **20**, 207–12.

Farrow, M.G. & Juberg, R.C. (1969). Genetics and laws prohibiting marriage in the United States. *Journal of the American Medical Association*, **209**, 534–8.

Fathzadeh, M., Bigi, M.A.B., Bazrgar, M. *et al.* (2008). Genetic counseling in Southern Iran: consanguinity and reason for referral. *Journal of Genetic Counseling*, **17**, 472–9.

FCP. (1987). *The Family Code of the Philippines*. Available at www.weddingsatwork.com/culture_laws_familycode01.shtml [26 April 2011].

Federal Authorities of the Swiss Confederation (2010). *SR311.0 Swiss Criminal Code.* Available at www.admin.ch./ [30 June 2011].

Fei, H.-T. (1939). *Peasant Life in China*. New York: Dutton.

Feinberg, A.P. (2007). Phenotypic plasticity and the epigenetics of human disease. *Nature*, **447**, 433–40.

Feinmesser, M., Tell, L. & Levi, H. (1989). Consanguinity among parents of hearing-impaired children in relation to ethnic groups in the Jewish population of Jerusalem. *Audiology*, **28**, 268–71.

Fenner, L., Egger, M. & Gagneux, S. (2009). Annie Darwin's death, the evolution of tuberculosis and the need for systems epidemiology. *International Journal of Epidemiology*, **38**, 1425–8.

Ferák, V., Genčik, A. & Genčikova, A. (1982). Population genetical aspects of primary congenital glaucoma. II. Fitness, parental consanguinity, founder effect. *Human Genetics*, **61**, 190–200.

Ferguson, N. (1999). *The House of Rothschild: Vol. 1 Money's Prophets 1798–1848*. London: Penguin Books.

Ferreiro-Barros, C.C., Tengan, C.H., Barros, M.H. *et al.* (2008). Neonatal mitochondrial encephaloneuromyopathy due to a defect of mitochondrial protein synthesis. *Journal of Neurological Science*, **275**, 128–32.

Ferris, S.D., Sage, R.D. & Wilson, A.C. (1982). Evidence from mtDNA that common strains of inbred mice are descended from a single female. *Nature*, **295**, 163–5.

Feuer, J. (2011). Relatively speaking: halachic and legal issues of gamete donation. *Medicine and Law*, **30**, 239–66.

Finkelhor, D. (1980). Sex among siblings: a survey on prevalence, variety, and effects. *Archives of Sexual Behavior*, **9**, 171–94.

Firdous, N. (2005). Prevention of thalasaemia and haemoglobinopathies in remote and isolated communities – The Maldives experience. *Annals of Human Biology*, **32**, 131–7.

Firdous, N., Gibbons, S. & Modell, B. (2011). Falling prevalence of beta-thalassaemia and eradication of malaria in the Maldives. *Journal of Community Genetics*, **2**, 173–9.

Firth, H.V. & Wright, C.F. (2011). The Deciphering Developmental Disorders (DDD) Study. *Developmental Medicine and Child Neurology*, **53**, 702–3.

Fisher, R.A., Hodges, M.D. & Newlands, E.S. (2004). Familial recurrent hydatidiform mole: a review. *Journal of Reproductive Medicine*, **49**, 595–601.

Fisloglu, H. (2001). Consanguineous marriage and marital adjustment in Turkey. *The Family Journal*, **9**, 215–22.

Flynn, J.R. (1987). Massive IQ gains in 14 countries: what IQ tests really measure. *Psychological Bulletin*, **101**, 171–91.

Flynn, M. (1986). Mortality, morbidity and marital features of travellers in the Irish Midlands. *Irish Medical Journal*, **79**, 308–10.

Fowler, J.H., Settle, J.E. & Christakis, N.A. (2011). Correlated genotypes in friendship networks. *Proceedings of the National Academy of Sciences of the United States of America*, **108**, 1993–7.

Fraser, F.C. & Biddle, C.J. (1976). Estimating the risks for offspring of first-cousin matings. An approach. *American Journal of Human Genetics*, **28**, 522–6.

Freire-Maia, N. (1963). The load of lethal mutations in White and Negro Brazilian populations. I. Second survey. *Acta Genetica et Statistica Medica, Basel*, **13**, 185–98.

Freire-Maia, N. (1968). Inbreeding levels in American and Canadian populations: a comparison with Latin America. *Eugenics Quarterly*, **15**, 22–3.

Freire-Maia, N., Chautard-Freire-Maia, E.A., Barbosa, C.A.A. & Krieger, H. (1983). Inbreeding effect on infant mortality. *American Journal of Medical Genetics*, **16**, 336–8.

Freire-Maia, N., Freire-Maia, A. & Quelce-Salgado, A. (1963). The load of lethal mutations in White and Negro Brazilian populations. I. First survey. *Acta Genetica et Statistica Medica, Basel*, **13**, 185–98.

Freire-Maia, N., Guaraciaba, M.A. & Quelce-Salgado, A. (1964). The genetical load in the Bauru Japanese isolate in Brazil. *Annals of Human Genetics*, **27**, 329–39.

Freire-Maia, N. & Krieger, H. (1963). A Jewish isolate in Southern Brazil. *Annals of Human Genetics*, **27**, 31–9.

Freire-Maia, N. & Takehara, N. (1977). Inbreeding effect on precocious mortality in Japanese communities of Brazil. *Annals of Human Genetics*, **41**, 99–102.

Fricke, T., Axinn, W.G. & Thornton, A. (1993). Marriage, social inequality, and women's contact with their natal families in alliance societies: two Tamang examples. *American Anthropologist*, **95**, 395–419.

Fricke, T. & Teachman, J.D. (1993). Writing the names: marriage style, living arrangements, and first birth interval in a Nepali society. *Demography*, **30**, 175–88.

Fried, K. & Kaufman, S. (1980). Congenital afibrinogenemia in 10 offspring of uncle–niece marriages. *Clinical Genetics*, **17**, 223–7.

Friedman, L.M. & Avraham, K.B. (2009). MicroRNAs and epigenetic regulation in the mammalian inner ear: implications for deafness. *Mammalian Genome*, **20**, 581–603.

Frishberg, Y., Ben-Neriah, Z., Suvanto, M. *et al.* (2007). Misleading findings of homozygosity mapping resulting from three novel mutations in *NPHS1* encoding nephrin in a highly inbred community. *Genetics in Medicine*, **9**, 180–4.

Frodsham, A.J. & Hill, A.V.S. (2004). Genetics of infectious diseases. *Human Molecular Genetics*, **13**, R187–94.

Frydman, M., Bonné-Tamir, B., Braude, E., Zamir, R. & Creter, D. (1986). Male fertility in factor XIII deficiency. *Fertility and Sterility*, **45**, 729–31.

Fuess, V.L.R., Benot, R.F. & da Silveira, J.A.M. (2002). Delay in maturation of the auditory pathway and its relationship to language acquisition disorders. *Ear, Nose and Throat Journal*, **81**, 706–12.

Fumagalli, M., Pozzoli, U., Cagliani, R. *et al.* (2010). Genome-wide identification of susceptibility alleles for viral infections through a population genetics approach. *PLoS Genetics*, **6**, e1000849.

Furuumi, H., Firdous, N., Inoue, T. *et al.* (1998). Molecular basis of beta-thalassemia in the Maldives. *Hemoglobin*, **22**, 141–51.

Fuster, V. (2003). Inbreeding pattern and reproductive success in a rural community from Galicia (Spain). *Journal of Biosocial Science*, **35**, 83–93.

Fuster, V., Jiménez, A.M. & Colantonio, S.E. (2001). Inbreeding in Gredos mountain range (Spain): contribution of multiple consanguinity and intervalley variation. *Human Biology*, **73**, 249–70.

Gadisseur, A., Hermans, C., Berneman, Z. *et al.* (2009). Laboratory diagnosis and molecular classification of von Willebrand disease. *Acta Haematologica*, **121**, 71–84.

Gallin, B. (1963). Cousin marriage in China. *Ethnology*, **2**, 104–8.

Galton, F. (1869). *Hereditary Genius*. London: Macmillan.

Gane, C.H.W. & Stoddart, C.N. (1988). *A Casebook on Scottish Criminal Law*, Edinburgh: Green, pp. 674–7.

Gardner, J. (1861). On the intermarriage of relations, as a cause of degeneracy of the offspring. *British Medical Journal*, **1**, 290–1.

Garrod, A.E. (1902). The incidence of alkaptonuria: a study in chemical individuality. *Lancet*, **2**, 1616–20.

Garrod, A.E. (1908). The Croonian Lectures on Inborn Errors of Metabolism. I. General and Introductory; II. Alkaptonuria; III. & IV. Cystinuria. *Lancet*, **2**, 1–7, 73–9, 142–8, 214–20.

Garshasbi, H., Hadavi, V., Habibi, H. *et al.* (2008). A defect in the TUSC3 gene is associated with autosomal recessive mental retardation. *American Journal of Human Genetics*, **82**, 1158–64.

Geddes, W.R. (1963). *Peasant Life in Communist China*. Ithaca, NY: Cornell University Press.

Geha, R.S., Malakian, A., LeFranc, G., Chayban, D. & Serre, J.L. (1976). Immunologic reconstitution in severe combined immunodeficiency following transplantation with parental bone marrow. *Pediatrics*, **58**, 451–5.

Genčik, A., Genčikova, A. & Ferák, V. (1982). Population genetical aspects of primary congenital glaucoma. I. Incidence, prevalence, gene frequency, and age of onset. *Human Genetics*, **61**, 193–7.

George, K., Vedamony, J., Idikulla, J. & Rao, P.S.S. (1992). The effect of consanguinity on pregnancy-induced hypertension. *Australian and New Zealand Journal of Obstetrics and Gynaecology*, **32**, 231–2.

George, S., Abel, R. & Miller, B.D. (1992). Female infanticide in rural South India. *Economic and Political Weekly*, **27**, 1153–6.

George, S.M. (1997). Female infanticide in Tamil Nadu, India: from recognition back to denial? *Reproductive Health Matters*, **5**, 124–132.

Georgiou, T., Funnell, C.L., Casssels-Brown, A. & O'Conor, R. (2004). Influence of ethnic origin on the incidence of keratoconus and associated atopic disease in Asians and white patients. *Eye*, **18**, 379–83.

Gerber, S., Rozet, J.M., Takezawa, S.I. *et al.* (2000). The photoreceptor cell-specific nuclear receptor gene (PNR) accounts for retinitis pigmentosa in the Crypto-Jews from Portugal (Marranos), survivors from the Spanish Inquisition. *Human Genetics*, **107**, 276–84.

Gershoni-Baruch, R., Shinawi, M., Shamaly, H., Katsinetz, L. & Brik, R. (2002). Familial Mediterranean Fever: the segregation of four different mutations in 13 individuals from one inbred family: genotype-phenotype correlation and intrafamilial variability. *American Journal of Medical Genetics*, **109**, 198–201.

Gev, D., Roguin, N. & Freundlich, E. (1986). Consanguinity and congenital heart disease in the rural Arab population in northern Israel. *Human Heredity*, **36**, 213–7.

Gherman, A., Chen, P.E., Teslovich, T.M. *et al.* (2007). Population bottlenecks as a potential major shaping force of human genome architecture. *PLoS Genetics*, **3**, e119.

Ghiasvand, N.M., Shirzad, E., Naghavi, M. & Mahdavi, M.R.V. (1998). High incidence of autosomal recessive nonsyndromal congential retinal nonattachment (NCRNA) in an Iranian founding population. *American Journal of Medical Genetics*, **78**, 226–32.

Ghurye, Y.E. (1936). A note on cross-cousin marriage and dual organization in Kathiawar. *Journal of the University of Bombay*, **5**, 88–90.

Gibson, J., Morton, N.E. & Collins, A. (2006). Extended tracts of homozygosity in outbred human populations. *Human Molecular Genetics*, **15**, 789–95.

Gilani, A.I., Jadoon, A.S., Qaiser, R. *et al.* (2007). Attitudes towards genetic diagnosis in Pakistan: a survey of medical and legal communities and parents of thalassemic children. *Community Genetics*, **10**, 140–6.

Gilani, G.M., Kamal, S. & Gilani, S.A.M. (2006). Risk factors for breast cancer for women in Punjab, Pakistan: results from a case-control study. *Pakistan Journal of Statistics and Operation Research*, **2**, 17–26.

Girimaji, S.R., Srinath, S. & Seshadri, S.P. (1994). A clinical study of infants presenting to a mental retardation clinic. *Indian Journal of Pediatrics*, **61**, 373–8.

Glendon, M.A. (1989). *The Transformation of Family Law. State, Law, and the Family in the United States and Western Europe*. Chicago: University of Chicago Press.

Glinka, J., Artaria, M.D. & Koesbardiati, T. (1996). On the relationship between cleft lip and cleft palate and consanguinity. *Homo*, **46**, 253–5.

Gnanalingam, M.G., Gnanalingam, K.K. & Singh, A. (1999). Congenital heart disease and parental consanguinity in South India. *Acta Paediatrica*, **88**, 473–4.

Gohh, R.Y., Morrissey, P., Madras, P.N. & Monaco, A.P. (2001). Controversies in organ donation: the altruistic living donor. *Nephrology, Dialysis and Transplantation*, **16**, 619–21.

Golalipour, M.J., Mirfazeli, A. & Behnampour, N. (2007). Birth prevalence of oral clefting in northern Iran. *Cleft Palate-Craniofacial Journal*, **44**, 378–80.

Goldberg, H. (1967). FBD marriage and demography among Tripolitan Jews in Israel. *Southwestern Journal of Anthropology*, **23**, 176–91.

Goldman, I.L. (2002). Raymond Pearl, smoking and longevity. *Genetics*, **162**, 997–1001.

Goldschmidt, E., Ronen, A. & Ronen, I. (1960). Changing marriage systems in the Jewish community of Israel. *Annals of Human Genetics*, **24**, 191–204.

Goll, S. (2011). *Marriage Between Cousins Faces Ban*. Available at www.newsinenglish.no/ [10 February 2011].

Golubovsky, M. (2008). Unexplained infertility in Charles Darwin's family: genetic aspect. *Human Reproduction*, **23**, 1237–8.

Gomaa, A. (2007). Genetic eye diseases and genetic counselling services in Egypt. *Community Eye Health*, **20**, 11.

Gonis, N. (2000). Incestuous twins in the city of Arsinoe. *Zeitschrfit für Papyrologie und Epigraphik*, **133**, 197–8.

Goodwin, J. & Gross, M. (1979). Pseudoseizures and incest. *American Journal of Psychiatry*, **136**, 1231.

Goodwin, J., Simms, M. & Bergman, R. (1979). Hysterical seizures: a sequel to incest. *American Journal of Orthopsychiatry*, **49**, 698–703.

Goody, J. (1985). *The Development of the Family and Marriage in Europe*. Cambridge: Cambridge University Press.

Gordon, C.R., Bar-Ziv, Y., Frydman, M., Zlotogora, J. & Gadoth, N. (1990). Mucolipidosis III and Bardet-Biedl syndrome in the same family: diagnostic pitfalls. *Brain and Development*, **12**, 403–7.

Gordon, M., Gorman, D.R., Hashem, S. & Stewart, D.G.T. (1991). The health of traveller's children in Northern Ireland. *Public Health*, **105**, 387–91.

Government of DPR Korea. (1994). *The Family Law of the Democratic People's Republic of Korea*. Pyongyang: Foreign Languages Publishing House.

Government of PR China. (1982). *The Marriage Law of the People's Republic of China*. Beijing: Foreign Languages Press.

Government of PR China. (2001). *Decision of the Standing Committee of the National People's Congress on the Amendment to the Marriage Law of the People's Republic of China* (Order of the President No. 51). Available at www.procedurallaw.cn [31 January 2011].

Govinda Reddy, P. (1988). Consanguineous marriages and marriage payment: a study among three South Indian caste groups. *Annals of Human Biology*, **15**, 263–8.

Gowri, V., Udayakumar, A.M., Bsiso, W., Al Farsi, Y. & Rao, K. (2011). Recurrent early pregnancy loss and consanguinity in Omani couples. *Acta Obstetrica et Gynecologica Scandinavica*, **90**, 1167–9.

Grant, J.C. & Bittles, A.H. (1997). The comparative role of consanguinity in infant and childhood mortality in Pakistan. *Annals of Human Genetics*, **16**, 143–9.

Gray, L.H. (1915). Iranian marriages. In *Encyclopaedia of Religion and Ethics*, vol. 8, J. Hastings, ed., Edinburgh: Clark, pp. 455–9.

Greenstein, M.A. & Bernstein, B.A. (1996). Jewish coulture in North America. In *Cultural and Ethnic Diversity – A Guide for Genetic Professionals*, N.L. Fisher, (ed.), Baltimore, MD: Johns Hopkins University Press, pp. 198–219.

Greenwald, E. & Leitenberg, H. (1989). Long-term effects of sexual experiences with siblings and nonsiblings during childhood. *Archives of Sexual Behavior*, **18**, 389–99.

Griffin, J.E., McPhaul, M.J., Russell, D.W. &Wilson, J.D. (1995). The androgen resistance syndromes: steroid 5α-reductase-2 deficiency, testicular feminization, and related disorders. In *The Metabolic and Molecular Basis of Disease*, 7th edn. C.R. Scriver, A.L. Beaudet, W.S. Sly, & D. Valle (eds.), New York: McGraw-Hill, pp. 2967–98.

Grjibovski, A.M., Magnus, P. & Stoltenberg, C. (2009). Decrease in consanguinity among parents of children born in Norway to women of Pakistani origin: A registry-based study. *Scandinavian Journal of Public Health*, **37**, 232–8.

Gross, M. (1979). Incestuous rape: a cause for hysterical seizures in four adolescent girls. *American Journal of Orthopsychiatry*, **49**, 704–8.

Grytten, N., Glad, S.B., Aarseth, J.H. *et al.* (2006). A 50-year follow-up of the incidence of multiple sclerosis in Hordaland County, Norway. *Neurology*, **66**, 182–6.

Gu, S-M., Kumaramanickavel, G., Srikumari, C.R., Denton, M.J. & Gal, A. (1999). Autosomal recessive retinitis pigmentosa locus RP28 maps between D2S1337 and D2S286 on chromosome 2p11-p15 in an Indian family. *Journal of Medical Genetics*, **36**, 705–7.

Gunaid, A.A., Hummad, A.N. & Tamim, K.A. (2004). Consanguineous marriage in the capital city Sana'a, Yemen. *Journal of Biosocial Science*, **36**, 111–21.

Gupta, P.K., Saxena, R., Adamtziki, E. *et al.* (2008). Genetic defects in von Willebrand disease type 3 in Indian and Greek patients. *Blood Cells, Molecules and Diseases*, **41**, 219–22.

Gurling, H.M., Kalsi, G., Brynjolfson, J. *et al.* (2001). Genomewide genetic linkage analysis confirms the presence if susceptibility loci for schizophrenia, on chromosomes 1q32.2, 5q33.2, and 8p21–22, and provides support for linkage to schizophrenia on chromosomes 11q23.3–24 and 20q12.1–11.23. *American Journal of Human Genetics*, **68**, 661–73.

Gustavson, K-H. (2005). Prevalence and aetiology of congential birth defects, infant mortality and mental retardation in Lahore, Pakistan: a prospective cohort study. *Acta Paediatrica*, **94**, 769–74.

Guy, R., Forsyth, J.M., Cooper, A. & Morton, R.E. (2001). Co-existence of lysosomal storage diseases in a consanguineous family. *Child: Care, Health and Development*, **27**, 173–81.

Güz, K., Dedeoglu, N. & Lûleci, G. (1989). The frequency and medical effects of consanguineous marriages in Antalya, Turkey. *Hereditas*, **111**, 79–83.

Haas, C.C. (1989). Bighorn lamb mortality: predation, inbreeding, and population effects. *Canadian Journal of Zoology*, **67**, 699–705.

Hadj Salem, I., Kamoun, F., Louhichi, N. *et al.* (2011). Mutations in LAMA2 and CPN3 genes associated with genetic and phenotypic heterogeneities within a single consanguineous family involving both congenital and progressive dystrophies. *Bioscience Reports*, **31**, 125–35.

Hafez, M., El-Tahan, H., Awafalla, M. *et al.* (1983). Consanguineous matings in the Egyptian population. *Journal of Medical Genetics*, **20**, 58–60.

Hajnal, J. (1963). Concepts of random mating and the frequency of consanguineous marriages. *Proceedings of the Royal Society, Series B*, **159**, 125–77.

Haldane, J.B.S. (1939). The spread of harmful autosomal recessive genes in human populations. *Annals of Eugenics*, **9**, 232–7.

Haldane, J.B.S. (1949). Disease and evolution. *La Ricerca Scientifica*, **19** Suppl. A, 68–76.

Halliday, J., Oke, K., Breheny, S., Algar, E. & Amor, D.J. (2004). Beckwith-Wiedemann syndrome and IVF: a case-control study. *American Journal of Human Genetics*, **75**, 526–8.

Hallmayer, J., Cleveland, S., Torres, A. *et al.* (2011). Genetic heritability and shared environmental factors among twin pairs with autism. *Archives of General Psychiatry*, **68**, 1095–102.

Hamamy, H., Al-Hait, S., Alwan, A. & Ajlouni, K. (2007). Jordan: communities and community genetics. *Community Genetics*, **10**, 52–60.

Hamamy, H., Antonarakis, S.E., Cavalli-Sfroza, L.L. *et al.* (2011). Consanguineous marriages, pearls and perils: Geneva International Consanguinity Workshop Report, *Genetics in Medicine*, **13**, 841–7.

Hamamy, H. & Bittles, A.H. (2009). Genetic clinics in Arab countries: meeting individual, family and community needs. *Public Health Genomics*, **12**, 30–40.

Hamamy, H., Jamhawi, L., Al-Darawsheh, J. & Ajlouni, K. (2005). Consanguineous marriages in Jordan: Why is the rate changing with time? *Clinical Genetics*, **76**, 511–6.

Hamamy, H.A., Al-Hakkak, Z.S. & Al-Taha, S. (1990). Consanguinity and the genetic control of Down syndrome. *Clinical Genetics*, **37**, 24–9.

Hamed, R.M. (2002). The spectrum of chronic renal failure among Jordanian children. *Journal of Nephrology*, **15**, 130–5.

Hamidieh, A.A., Ostadali, M., Houseini, A. *et al.* (2011). Using other-related donors for allogeneic blood and marrow transplantation in Iran. Paris: European Group for Blood and Bone Marrow Transplantation, 37th Annual Meeting. Downloaded 13 July 2011.

Hamilton, W.D. (1963). The evolution of altruistic behaviour. *The American Naturalist*, **97**, 354–6.

Hamilton, W.D. (1964). The genetical evolution of social behaviour. *Journal of Theoretical Biology*, **7**, 1–16.

Hammami, A., Elgazzeh, M., Chalbi, N. & Mansour, B.A. (2005). Endogamy and consanguinity in Mauritania. *Tunis Medicale*, **83**, 38–42. [In French.]

Hammel, E.A., McDaniel, C.K. & Wachter, K.W. (1979). Demographic consequences of incest tabus: a microsimulation analysis. *Science*, **205**, 972 7.

Hampshire, K.R. & Smith, M.T. (2001). Consanguineous marriage among the Fulani. *Human Biology*, **73**, 597–603.

Hann, K.L. (1985). Inbreeding and fertility in a South Indian population. *Annals of Human Biology*, **12**, 267–74.

Harlap, S., Kleinhaus, K., Perrin, M.C. *et al.* (2008). Consanguinity and birth defects in the Jerusalem perinatal study cohort. *Human Heredity*, **66**, 180–9.

Harpending, H.C., Batzer, M.A., Gurven, M. *et al.* (1998). Genetic traces of ancient demography. *Proceedings of the National Academy of Sciences of the United States of America*, **95**, 1961–7.

Hart, T.C., Hart, P.S., Michalec, M.D. *et al.* (2000). Localisation of a gene for prepubertal periodontitis to chromosome 11q14 and identification of a cathepsin C gene mutation. *Journal of Medical Genetics*, **37**, 95–101.

Hashimoto, S., Boissel, S., Zarhrate, M. *et al.* (2011). MED23 mutation links intellectual disability to dysregulation of immediate early gene expression. *Science*, **333**, 1161–3.

Hashmi, M.A. (1997). Frequency of consanguinity and its effect on congenital malformation – a hospital based study. *Journal of the Pakistan Medical Association*, **47**, 75–8.

Hass, C. (1989). Bighorn sheep mortality: predation, inbreeding and population effects. *Canadian Journal of Zoology*, **67**, 699–705.

Hassan, I., Haleem, A.A. & Bhutta, Z.A. (1997). Profile and risk factors for congenital heart disease. *Journal of the Pakistan Medical Association*, **47**, 78–81.

Hassan, M.O., Albarwani, S., Al Yahyaee, S. *et al.* (2005). A family study in Oman: large consanguineous, polygamous Omani Arab pedigrees. *Community Genetics*, **8**, 56–60.

Hassan, M.O., Al Kharusi, W. & Ziada, A. (2001). Blood pressure and its reactivity in the offspring of first cousin hypertensive and first cousin normotensive parents: a preliminary report. *Journal of Human Hypertension*, **15**, 869–72.

Hassan, S.M., Hamza, N., Al-Lawatiya, F.J. *et al.* (2010). Extended molecular spectrum of β- and α-thalassemia in Oman. *Hemoglobin*, **34**, 127–34.

Havlicek, J. & Roberts, S.C. (2009). MHC-correlated mate choice in humans: a review. *Psychoneuroendocrinology*, **34**, 497–512.

Hawass, Z., Gad, Y.Z., Ismail, S. *et al.* (2010). Ancestry and pathology in King Tutankhamun's family. *Journal of the American Medical Association*, **303**, 638–47.

Hawks, J., Wang, E.T., Cochran, G.M., Harpending, H.C. & Moyzis, R.K. (2007). Recent acceleration of human adaptive selection. *Proceedings of the National Academy of Sciences of the United States of America*, **104**, 20753–8.

Hayakawa, M., Fujiki, K., Kanai, A. *et al.* (1997a). Multicenter genetic study of retinitis pigmentosa in Japan: 1. Genetic heterogeneity in typical retinitis pigmentosa. *Japanese Journal of Ophthalmology*, **41**, 1–6.

Hayakawa, M., Fujiki, K., Kanai, A. *et al.* (1997b). Multicenter genetic study of retinitis pigmentosa in Japan: 1. Prevalence of autosomal recessive retinitis pigmentosa. *Japanese Journal of Ophthalmology*, **41**, 7–11.

Hayes, C.E. (2000). Vitamin D: a natural inhibitor of multiple sclerosis. *Proceedings of the Nutritional Society*, **59**, 531–5.

Hayward, B.E., De Vos, M., Talati, N. *et al.* (2009). Genetic and epigenetic analysis of hydatidiform mole. *Human Mutation*, **30**, E629–39.

Hecht, F. & Sandberg, A.A. (1987). Genetic history: I. The Schneersons of Lubavich. *Clinical Genetics*, **32**, 70–4.

Hedayat, K.M., Shooshtarizadeh, P. & Raza, M. (2006). Therapeutic abortion in Islam: contemporary views of Muslim Shiite scholars and effect of recent Iranian legislation. *Journal of Medical Ethics*, **32**, 652–7.

Hedrick, P.W. (2011). Selection and mutation for α thalassemia in nonmalarial and malarial environments. *Annals of Human Genetics*, **75**, 468–74.

Hedrick, P.W. & Black, F.L. (1997). HLA and mate selection: no evidence in South Americans. *American Journal of Human Genetics*, **61**, 505–11.

Heemskerk, M.B.A., van Walraven, S.M., Cornelissen, J.J. *et al.* (2005). How to improve the search for an unrelated haematopoietic stem cell donor. Faster is better than more! *Bone Marrow Transplantation*, **35**, 645–52.

Heinisch, U., Zlotogora, J., Kafert, S. & Gieslemann, V. (1995). Multiple mutations are responsible for the high frequency of metachromatic leukodystrophy in a small geographic area. *American Journal of Human Genetics*, **56**, 51–7.

Helgason, A., Pálsson, S., Gudbjartsson, D.F., Kristjánsson, T. & Stefánsson, K. (2008). An association between the kinship and fertility of human couples. *Science*, **319**, 813–6.

Helgason, A., Yngvadóttir, B., Hrafnkelsson, B., Gulcher, J. & Stefánsson, K. (2005). An Icelandic example of the impact of population structure on association studies. *Nature Genetics*, **37**, 90–5.

Hellani, A., Schuchman, E.H., Al-Odaib, A. *et al.* (2004). Genetic diagnosis for Niemann-Pick disease type B. *Prenatal Diagnosis*, **24**, 943–8.

Henderson, J. (1983). Is incest harmful? *Canadian Journal of Psychiatry*, **28**, 34–40.

Herman, J. (1981). Father–daughter incest. *Professional Psychology*, **12**, 76–80.

Hessini, L. (2007). Abortion and Islam: policies and practice in the Middle East and North Africa. *Reproductive Health Matters*, **15**, 1–10.

Hickman, M., Modell, B., Greengross, P. *et al.* (1999). Mapping the prevalence of sickle cell and beta thalassaemia in England: estimating and validating ethnic-specific rates. *British Journal of Haematology*, **104**, 860–7.

Higgins, J.J., Pucilowska, J., Lombardi, R.Q. & Rooney, J.P. (2004). A mutation in a novel ATP-dependent Lon protease gene in a kindred with mild mental retardation. *Neurology*, **63**, 1927–31.

Hmani-Aifa, M., Benzina, Z., Zulfiqar, F. *et al.* (2009). Identification of two new mutations in the *GPR98* and the *PDE6B* genes segregating in a Tunisian family. *European Journal of Human Genetics*, **17**, 474–82.

HMSO (1870). *British Parliamentary Papers, Parliamentary Debates*, Third Series, **203**, London: Her Majesty's Stationary Office, pp. 817–8, 1006–10.

Ho, H.N., Gill, T.J., Nsieh, R.-P., Hsieh, H.-J. & Lee, T.-Y. (1990). Sharing of human leukocyte antigens in primary and secondary recurrent spontaneous abortions. *American Journal of Obstetrics and Gynecology*, **163**, 178–88.

Hoben, A.D., Buunk, A.P., Fincher, C.L., Thornhill, R. & Schaller, M. (2010). On the adaptive origins and maladaptive consequences of human inbreeding: parasite prevalence, immune functioning, and consanguineous marriage. *Evolutionary Psychology*, **8**, 658–76.

Hoggart, C.J., Parra, E.J., Shriver, M.D. *et al.* (2003). Control of confounding of genetic associations in stratified populations. *American Journal of Human Genetics*, **72**, 1492–504.

Holliday, E.G., Nyholt, D.R., Tirupati, S. *et al.* (2009). Strong evidence for a novel schizophrenia risk locus on chromosome 1p31.1 in homogeneous pedigrees from Tamil Nadu, India. *American Journal of Psychiatry*, **166**, 206–15.

Hom, J., Turner, M.B., Risser, R., Bonte, F.J. & Tintner, R. (1994). Cognitive deficits in asymptomatic first-degree relatives of Alzheimer's disease patients. *Journal of Clinical and Experimental Neuropsychology*, **16**, 568–76.

Honeyman, M.M., Bahl, L., Marshall, T. & Wharton, B.A. (1987). Consanguinity and fetal growth in Pakistani Moslems. *Archives of Disease in Childhood*, **62**, 231–5.

Hoodfar, E. & Teebi, A.S. (1996). Genetic referrals of Middle Eastern origin in a western city: inbreeding and disease profile. *Journal of Medical Genetics*, **33**, 212–5.

Hopewell, S., Loudon, K., Clarke, M.J., Oxman, A.D. & Dickersin, K. (2009). Publication bias in clinical trials due to statistical significance or direction of trial results. *Cochrane Database of Systematic Reviews*, Issue 1. Art. No.: MR000006. DOI: 10.1002/14651858.MR000006.pub3.

Hopkins, K. (1980). Brother–sister marriage in Roman Egypt. *Comparative Studies in Social History*, **22**, 303–54.

Hoppenbrouwers, I.A., Cortes, L.M., Aulchenko, Y.S. *et al.* (2007). Familial clustering of multiple sclerosis in a Dutch genetic isolate. *Multiple Sclerosis*, **13**, 17–24.

Horder, T.J. (1933). Eugenics – and the doctor. *British Medical Journal*, **2**, 1057–60.

Hornby, S.J., Dandona, L., Foster, A., Jones, R.B. & Gilbert, C.E. (2001). Clinical findings, consanguinity, and pedigrees in children with anophthalmos in southern India. *Developmental Medicine and Child Neurology*, **43**, 392–8.

Hosoda, Y., Fujiki, K. & Nakajima, A. (1983). A research of consanguinity in Japanese young couples. *Japanese Journal of Human Genetics*, **28**, 205–7.

Hsu, F.L.K. (1945). Observations on cross-cousin marriage in China. *American Anthropologist*, **47**, 83–103.

Hurley, C.K., Wade, J.A., Oudshoorn, M. *et al.* (1999). A special report: histocompatibility testing guidelines for hematopoietic stem cell transplantation using volunteer donors. *Human Immunology*, **60**, 347–60.

Hurwich, B.J., Rosner, B., Nubani, N., Kass, E.H. & Lewitter, F.I. (1982). Familial aggregation of blood pressure in a highly inbred community, Abu Ghosh, Israel. *American Journal of Epidemiology*, **115**, 646–56.

Husain, A. & Chapel, J.L. (1983). History of incest in girls admitted to a psychiatric hospital. *American Journal of Psychiatry*, **140**, 591–3.

Hussain, R. (1998). The role of consanguinity and inbreeding as a determinant of spontaneous abortion in Karachi, Pakistan. *Annals of Human Genetics*, **62**, 147–57.

Hussain, R. (1999). Community perceptions of reasons for preference for consanguineous marriages in Pakistan. *Journal of Biosocial Science*, **31**, 449–61.

Hussain, R. (2002). Lay perceptions of genetic risks attributable to inbreeding in Pakistan. *American Journal of Human Biology*, **14**, 264–74.

Hussain, R. (2005). The effect of religious, cultural and social identity on population genetic structure among Muslims in Pakistan. *Annals of Human Biology*, **32**, 145–53.

Hussain, R. & Bittles, A.H. (1998). The prevalence and demographic characteristics of consanguineous marriages in Pakistan. *Journal of Biosocial Science*, **30**, 261–79.

Hussain, R. & Bittles, A.H. (1999). Consanguinity and differentials in age at marriage, contraceptive use and fertility in Pakistan. *Journal of Biosocial Science*, **31**, 121–38.

Hussain, R. & Bittles, A.H. (2000). Sociodemographic correlates of consanguineous marriage in the Muslim population of India. *Journal of Biosocial Science*, **32**, 433–42.

Hussain, R., Fikree, F.F. & Berendes, H.W. (2000). The role of son preference in reproductive behaviour in Pakistan. *Bulletin of the World Health Organization*, **78**, 379–88.

Husseini, A. & Akkawi, M. (2005). Maternal serum screening of Palestinian women in the West Bank. *Eastern Mediterranean Health Journal*, **11**, 824–7.

Hussels, I. (1969). Genetic structure of Laas, a Swiss isolate. *Human Biology*, **41**, 469–79.

Hussien, F.H. (1971). Endogamy in Egyptian Nubia. *Journal of Biosocial Science*, **3**, 251–7.

Huth, A.H. (1875). *The Marriage of Near Kin, considered with respect to the laws of nations, the results of experience, and the teachings of biology.* London: Churchill.

Huth, A.H. (1877). Cross-fertilization of plants, and consanguineous marriage. *Westminster Review*, New Series, **3**, 466–85.

Ihara, Y., Aoki, K., Tokunaga, K., Takahashi, K. & Juji, T. (2000). HLA and human mate choice: tests on Japanese couples. *Anthropological Science*, **108**, 199–214.

IIPS. (1995). *National Family Health Survey, 1992–93.* Bombay: International Institute for Population Sciences.

Imaizumi, Y. (1986a). A recent survey of consanguineous marriages in Japan: religion and socioeconomic class effects. *Annals of Human Biology*, **13**, 317–30.

Imaizumi, Y. (1986b). Factors influencing the frequency of consanguineous marriages in Japan: marital distance and opportunity of encounter. *Human Heredity*, **36**, 304–9.

Imaizumi, Y. (1987). Reasons for consanguineous marriage in Japan. *Journal of Biosocial Science*, **19**, 97–106.

Imaizumi, Y. (1988). Familial aggregation of consanguineous marriages in Japan. *Journal of Biosocial Science*, **20**, 99–109.

Imaizumi, Y. & Shinozaki, N. (1984). Frequency of consanguineous marriages in Japan: geographical variations. *Japanese Journal of Human Genetics*, **29**, 381–5.

Imaizumi, Y., Shinozaki, N. & Aoki, H. (1975). Inbreeding in Japan: results of a nation-wide study. *Japanese Journal of Human Genetics*, **20**, 91–107.

IMSGC & WTCCC (2011). Genetic risk and a primary role for cell-mediated immune mechanisms in multiple sclerosis. *Nature*, **476**, 214–8.

Indian Genome Variation Consortium. (2008). Genetic landscape of the people of India: a canvas for disease gene exploration. *Journal of Genetics*, **87**, 3–20.

Inhorn, M.C., Kobeissi, L., Nassar, Z., Lakkis, D. & Fakih, M.H. (2009). Consanguinity and family clustering of male factor infertility in Lebanon. *Fertility & Sterility*, **91**, 1104–9.

Irmansyah, Schwab, S.G., Heriani *et al.* (2008). Genome-wide scan in 124 Indonesian sib-pair families with schizophrenia reveals genome-wide significant linkage to a locus on chromosome 3p26–21. *American Journal of Medical Genetics Part B, Neuropsychiatric Genetics*, **147B**, 1245–52.

Ishikuni, N., Nemoto, H., Neel, J.V., Drew, A.L., Yanase, T. & Matsumoto, Y.S. (1960). Hosojima. *American Journal of Human Genetics*, **12**, 67–75.

Ismail, J., Jafar, T.H., Jafary, F.H., White, F., Faruqui, A.M. & Chaturvedi, N. (2004). Risk factors for non-fatal myocardial infarction in young South Asian adults. *Heart*, **90**, 259–63.

Ito, K. (1972). Genetic studies in some Japanese populations. V. Multivariate regression analyses of anthropometric measurements on the effect of consanguinity in Shizuoka children. *Japanese Journal of Human Genetics*, **17**, 180–92.

Iyer, S. (2002). *Demography and Religion in India*. New Delhi: Oxford University Press.

Jaber, L., Halpern, G.J. & Shohat, T. (2000). Trends in the frequencies of consanguineous marriages in the Israeli Arab community. *Clinical Genetics*, **58**, 106–10.

Jaber, L., Merlob, P., Bu, X., Trotter, J. & Shohat, M. (1992). Marked parental consanguinity as a cause for increased major malformations in an Israeli Arab community. *American Journal of Medical Genetics*, **44**, 1–6.

Jaber, L., Merlob, P., Gabriel, R. & Shohat, M. (1997a). Effects of consanguineous marriage on reproductive outcome in an Arab community in Israel. *Journal of Medical Genetics*, **34**, 1000–2.

Jaber, L., Nahmani, A., Halpern, G.J. & Shohat, M. (2002). Facial clefting in an Arab town in Israel. *Clinical Genetics*, **61**, 448–53.

Jaber, L., Nahmani, A. & Shohat, M. (1997b). Speech disorders in Israeli Arab children. *Israeli Journal of Medical Science*, **33**, 663–5.

Jaber, L., Shohat, M. & Halpern, G.J. (1996). Demographic characteristics of the Israeli Arab community in connection with consanguinity. *Israeli Journal of Medicine*, **32**, 1286–9.

Jaber, L., Shohat, T., Rotter, J.I. & Shohat, M. (1997c). Consanguinity and common adult diseases in Israeli Arab communities. *American Journal of Medical Genetics*, **70**, 346–8.

Jacob, S., McClintock, M.K., Zelano, B. & Ober, C. (2002). Paternally inherited HLA alleles are associated with women's choice of body odor. *Nature Genetics*, **30**, 175–9.

Jacob, S.M. (1911). Inbreeding in a stable simple Mendelian population with special reference to cousin marriage. *Proceedings of the Royal Society Series B*, **84**, 23–42.

Jacob John, T. & Jayabal, P. (1971). Effects of consanguineous marriage on reproductive outcome in an Arab community in Israel. *Indian Journal of Medical Research*, **59**, 1050–3.

Jacoby, H.G. & Mansuri, G. (2007). *Watta satta*: bride exchange and women's welfare in rural Pakistan. *World Bank Policy Research Working Paper 4126*. Washington, DC, The World Bank.

Jacoby, H.G. & Mansuri, G. (2010). *Watta satta*: bride exchange and women's welfare in rural Pakistan. *American Economic Review*, **100**, 1804–25.

Jacquard, A. (1975). Inbreeding: one word, several meanings. *Theoretical Population Biology*, **7**, 338–63.

Jain, V.K., Nalini, P., Chandra, R. & Srinivasan, S. (1993). Congenital malformations, reproductive wastage and consanguineous mating. *Australian and New Zealand Journal of Obstetrics and Gynaecology*, **33**, 33–6.

Jamilian, A., Nayeri, F. & Babayan, A. (2007). Incidence of cleft lip and palate in Iran. *Journal of the Indian Society of Pedodontics and Preventive Dentistry*, **25**, 174–6.

Jancar, J. & Johnston, S.J. (1990). Incest and mental handicap. *Journal of Mental Deficiency Research*, **34**, 483–90.

Janson, S., Jayakoddy, A., Abulaban, A. & Gustavson, K.H. (1990). Severe mental retardation in Jordanian children. A retrospective study. *Acta Paediatrica Scandinavica*, **79**, 1099–104.

Janssens, P.M. (2003). No reason for a reduction in the number of offspring per sperm donor because of possible transmission of autosomal dominant diseases. *Human Reproduction*, **28**, 669–71.

Janssens, P.M., Nap, A.W. & Bancsi, L.F. (2011). Reconsidering the number of offspring per gamete donor in the Dutch open-identity system. *Human Fertility*, **14**, 106–14.

Janssens, P.M., Simons, A.H., van Kooij, R.J., Blokzijl, E. & Dunselman, G.A. (2006). A new Dutch Law regulating provision of identifying information of donors to offspring: background, content and impact. *Human Reproduction*, **21**, 852–6.

Jarisch, A., Giunta, C., Zielen, S., König, R. & Steinmann, B. (1998). Sibs affected with both Ehlers-Danlos syndrome type VI and cystic fibrosis. *American Journal of Medical Genetics*, **78**, 455–60.

Jazayeri, R., Saberi, S.H. & Soleymanzadeh, N. (2010). Etiological characteristics of people with intellectual disability in Iran. *Neurosciences (Riyadh)*, **15**, 258–61.

Jeffrey, R., Jeffrey, P. & Lyon, A. (1984). Female infanticide and amniocentesis. *Social Science and Medicine*, **19**, 1207–12.

Jeganathan, D., Chodhari, R., Meeks, M. *et al.* (2004). Loci for primary ciliary dyskinesia map to chromosome 16p12.1–12.2 and 15q13.1–15.1 in Faroe Islands and Israeli Druze genetic isolates. *Journal of Medical Genetics*, **41**, 233–40.

Jenkins, W.G. (1891). Heredity in relation to deafness. *American Annals of the Deaf*, **36**, 97–111.

Jha, P., Kumar, R., Vasa, P. *et al.* (2006). Low female-to-male sex ratio of children born in India: national survey of 1.1 million households. *Lancet*, **367**, 211–8.

Jin, K., Ho, H.-N., Speed, T. & Gill, T.J. (1995b). Reproductive failure and the major histocompatibility complex. *American Journal of Human Genetics*, **56**, 1456–67.

Jin, K., Speed, T.P. & Thomson, G. (1995a). Tests of random mating for a highly polymorphic locus: applications to HLA data. *Biometrics*, **51**, 1064–76.

Johnson, P.M., Chia, K.V., Risk, J.M., Barnes, R.M.R. & Woodrow, J.C. (1988). Immunological and immunogenetic investigation of recurrent spontaneous abortion. *Disease Markers*, **6**, 163–71.

Johnstone, F. & Inglis, L. (1974). Familial trends in low birth weight. *British Medical Journal*, **3**, 659–61.

Joseph, M., Zoubeidi, T., Al-Dhaheri, S.M. *et al.* (2009). Paternal asthma is a predictor for childhood asthma in the consanguineous families from the United Arab Emirates. *Journal of Asthma*, **46**, 175–8.

Joseph, S.E. (2007). "Kissing Cousins": consanguineous marriage and early mortality in a reproductive isolate. *Current Anthropology*, **48**, 756–64.

Joshi, S., Iyer, S. & Do, Q-T. (2009). *Why Marry a Cousin? Insights from Bangladesh.* Washington, DC: The World Bank.

Joshi, S.N. & Venugopalan, P. (2007). Clinical characteristics of neonates with inborn errors of metabolism detected by Tandem MS analysis in Oman. *Brain and Development*, **29**, 543–6.

Jurdi, R. & Saxena, P.C. (2003). The prevalence and correlates of consanguineous marriages in Yemen: similarities and contrasts with other Arab countries. *Journal of Biosocial Science*, **35**, 1–13.

Jursić, A. & Skrinjarić, I. (1988). The inheritance of hypodontia in families – the segregational analysis. *Acta Stomatologica Croatica*, **22**, 261–9. [In Croatian.]

Jutla, R.K. & Heimbach, D. (2004). Love burns: an essay about bride burning in India. *Journal of Burn Care & Rehabilitation*, **25**, 165–70.

Kabarity, A., Al-Awadi, S.A., Farag, T.I. & Mallalah, G. (1981). Autosomal recessive "uncomplicated" profound childhood deafness in an Arabic family with high consanguinity. *Human Genetics*, **57**, 444–6.

Kabré, A., Badiane, S.B., Sakho, Y., Ba, M.C. & Gueye, M. (1994). Genetic and etiologic aspects of spina bifida in Senegal. Apropos of 211 cases collected at the neurological clinic UHC of Fann. *Dakar médical*, **39**, 113–9. [In French.]

Kaeuffer, R., Coltman, D.W., Chapuis, J.L., Pontier, D. & Réale, D. (2007). Unexpected heterozyosity in an island mouflon population founded by a single pair of individuals. *Proceedings of the Royal Society Series B*, **274**, 527–33.

Kalaydjieva, L., Hallmayer, J., Chandler, D. *et al.* (1996). Gene mapping in Gypsies identifies a novel demyelinating neuropathy on chromosome 8q24. *Nature Genetics*, **14**, 214–7.

Kamin, L.J. (1980). Inbreeding depression and IQ. *Psychological Bulletin*, **87**, 469–78.

Kanaan, Z.M., Mahfouz, R. & Tamim, H. (2008). The prevalence of consanguineous marriages in an underserved area in Lebanon and its association with congenital anomalies. *Genetic Testing*, **12**, 367–72.

Kapadia, K.M. (1958). *Marriage and Family in India*, 2nd edn. Calcutta: Oxford University Press, pp. 117–37.

Kar, B., John, S. & Kumaramanickavel, G. (1995). Retinitis pigmentosa in India: a genetic and segregation analysis. *Clinical Genetics*, **47**, 75–9.

Karimi, M., Bonyadi, M., Galehdari, M.R. & Zareifar, S. (2008). Termination of pregnancy due to thalassaemia major, haemophilia, and Down's syndrome: the view of Iranian physicians. *BMC Medical Ethics*, **9**, 19.

Karimi, M., Johari, S. & Cohan, N. (2010). Attitude towards prenatal diagnosis for beta-thalassaemia major and medical abortion in southern Iran. *Hemoglobin*, **34**, 49–54.

Karnib, H.H., Gharavi, A.G., Aftimos, G. *et al.* (2010). A 5-year survey of biopsy proven kidney diseases in Lebanon: significant variation in prevalence of primary glomerular diseases by age, population structure and consanguinity. *Nephrology, Dialysis, Transplantation*, **25**, 3962–9.

Katz-Sidlow, R.J. (1998). In the Darwin family tradition: another look at Charles Darwin's ill health. *Journal of the Royal Society of Medicine*, **91**, 484–8.

Kaufman, L., Ayub, M. & Vincent, J.B. (2010). The genetic basis of non-syndromic intellectual disability: a review. *Journal of Neurodevelopmental Disorders*, **2**, 182–209.

Kaur, M. & Balgir, P.P. (2005). APOE2 and consanguinity: a risky combination for Alzheimer's disease. *Journal of Alzheimer's Disease*, **8**, 293–7.

Kazaura, M.R. & Lie, R.T. (2002). Down's syndrome and paternal age in Norway. *Paediatric and Perinatal Epidemiology*, **16**, 314–9.

Kazaura, M.R., Lie, R.T. & Skjaerven, R.S. (2006). Grandparent's age and the risk of Down's syndrome in Norway. *Acta Obstetrica et Gynecologica Scandinavica*, **85**, 236–40.

Kende, G., Toren, A., Mandel, M., Numann, Y. *et al.* (1994). Familial leukaemia: description of two kindreds and a review of the genetic aspects of the disease. *Acta Haematologica*, **92**, 208–11.

Kerkeni, E., Monastri, K., Seket, B., Guediche, M.N. & Ben Chiekh, H. (2007). Interplay of socio-economic factors, consanguinity, fertility, and offspring mortality in Monastir, Tunisia. *Croatian Medical Journal*, **48**, 701–7.

Keskin, A., Turk, T., Polat, A., Koyuncu, H. & Saracoglu, B. (2000). Premarital screening of beta-thalassemia in the province of Denizli, Turkey. *Acta Haematologica*, **104**, 31–3.

Khabori, M.A. & Patton, M.A. (2008). Consanguinity and deafness in Omani children. *International Journal of Audiology*, **47**, 30–3.

Khan, A.K., Rafiq, M.A., Noor, A. *et al.* (2011). A novel deletion mutation in the *TUSC3* gene in a consanguineous Pakistani family with autosomal recessive nonsyndromic intellectual disability. *BMC Medical Genetics*, **12**, 56.

Khan, A.O., Al-Abdi, L., Mohamed, J.Y., Aldahmesh, M.A. & Alkuraya, F.S. (2011). Familial juvenile glaucoma with underlying homozygous p.G61E CYP1B1 mutations. *Journal of the American Association for Pediatric Ophthalmology and Starbismus*, **15**, 198–9.

Khandekar, R., Khabori, M., Jaffer Mohammed, A. & Gupta, R. (2006). Neonatal screening for hearing impairment – the Oman experience. *International Journal of Pediatric Otorhinolaryngology*, **70**, 663–70.

Khatri, M.L., Bemghazil, M., Shafi, M. & Machina, A. (1999). Xeroderma pigmentosa in Libya. *International Journal of Dermatology*, **38**, 520–4.

Khawaja, M. & Hammoury, N. (2008). Coerced sexual intercourse within marriage: a clinic-based study of pregnant Palestinian refugees in Lebanon. *Journal of Midwifery and Women's Health*, **53**, 150–4.

Khawaja, M & Tewtel-Salem, M. (2004). Agreement between husband and wife reports of domestic violence: evidence from poor refugee communities in Lebanon. *International Journal of Epidemiology*, **33**, 526–33.

Khlat, M. (1985). Consanguineous marriages in Beirut: time trends, spatial distribution. *Social Biology*, **35**, 324–30.

Khlat, M. (1988). Consanguineous marriage and reproduction in Beirut, Lebanon. *American Journal of Human Genetics*, **43**, 188–96.

Khlat, M. (1989). Inbreeding effects on fetal growth in Beirut, Lebanon. *American Journal of Physical Anthropology*, **80**, 481–4.

Khlat, M. (1997). Endogamy in the Arab world. In *Genetic Disorders among Arab Populations*, A.S. Teebi & T.I. Farag, eds., New York: Oxford University Press, pp. 63–80.

Khlat, M., Halabi, S., Khudr, A. & De Kaloustian, V.M. (1986). Perception of consanguineous marriages and their genetic effects among a sample of couples from Beirut. *American Journal of Medical Genetics*, **25**, 299–306.

Khoury, M.J., Cohen, B.H., Diamond, E.L., Chase, G.A. & McKusick, V.A. (1987). Inbreeding and prereproductive mortality in the Old Order Amish. III. Direct and indirect effects of inbreeding. *American Journal of Epidemiology*, **125**, 473–83.

Khoury, S.A. & Massad, D. (1992). Consanguineous marriage in Jordan. *American Journal of Medical Genetics*, **43**, 769–75.

Khoury, S.A. & Massad, D.F. (2000). Consanguinity, fertility, reproductive wastage, infant mortality and congenital malformations in Jordan. *Saudi Medical Journal*, **21**, 150–4.

Khrouf, N., Spång, R., Podgorna, T. *et al.* (1986). Malformations in 10,000 consecutive births in Tunis. *Acta Paediatrica Scandinavica*, **75**, 534–9.

Khuri, F.I. (1970). Parallel cousin marriage reconsidered: a Middle Eastern practice that nullifies the effects of marriage on the intensity of family relationships. *Man*, **5**, 597–618.

Kilpatrick, D.C. (1984). A case of materno-fetal histocompatability – implications for leucocyte transfusion treatment for recurrent abortions. *Scottish Medical Journal*, **29**, 110–2.

Kimmel, G., Jordan, M.I., Halperin, E., Shamir, R. & Karp, R.M. (2007). A randomization test for controlling population stratification in whole-genome association studies. *American Journal of Human Genetics*, **81**, 895–905.

Kirin M., McQuillan, R., Franklin, C.S. *et al.* (2010). Genomic runs of homozygosity record population history and consanguinity. *PLoS One*, **5**, e13996.

Kisch, S. (2008). "Deaf discourse": the social construction of deafness in a Bedouin community. *Medical Anthropology*, **27**, 283–313.

Kishimoto, K. (1962). Preliminary report of activities of the Consanguinity Study Group of the Science Council of Japan. *Eugenics Quarterly*, **9**, 5–13.

Kishore, R.R. (2005). Human organs, scarcities, and sale: morality revisited. *Journal of Medical Ethics*, **31**, 362–5.

Kitchens, C.S. & Newcomb, T.F. (1979). Factor XIII. *Medicine*, **58**, 413–29.

Klebanoff, M.A., Graubard, B.I., Kessel, S.S. & Berendes, H.W. (1984). Low birth weight across generations. *Journal of the American Medical Association*, **252**, 2423–7.

Klein, T., Yaniv, I., Stein, J. *et al.* (2005). Extended family studies for the identification of allogeneic stem cell transplant donors in Jewish and Arabic patients in Israel. *Pediatric Transplantation*, **9**, 52–5.

Knight, K. (2003). Consanguinity in Canon Law. In *The Catholic Encyclopedia*, Online edn., Vol. IV. Available at www.newadvent.org [25 February 2004].

Kobbé, C.W. (1987). *The Definitive Kobbe's Opera Book*, edited, revised and updated by The Earl of Harewood. New York: Putnam, pp. 1356–8.

Kobeissi, L. & Inhorn, M.C. (2007). Health issues in the Arab American community. Male infertility in Lebanon: a case-controlled study. *Ethnicity and Disease*, **17**, S3–33–8.

Koç, I. (2008). Prevalence and sociodemographic correlates of consanguineous marriages in Turkey. *Journal of Biosocial Science*, **40**, 137–48.

Komai, T. & Tanaka, K. (1972). Genetic studies in some Japanese populations. VIII. Effects of inbreeding on physical development in Shizuoka school-children. Conclusions and remarks. *Japanese Journal of Human Genetics*, **17**, 209–16.

Komlos, L., Zamir, R., Joshua, H. & Halbrecht, I. (1977). Common HLA antigens in couples with repeated abortions. *Clinical Immunology & Immunopathology*, **7**, 330–5.

Korea Times (2009). *Family Visa Denied for Businessman's Wife*. Available at http://www.koreatimes.co.kr/www/news/special/2009/01/177_8686.html [31 January 2011].

Korotayev, A. (2000). Parallel-cousin (FBD) marriage, Islamization, and Arabization. *Ethnology*, **39**, 395–407.

Korver, A.M., Admiraal, R.J., Kant, S.G. *et al.* (2011). Causes of permanent childhood hearing impairment. *Laryngoscope*, **121**, 409–16.

Kostyu, D.D., Dawson, D.V., Elias, S. & Ober, C. (1993). Deficit of HLA homozygotes in a Caucasian isolate. *Human Immunology*, **37**, 135–42.

Krieger, H. (1969). Inbreeding effects on metrical traits in Northeastern Brazil. *American Journal of Human Genetics* **21**, 537–46.

Krishnamoorthy, S. & Audinarayana, N. (2001). Trends in consanguinity in South India. *Journal of Biosocial Science*, **33**, 185–97.

Krishnan, G. (1986). Effect of parental consanguinity on anthropometric measurements among the Sheikh Sunni Muslim boys of Delhi. *American Journal of Physical Anthropology*, **70**, 69–73.

Kromberg, J.G.R. & Jenkins T. (1982). Prevalence of albinism in the South African Negro. *South African Medical Journal*, **61**, 383–6.

Kudo, A., Ito, K. & Tanaka, K. (1972). Genetic studies in some Japanese populations. X. The effects of parental consanguinity on psychometric measurements, school performance and school attendance in Shizuoka school-children. *Japanese Journal of Human Genetics*, **17**, 231–48.

Kulkarni, M.L. & Kurian, M. (1990). Consanguinity and its effect on fetal growth and development: a south Indian study. *Journal of Medical Genetics*, **27**, 348–52.

Kulkarni, M.L., Mathew, M.A. & Reddy, V. (1989). The range of neural tube defects in southern India. *Journal of Medical Genetics*, **64**, 201–4.

Kumar, A. & Agarwal, M. (1988). Incest and anorexia nervosa. *British Journal of Psychiatry*, **152**, 713–4.

Kumar, V. (2004). Poisoning deaths in married women. *Journal of Clinical Forensic Medicine*, **11**, 2–5.

Kumar, V. & Tripathi, C.B. (2004). Burnt wives: a study of homicides. *Medicine Science & Law*, **44**, 55–60.

Kumaramanickavel, G., Joseph, B., Vidhya, A., Arokiasamy, T. & Sridhara Shetty, N. (2002). Consanguinity and ocular genetic diseases in South India: analysis of a five-year study. *Community Genetics*, **5**, 182–5.

Kumari, R. (1989). *Brides Are Not for Burning*. New Delhi, Radiant Publishers.

Kuper, A. (2009). Commentary: A Darwin family concern. *International Journal of Epidemiology*, **38**, 1439–42.

Kuss, A.W., Garshabi, M., Kahrizi, K. *et al.* (2011). Autosomal recessive mental retardation: homozygosity mapping identifies 27 single linkage intervals, at least 14 novel loci and several mutation hotspots. *Human Genetics*, **129**, 141–8.

Lagiou, P. & Trichopoulos, D. (2008). Birth size and the pathogenesis of breast cancer. *PLoS Medicine*, **5**, e194.

Laitenen, T., Koskimies, S. & Westman, P. (1993). Foeto-maternal compatibility in HLA-DR, DQ, and DP loci in Finnish couples suffering from recurrent spontaneous abortions. *European Journal of Immunogenetics*, **20**, 249–58.

Lakeman, P., Plass, A.M.C., Henneman, L. *et al.* (2009). Preconceptional ancestry-based carrier couple screening for cystic fibrosis and haemoglobinopathies: What determines the intention to participate or not and actual participation? *European Journal of Human Genetics*, **17**, 999–1009.

Lamdouar Bouazzaoui, N. (1994). Consanguinité et santé publique au Maroc. *Bulletin de l'Academie Nationale de Médecine*, **178**, 1013–27. [In French.]

Langhoff-Roos, J., Lindmark, G., Gustavson, K-H., Gebre-Medhin, M. & Meirik, O. (1987). Relative effect of parental birth weight on infant birth weight at term. *Clinical Genetics*, **32**, 240–8.

Larson, C.A. (1957). Some aspects of kin matings with mentally defective offspring. *Acta Genetica et Statistica Medica*, **7**, 382–5.

Lass-Hennemann, J., Deuter, C.E., Kuehl, L.K. *et al.* (2010). Effects of stress on human mating preferences: stressed individuals prefer dissimilar mates. *Proceedings of the Royal Society of London Series B*, **277**, 2175–83.

Lauc, T. (2003). Orofacial analysis on the Adriatic islands: an epidemiological study of malocclusions on Hvar island. *European Journal of Orthodontics*, **25**, 273–8.

Lauc, T., Rudan, P., Rudan, I. & Campbell, H. (2003). Effect of inbreeding and endogamy on occlusal traits in human isolates. *Journal of Orthodontics*, **30**, 301–8.

Laurier, V., Stoetzel, C., Muller, J. *et al.* (2006). Pitfalls of homozygosity mapping: an extended consanguineous Bardet-Biedel syndrome family with two mutant genes (*BBS*2, *BBS*10), three mutations, but no triallelism. *European Journal of Human Genetics*, **14**, 1195–203.

Layouni, S., Salzmann, A., Guiponni, M. *et al.* (2010). Genetic linkage study of an autosomal recessive form of juvenile myoclonic epilepsy in a consanguineous Tunisian family. *Epilepsy Research*, **90**, 33–8.

Leavitt, G.C. (2007). The incest taboo? A reconsideration of Westermarck. *Anthropological Theory*, **7**, 393–419.

Lebel, R.R. & Gallagher, W.B. (1989). Wisconsin consanguinity studies. II. Familial adenocarcinomatosis. *American Journal of Medical Genetics*, **33**, 1–6.

Le Lannou, D., Thépot, F. & Jouannet, P. (1998). Multicentre approaches to donor insemination in the French CECOS Federation: Nationwide evaluation, donor matching, screening for genetic diseases and consanguinity. *Human Reproduction*, **13**, 35–54.

Lencz, T., Lambert, C., DeRosse, P. *et al.* (2007). Runs of homozygosity reveal highly penetrant recessive loci in schizophrenia. *Proceedings of the National Academy of Sciences of the United States of America*, **104**, 19942–7.

Lentner, C. (1982). *Geigy Scientific Tables*, 8th edn., vol. 2, Basel: Ciba-Geigy, p. 229.

Le Roy Ladurie, E. (1980). *Montaillou: Cathars and Catholics in a French Village 1294–1324*, London: Penguin, pp. 36, 52, 179 sqq.

Leslie, P.W., MacCluer, J.W. & Dyke, B. (1978). Consanguinity avoidance and genotype frequencies in human populations. *Human Biology*, **50**, 281–99.

Leslie, P.W., Morrill, W.T. & Dyke, B. (1981). Genetic implications of mating structure in a Caribbean isolate. *American Journal of Human Genetics*, **33**, 90–104.

Leutenegger, A.L., Prum, B., Génin, E. *et al.* (2003). Estimation of the inbreeding coefficient through use of genomic data. *American Journal of Human Genetics*, **73**, 516–23.

Leutenegger, A.L., Sahbatou, M., Gazal, S., Cann, H. & Génin, E. (2011). Consanguinity around the world: What do the genomic data of the HGDP-CEPH diversity panel tell us? *European Journal of Human Genetics*, **19**, 583–7.

Levy, S., Sutton, G., Ng, P.C. *et al.* (2007). The diploid genome sequence of an individual human. *PloS Biology*, **5**, e254.

Lewando-Hundt, G., Shoham-Varid, I., Beckerleg, S. *et al.* (2001). Knowledge, action and resistance: the selective use of pre-natal screening among Bedouin women of the Negev, Israel. *Social Science and Medicine*, **52**, 561–9.

Lewontin, R., Kirk, D. & Crow, J. (1967). Selective mating, assortative mating, and inbreeding: definitions and implications. *Eugenics Quarterly*, **15**, 141–3.

Lezirovitz, K., Nicastro, F.S., Pardono, E. *et al.* (2006). Is autosomal recessive deafness associated with oculocutaneous albinism a "coincidence syndrome"? *Journal of Human Genetics*, **51**, 716–20.

Lezirovitz, K., Pardono, E., de Mello Auricchio, M.T. *et al.* (2008). Unexpected genetic heterogeneity in a large consanguineous Brazilian pedigree presenting deafness. *European Journal of Human Genetics*, **16**, 89–96.

Li, C.C. (1963a). Decrease of population fitness upon inbreeding. *Proceedings of the National Academy of Sciences of the United States of America*, **49**, 439–45.

Li, C.C. (1963b). Genetic load: three views. 3. The way the load ratio works. *American Journal of Human Genetics*, **115**, 316–21.

Liascovich, R., Rittler, M. & Castilla, E.E. (2001). Consanguinity in South America: demographic aspects. *Human Heredity*, **51**, 27–34.

Lichenstein, P., Yip, B.H., Björk, C. *et al.* (2009). Common genetic determinants of schizophrenia and bipolar disorder in Swedish families: a population-based study. *Lancet*, **373**, 234–9.

Lie, R.T., Rasmussen, S., Brunborg, H. *et al.* (1998). Fetal and maternal contributions to risk of pre-eclampsia: population base study. *British Medical Journal*, **316**, 1343–7.

Lieberman, D., Tooby, J. & Cosmides, L. (2003). Does morality have a biological basis? An empirical test of the factors governing moral sentiments relating to incest. *Proceedings of the Royal Society of London Series B*, **270**, 819–26.

Liede, A., Malik, I.A., Aziz, Z. *et al.* (2002). Contribution of BRAC1 and BRAC2 mutations of breast and ovarian cancer in Pakistan. *American Journal of Human Genetics*, **71**, 595–606.

Lindner, M., Abdoh, G., Fang-Hoffman, J. *et al.* (2007). Implementation of extended neonatal screening and a metabolic unit in the State of Qatar: developing and optimizing strategies in cooperation with the Neonatal Screening Center in Heidelberg. *Journal of Inherited Metabolic Disease*, **30**, 522–9.

Lipsitch, M., Bergstrom, C.T. & Antia, R. (2003). The effect of human leukocyte antigen heterozygosity on infectious disease outcome: the need for allele-specific measures. *BMC Medical Genetics*, **4**, 2.

Liu, F., Elefante, S., van Djuin, C.M. & Aulchenko, Y.S. (2006b). Ignoring distant genealogical loops leads to false-positives in homozygosity mapping. *Annals of Human Genetics*, **70**, 965–70.

Liu, H., Prugnolle, F., Manica, A. & Balloux, F. (2006a). A geographically explicit genetic map of worldwide human-settlement history. *American Journal of Human Genetics*, **79**, 230–7.

Liu, X., Xu, L., Zhang, S. & Xu, Y. (1994). Epidemiological and genetic studies of congenital profound deafness in the general population of Sichuan, China. *American Journal of Medical Genetics*, **53**, 192–5.

Loder, E. (2011). Editorial. A theme issue in 2011 on unpublished evidence. *British Medical Journal*, **342**, d2627.

López, N.J. (1992). Clinical, laboratory, and immunological studies of a family with a high prevalence of generalized prepubertal and juvenile periodontitis. *Journal of Periodontology*, **63**, 457–68.

Lowry, R.B., Johnson, C.Y., Gagnon, F. & Little, J. (2009). Segregation analysis of cleft lip with or without cleft palate in the First Nations (Amerindian) people of British Columbia and review of isolated cleft palate etiologies. *Birth Defects Research Part A: Clinical and Molecular Teratology*, **85**, 568–73.

LRB (2001). *Sex Crimes and Penalties in Wisconsin*. Part II, Section IV, Crimes against sexual morality. Informational Bulletin 01–1, January 2001. Madison, WI: State of Wisconsin Legislative Reference Bureau.

Ludovici, A.M. (1933–4). Eugenics and consanguineous marriages. *Eugenics Review*, **25**, 147–55.

Lukianowicz, N. (1972a). Incest I: Paternal incest. *British Journal of Psychiatry*, **120**, 301–8.

Lukianowicz, N. (1972b). Incest II: Other types of incest. *British Journal of Psychiatry*, **120**, 308–13.

Lumsden, C.J. & Wilson, E.O. (1980). Gene-culture translation in the avoidance of sibling incest. *Proceedings of the National Academy of Sciences of the United States of America*, **77**, 6248–50.

Lund, P.M. (1996). Distribution of oculocutaneous albinism in Zimbawbe. *Journal of Medical Genetics*, **33**, 641–4.

Lunde, A., Melve, K.K., Gjessing, H.K., Skjaerven, R. & Irgens, L.M. (2007). Genetic and environmental influences on birth weight, birth length, head circumference, and gestational age by use of population-based parent–offspring data. *American Journal of Epidemiology*, **165**, 734–41.

Lyons, E.J., Amos, W., Berkley, J.A. *et al.* (2009b). Homozygosity and risk of childhood death due to invasive bacterial disease. *BMC Medical Genetics*, **10**, 55.

Lyons, E.J., Frodsham, A.J., Zhang, L., Hill, A.V.S. & Amos, W. (2009a). Consanguinity and susceptibility to infectious diseases in humans. *Biology Letters*, **5**, 574–6.

Maatouk, F., Laamri, D., Argoubi, K. & Ghedra, H. (1995). Dental manifestations of inbreeding. *Journal of Clinical and Pediatric Dentistry*, **19**, 305–6.

Macpherson, A.G. (1968). An old Highland parish register: survivals of clanship and social change in Laggan, Inverness-shire, 1775–1854. *Scottish Studies*, **12**, 81–111.

Madani, K., Otoukesh, H., Rastegar, A. & Van Why S. (2001). Chronic renal failure in Iranian children. *Pediatric Nephrology*, **16**, 140–4.

Madhavan, T. & Narayan, J. (1991). Consanguinity and mental retardation. *Journal of Mental Deficiency Research*, **35** (Pt 2), 133–9.

Madrigal, L. & Ware, B. (1997). Inbreeding in Escazú, Costa Rica (1800–1840, 1850–1899): Isonymy and ecclesiastical dispensations. *Human Biology*, **69**, 703–14.

Magalhães, J.C.M. & Arce-Gomez, B. (1987). Study on a Brazilian isolate. *Human Heredity*, **37**, 278–84.

Magnus, P. (1984). Causes of variation in birth weight: a study of offspring of twins. *Clinical Genetics*, **25**, 15–24.

Magnus, P., Bakketeig, L.S. & Skjaerven, R. (1993). Correlations of birth weight and gestational age across generations. *Annals of Human Biology*, **20**, 231–8.

Magnus, P., Berg, K. & Bjerkedal, T. (1985). Association of parental consanguinity with decreased birth weight and increased rate of early death and congenital malformations. *Clinical Genetics*, **28**, 335–42.

Mahadevan, B. & Bhat, B.V. (2005). Neural tube defects in Pondicherry. *Indian Journal of Pediatrics*, **72**, 557–9.

Mahdi, A.H. (1991). Genetically determined neurodegenerative disorders: experiences in Saudi Arabia. *Annals of Tropical Paediatrics*, **11**, 17–23.

Mahdieh, N., Rabbani, B., Shirkavand, A. *et al.* (2011). Impact of consanguineous marriages in GJB2-related hearing loss in Iranian population: a report of a novel variant. *Genetic Testing and Molecular Biomarkers*, **15**, 489–93.

Maheshwari, M., Vijaya, R., Ghosh, M. *et al.* (2003). Screening of families with autosomal recessive non-syndromic hearing impairment (ARNSHI) for mutations in GJB2 gene: Indian scenario. *American Journal of Medical Genetics*, **120A**, 180–4.

Mahmood, S., Ahmad, W. & Hassan, M.J. (2011). Autosomal recessive primary microcephaly (MCPH): clinical manifestations, genetic heterogeneity and mutation continuum. *Orphanet Journal of Rare Diseases*, **6**, 39.

Makov, U. & Bittles, A.H. (1986). On the choice of mathematical models for the estimation of lethal gene equivalents in man. *Heredity*, **57**, 377–80.

Malhotra, K.C., Ahamadi, K., Kazi, R.B. & Bhosale, N. (1977). Consanguineous marriages among five Muslim isolates. *Journal of the Indian Anthropological Society*, **12**, 207–12.

Malik, S., Arshad, M., Amin-ud-Din, M. *et al.* (2004). A novel type of autosomal recessive syndactly: clinical and molecular studies in a family of Pakistani origin. *American Journal of Medical Genetics*, **126A**, 61–7.

Malini, S.S. & Ramachandra, N.B. (2006). Influence of advanced age of maternal grandmothers on Down syndrome. *BMC Medical Genetics*, **14**, 4.

Mange, A.P. (1964). Growth and inbreeding of a human isolate. *Human Biology*, **36**, 104–33.

Manica, A., Amos, W., Balloux, F. & Hanihara, T. (2007). The effect of ancient population bottlenecks on human phenotypic variation. *Nature*, **448**, 346–8.

Mannucci, P.M., Duga, S. & Peyvandi, P. (2004). Recessively inherited coagulation disorders. *Blood*, **104**, 1243–52.

Maniolo, T.A., Collins, F.S., Cox, N.J. *et al.* (2009). Finding the missing heritability of complex diseases. *Nature*, **461**, 747–53.

Manson, J.M. & Carr, M.C. (2003). Molecular epidemiology of hypospadias: review of genetic and environmental risk factors. *Birth Defects Research (Part A)*, **67**, 825–36.

Mansour, H., Fathi, W., Klei, L. *et al.* (2010). Consanguinity and increased risk for schizophrenia in Egypt. *Schizophrenia Research*, **120**, 108–12.

Mansour, H., Klei, L., Wood, J. *et al.* (2009). Consanguinity associated with increased risk for bipolar I disorder in Egypt. *American Journal of Medical Genetics Part B: Neuropsychiatric Genetics*, **150B**, 879–85.

Marazita, M.L., Field, L.L., Tunçbilek, G. *et al.* (2004). Genome-scan for loci involved in cleft lip with or without cleft palate in consanguineous families from Turkey. *American Journal of Medical Genetics*, **126A**, 111–22.

Marçallo, F.A., Freire-Maia, N., Azevedo, J.B.C. & Simões, I.A. (1964). Inbreeding effect on mortality and morbidity in South Brazilian populations. *Annals of Human Genetics*, **27**, 203–18.

Margetts, B.M., Yusof, S.M., Al Dallal, Z. & Jackson, A.A. (2002). Persistence of lower birth weight in second generation South Asian babies born in the United Kingdom. *Journal of Epidemiology and Community Health*, **56**, 684–7.

Marshall, C.R., Noor, A., Vincent, J.B. *et al.* (2008). Structural variation of chromosomes in autism spectrum disorder. *American Journal of Human Genetics*, **82**, 477–88.

Martelli-Junior, H., Bonan, P.R., Dos Santos, L.A. *et al.* (2008). Case reports of a new syndrome associating gingival fibromatosis and dental abnormalities in a consanguineous family. *Journal of Periodontology*, **79**, 1287–96.

Martin, A.O., Kurczynski, T.W. & Steinberg, A.G. (1973). Familial studies of medical and anthropometric variables in a human isolate. *American Journal of Human Genetics*, **25**, 581–93.

Martin, E. & Gamella, J.F. (2005). Marriage practices and ethnic differentiation: The case of the Spanish Gypsies (1870–2000). *History of the Family*, **10**, 45–63.

Martin, S.N., Sutherland, J., Levin, A.V. *et al.* (2000). Molecular characterisation of congenital glaucoma in a consanguineous Canadian community: a step towards preventing glaucoma related blindness. *Journal of Medical Genetics*, **37**, 422–7.

Martínez-Frías, M.L., Castilla, E.E., Bermejo, E., Prieto, L. & Orioli, I.M. (2000). Isolated small intestinal atresias in Latin America and Spain: epidemiological analysis. *American Journal of Medical Genetics*, **93**, 355–9.

Martin-Villa, J.M., De Juan, D., Vicario, J.L. *et al.* (1993). HLA class I, class II, and class III antigen sharing is not found in couples with unexplained infertility. *International Journal of Fertility*, **38**, 280–8.

Masri, A., Hamamy, H. & Khreisat, A. (2010). Profile of developmental delay in children under five years of age in a highly consanguineous community: A hospital-based study – Jordan. *Brain and Development*, **33**, 810–5.

Matessi, C. & Karlin, S. (1984). On the evolution of altruism by kin selection. *Proceedings of the National Academy of Sciences of the United States of America*, **81**, 1754–8.

Mathias, R.A., Bickel, C.A., Beaty, T.H. *et al.* (2000). A study of contemporary levels and temporal trends in inbreeding in the Tangier Island, Virginia, population using pedigree data and isonymy. *American Journal of Physical Anthropology*, **112**, 29–38.

Mathur, P.R.G. (1997–1998). Social stratification among the Muslims of Kerala. *Bulletin of the International Committee on Urgent Anthropological and Ethnological Research (IUAES)* No. 39, pp. 51–71.

Max Planck Institute (2011a). *Limits on the Protection of Legal Interests in the Criminalization of Incest. I. Criminal Prohibition of Incest in International Legal Comparison.* Available at http://www.mpicc.de/ww/en/pub/forschung/ forschungsarbeit/gemeinsame_projekte/inzest/inzeststrafbarkeit.htm [30 June 2011].

Max Planck Institute (2011b). *Limits on the Protection of Legal Interests in the Criminalization of Incest. II. Incest in Criminology and Genetics.* Available at http://www.mpicc.de/ww/en/pub/forschung/forschungsarbeit/gemeinsame_ projekte/inzest/inzeststra_krim.htm [30 June 2011].

Maziak, W. & Asfar, T. (2003). Physical abuse in low-income women in Aleppo, Syria. *Health Care for Women International*, **24**, 313–26.

McCormick, M.C. (1985). The contribution of low birth weight to infant mortality and childhood morbidity. *New England Journal of Medicine*, **312**, 82–90.

McCullough, J.M. & O'Rourke, D.H. (1986). Geographic distribution of consanguinity in Europe. *Annals of Human Biology*, **13**, 359–68.

McDowell, G.A., Mules, E.H., Fabacher, P., Shapira, E. & Blitzer, M.G. (1992). The presence of two different infantile Tay-Sachs disease mutations in a Cajun population. *American Journal of Human Genetics*, **51**, 1071–7.

McEvoy, B.P., Montgomery, G.W., McRae, A.F. *et al.* (2009). Geographical structure and differential natural selection among North European populations. *Genome Research*, **19**, 804–14.

McEwan, I. (1980). *The Cement Garden.* London: Jonathan Cape.

McGregor, T.L., Misri, A., Bartlett, J. *et al.* (2010). Consanguinity mapping of congenital heart disease in a South Indian population. *PLoS One*, **5**, e10286.

McIntyre, J.A., Faulk, W.P., Nichols-Johnson, V.R. & Taylor, C.G. (1986). Immunologic testing and immunotherapy in recurrent spontaneous abortion. *Obstetrics and Gynecology*, **67**, 169–75.

McQuillan, R., Leutenegger, A-L., Abdel-Rahman, R. *et al.* (2008). Runs of homozygosity in European populations. *American Journal of Human Genetics*, **83**, 359–72.

McWhirter, R.E., McQuillan, R., Visser, E., Counsell, C. & Wilson, J.F. (2012). Genome-wide homozygosity and multiple sclerosis in Orkney and Shetland Islanders. *European Journal of Human Genetics*, **20**, 198–202.

Medlej-Hashim, M., Chouery, E., Salem, N. *et al.* (2011). Familial Mediterranean fever in a large Lebanese family: multiple MEFV mutations and evidence for a Founder effect on the p.{M6941] mutation. *European Journal of Medical Genetics*, **54**, 50–4.

Mégarbané, A., Noujeim, Z., Fabre, M. & Der Kaloustian, V.M. (1998). New form of hidrotic ectodermal dysplasia in a Lebanese family. *American Jorunal of Medical Genetics*, **75**, 196–9.

Meijer, M.J. (1971). *Marriage Law and Policy in the Chinese People's Republic*. Hong Kong: Hong Kong University Press.

Meijer, M.J. (1978). Marriage law and policy in the Chinese People's Republic of China. In *Chinese Family Law and Social Change*, D.C. Buxbaum, ed., Seattle: University of Washington Press, pp. 436–83.

Meijer, R.P. & Groeneveld, A.E. (2007). Intersex: four cases in one family. *Journal of Pediatric Urology*, **3**, 137–41.

Memish, Z.A. & Saeedi, M.Y. (2011). Six-year outcome of the national premarital screening and genetic counselling program for sickle cell disease and β-thalassaemia in Saudi Arabia. *Annals of Saudi Medicine*, **31**, 229–35.

Merzario, R. (1990). Land, kinship and consanguineous marriage in Italy from the seventeenth to the nineteenth centuries. *Journal of Family History*, **15**, 529–46.

Meyer, B. (2005). Strategies for the prevention of hereditary diseases in a highly consanguineous population. *Annals of Human Biology*, **32**, 174–9.

Meyer, E., Lim, D., Pasha, S. *et al.* (2009). Germline mutation in *NLRP2* (*NALP2*) in a familial imprinting disorder (Beckwith–Wiedemann syndrome). *PLoS Genetics*, **5**, e1000423.

Micali, G., Nasca, M.R., Innocenzi, D. *et al.* (2005). Association of palmoplantar keratoderma, cutaneous squamous cell carcinoma, dental anomalies, and hypogenitalism in four siblings with 46,XX karotype: a new syndrome. *Journal of the American Academy of Dermatology*, **53**, S234–9.

Michalopolous, A., Tzelepis, G. & Geroulanos, S. (2003). Morbid obesity and hyper-somnolence in several members of an ancient royal family. *Thorax*, **58**, 281–2.

Middleton, R. (1962). Brother–sister and father–daughter marriage in Ancient Egypt. *American Sociology Review*, **27**, 603–11.

Migicovsky, Z. & Kovalchuk, I. (2011). Epigenetic memory in mammals. *Frontiers in Genetics*, **2**, 28.

Milan, M., Astolfi, G., Volpato, S. *et al.* (1994). 766 cases of oral cleft in Italy. *European Journal of Epidemiology*, **10**, 317–24.

Milinski, M. (2006). The major histocompatibility complex, sexual selection and mate choice. *Annual Review of Ecological and Evolutionary Systems*, **37**, 159–86.

Miller, J.F., Williamson, E., Glue, J. *et al.* (1980). Fetal loss after implantation. *Lancet*, **2**, 554–6.

Milo, R. & Kahana, E. (2010). Multiple sclerosis: geoepidemiology, genetics and the environment. *Autoimmunity Reviews*, **9**, A387–94.

Mingroni, M.A. (2004). The secular rise in IQ: giving heterosis a closer look. *Intelligence*, **32**, 65–83.

Ministry of Health, PRC (1994). *Law of the People's Republic of China on Maternal and Infant Health Care*. Beijing: Legislative Affairs Commission of the Standing Committee of the National People's Congress of the People's Republic of China.

Mir, A., Kaufman, L., Noor, A. *et al.* (2009). Identification of mutations in TRAPPC9, which encodes the NIK- and IKK-beta-binding protein, in nonsyndromic autosomal-recessive mental retardation. *American Journal of Human Genetics*, **85**, 909–15.

Mitchell, A.L., Dwyer, A., Pitteloud, N. & Quinton, R. (2011). Genetic basis and variable phenotypic expression of Kallmann syndrome: towards a unifying theory. *Trends in Endocrinology and Metabolism*, **22**, 249–58.

Mitchell, A.M. (1862). Blood-relationship in marriage considered in its influence upon the offspring. *Memoirs of the Anthropological Society of London*, **2**, 402–56.

Mitchell, A.M. (1864–5). On the influence which consanguinity in the parentage exercises upon the offspring. *Edinburgh Medical Journal*, **10**, 781–94, 894–913, 1074–85.

Mochida, G.H., Mahajnah, M., Hill, A.D. *et al.* (2009). A truncating mutation of TRAPPC9 is associated with autosomal-recessive intellectual disability and postnatal microcephaly. *American Journal of Human Genetics*, **85**, 897–902.

Modell, B., Darlison, M., Khan, M. & Harris, R. (2000). Role of genetic diagnosis registers in ongoing consultation with the community. *Community Genetics*, **3**, 144–7.

Modell, B. & Darr, A. (2002). Genetic counseling and customary consanguineous marriage. *Nature Reviews. Genetics*, **3**, 225–9.

Modell, B., Khan, M., Darlison, M. *et al.* (2001). A national register for surveillance of inherited disorders: β thalassaemia in the United Kingdom. *Bulletin of the World Health Organization*, **79**, 1006–13.

Modell, B., Petrou, M., Layton, M. *et al.* (1997). Audit of prenatal diagnosis for haemoglobin disorders in the United Kingdom: the first 20 years. *British Medical Journal*, **315**, 779–84.

Modin, H., Masterman, T., Thorlacius, T. *et al.* (2003). Genome-wide linkage screen of a consanguineous multiple sclerosis kinship. *Multiple Sclerosis*, **9**, 128–34.

Moghraby, J.S., Tamim, H., Anacan, V., Al Khalaf, H. & Moghraby, S.A. (2010). HLA sharing among couples appears unrelated to idiopathic recurrent fetal loss in Saudi Arabia. *Human Reproduction*, **25**, 1900–5.

Mohamed, M.S. (1995). An epidemiological study on consanguineous marriage among urban population in Alexandria. *Journal of the Egypt Public Health Association*, **70**, 293–305.

Mohan Reddy, B. & Malhotra, K.C. (1991). Relationship between birth order of spouses with different degrees of consanguineous relationship. *Human Biology*, **63**, 489–98.

Mokhtar, M.M. & Abdel-Fattah, M.M. (2001). Consanguinity maternal age as risk factors for reproductive losses in Alexandria, Egypt. *European Journal of Epidemiology*, **17**, 559–65.

Molinari, F., Rio, M., Meskenaite, V. *et al.* (2002). Truncating neurotrypsin mutation in autosomal recessive nonsyndromic mental retardation. *Science*, **298**, 1779–81.

Moolhuijzen, P., Black, M.L., Bellgard, M. & Bittles, A.H. (2009). A bioinformatics approach to the study of autosomal recessive non-syndromic intellectual disability. *Twin Research and Human Genetics*, **12**, 225.

Moore, B., Hu, H., Singleton, M. *et al.* (2011). Global analysis of disease-related DNA sequence variation in 10 healthy individuals: Implications for whole genome-based clinical diagnostics. *Genetics in Medicine*, **13**, 210–7.

Moore, M.J. (1987). Inbreeding and reproductive parameters among Mennonites in Kansas. *Social Biology*, **34**, 180–6.

Moreno-Fuenmayor, H., Champin, J., Alvarez-Arratia, M. & Sanchéz, O. (1993). Epidemiology of congenital malformations in Bolivar City, Venezuela. Analysis of the consanguinity factor. *Investigaciónes Clinica*, **34**, 5–14. [In Spanish.]

Morgan, L.H. (1871). *Systems of Consanguinity and Affinity of the Human Family*. Washington, DC: Smithsonian Institution.

Morris, J.K. & Alberman, E. (2009). Trends in Down's syndrome live births and antenatal diagnoses in England and Wales from 1989 to 2008: analysis of data from the National Down Syndrome Cytogenetic Register. *British Medical Journal*, **339**, b3794.

Morris, P. (1991). Incest or survival strategy? Plebian marriage within the prohibited degrees in Somerset, 1730–1835. *Journal of the History of Sexuality*, **2**, 235–65.

Morrow, E.M., Yoo, S.Y., Flavell, S.W. *et al.* (2008). Identifying autism loci by tracing recent shared ancestry. *Science*, **321**, 218–23.

Morton, N.E. (1958). Empirical risks in consanguineous marriages: birth weight, gestation time, and measurements of infants. *American Journal of Human Genetics*, **10**, 344–9.

Morton, N.E. (1978). Effect of inbreeding on IQ and mental retardation. *Proceedings of the Naional Academy of Sciences of the United States of America*, **75**, 3906–8.

Morton, N.E., Crow, J.F. & Muller, H.J. (1956). An estimate of the mutational damage in man from data on consanguineous marriages. *Proceedings of the National Academy of Sciences of the United States of America*, **42**, 855–63.

Morton, N.E., Klein, D., Hussels, I.E. *et al.* (1973). Genetic structure of Switzerland. *American Journal of Human Genetics*, **25**, 347–61.

Mosayebi, Z. & Movahedian, A.H. (2007). Pattern of congential malformations in consanguineous versus nonconsanguineous marriages in Kashan, Islamic Republic of Iran. *East Mediterrranean Health Journal*, **13**, 868–75.

Moslemi, A-R. & Darin, N. (2007). Molecular genetic and clinical aspects of mitochondrial disorders in childhood. *Mitochondrion*, **7**, 241–52.

Motazacker, M.M., Rost, B.R., Hucho, T. *et al.* (2007). A defect in the ionotropic glutamate receptor 6 gene (GRIK2) is associated with autosomal recessive mental retardation. *American Journal of Human Genetics*, **81**, 792–8.

Motlagh, M.G., Seddigh, A., Dashti, B. Leckman, J.F. & Alaghband-Rad, J. (2008). Consanguineous Iranian kindreds with severe Tourette syndrome. *Movement Disorders*, **23**, 2079–83.

Movahedi, M., Aghamohammadi, A., Rezaei, N. *et al.* (2004). Chronic granulomatous disease: a clinical survey of 41 patients from the Iranian Primary Immunodeficiency Registry. *International Archives of Allergy and Immunology*, **134**, 253–9.

Mowbray, J.F., Gibbings, C., Liddell, H. *et al.* (1985). Controlled trial of treatment of recurrent spontaneous abortion by immunisation with paternal cells. *Lancet*, **1**, 941–3.

Mowbray, J.F., Underwood, J. & Gill, T.J. (1991). Familial recurrent spontaneous abortion. *American Journal of Reproductive Immunology*, **26**, 17–8.

Mroz, J. (2011). *One sperm donor, 150 offspring*. Available at http://www.nytimes.com/2011/09/06/health/06donor.html [07 September 2011].

Müller-Freienfels, W. (1978). Soviet family law and comparative Chinese developments. In *Chinese Family Law and Social Change*, D.C. Buxbaum, ed., Seattle: University of Washington Press, pp. 323–99.

Mumtaz, G., Nassar, A.H., Mahfoud, Z. *et al.* (2010). Consanguinity: a risk factor for preterm birth at less than 33 weeks' gestation. *American Journal of Epidemiology*, **172**, 1424–30.

Mumtaz, G., Tamin, H., Kanaan, M. *et al.* (2007). Effect of consanguinity on birth weight for gestational age in a developing country. *American Journal of Epidemiology*, **165**, 742–52.

Murdock, G.P. (1967). *Ethnographic Atlas*. Pittsburgh, PA: University of Pittsburgh Press.

Murphy, M., McHugh, B., Tighe, O. *et al.* (1999). Genetic basis of transferase-deficient galactosaemia in Ireland and the population history of the Irish Travellers. *European Journal of Human Genetics*, **7**, 549–54.

Murphy, R.F. & Kasdan, L. (1959). The structure of parallel cousin marriage. *American Anthropologist*, **61**, 17–29.

Murray, M.A. (1934). Marriage in ancient Egypt. *Proceedings of the International Congress of Anthropological and Ethnological Sciences*, London: Royal Institute of Anthropology, p. 282.

Murshid, W.R. (2000). Spina bifida in Saudi Arabia: Is consanguinity among the parents a risk factor? *Pediatric Neurosurgery*, **32**, 10–2.

Murshid, W.R., Jarallah, J.S. & Dad, M.I. (2000). Epidemiology of infantile hydrocephalus in Saudi Arabia: birth prevalence and associated factors. *Pediatric Neurosurgery*, **32**, 119–23.

Musani, M.A., Rauf, A., Ahsan, M. & Khan, F.A. (2011). Frequency and causes of hearing impairment in tertiary care center. *Journal of the Pakistan Medical Association*, **61**, 141–4.

Myers, W.C. & Brasington, S.J. (2002). A father marries his daughters: a case of incestuous polygamy. *Journal of Forensic Science*, **47**, 1112–6.

Nabulsi, A. (1995). Mating patterns of the Abbad tribe in Jordan. *Social Biology*, **42**, 162–74.

Nabulsi, M.M., Tamim, H., Sabbagh, M. *et al.* (2003). Parental consanguinity and congenital heart malformations in a developing country. *American Journal of Medical Genetics*, **116A**, 342–7.

Naderi, S. (1979). Congenital abnormalities in newborns of consanguineous and nonconsanguineous parents. *Obstetrics and Gynecology*, **53**, 195–9.

Naidu, J.M., Babu, B.V., Kusuma, Y.S., Yasmin & Sachi Devi, S. (1995). Inbreeding effects on reproductive outcome among seven tribes of Andhra Pradesh, India. *American Journal of Human Biology*, **7**, 589–95.

Nair, R.R. & Murty, J.S. (1985). ABO blood group incompatibility and inbreeding effects: evidence for an interaction. *Human Genetics*, **69**, 147–50.

Nair, R.R. & Thomas, S.V. (2004). Genetic liability to epilepsy in Kerala State, India. *Epilepsy Research*, **62**, 157–64.

Najmabadi, H., Ghamari, A., Sahebjam, F. *et al.* (2006). Fourteen-year experience of prenatal diagnosis of thalassaemia in Iran. *Community Genetics*, **9**, 93–7.

Najmabadi, H., Hu, H., Garshabi, M. *et al.* (2011). Deep sequencing reveals 50 novel genes for recessive cognitive disorders. *Nature*, **478**, 57–63.

Najmabadi, H., Motazacher, M.M., Garshasbi, M. *et al.* (2007). Homozygosity mapping in consanguineous families reveals extreme heterogeneity of non-syndromic autosomal recessive mental retardation and identifies 8 novel gene loci. *Human Genetics*, **121**, 43–8.

Nakashima, I.I. & Zakus, G. (1979). Incestuous families. *Pediatric Annals*, **8**, 29–42.

Nakazawa, S., Shiokawa, T., Ishikawa, T., Tanaka, K. & Katui, A. (1972). Genetic studies in some Japanese populations. VII. Psychometric data of school-children in Shizuoka. *Japanese Journal of Human Genetics*, **17**, 219–30.

Nalls, M.A., Guerreiro, R.J., Simon-Sanchez, J. *et al.* (2009a). Extended tracts of homozygosity identify novel candidate genes associated with late-onset Alzheimer's disease. *Neurogenetics*, **10**, 183–90.

Nalls, M.A., Simon-Sanchez, J., Gibbs, J.R. *et al.* (2009). Measures of autozygosity in decline: globalization, urbanization, and its implications for medical genetics. *PLoS Genetics*, **5**, e1000415.

Nanda, A., Al-Ateeqi, W.A., Al-Khawari, M.A., Alsaleh, Q.A. & Anim, J.T. (2010). GAPO syndrome: a report of two siblings and a review of the literature. *Pediatric Dermatology*, **27**, 156–61.

National Conference of Commissioners (1970). *Handbook on Uniform State Laws and Proceedings of the Annual Conference Meeting in its Seventy-Ninth Year*. Baltimore, MD: Port City Press.

Nebert, D.W., Galvez-Peralta, M., Shi, Z. & Dragin, N. (2010). Inbreeding and epigenetics: beneficial as well as deleterious effects. *Nature Reviews. Genetics*, **11**, 662.

Neel, J.V & Schull, W.J. (1962). The effect of inbreeding on mortality and morbidity in two Japanese cities. *Proceedings of the National Academy of Sciences of the United States of America*, **48**, 573–82.

Neel, J.V., Schull, W.J., Kimura, T. *et al.* (1970b). The effects of parental consanguinity and inbreeding in Hirado, Japan. III. Vision and hearing. *Human Heredity*, **20**, 129–55.

Neel, J.V., Schull, W.J., Yamamoto, M. *et al.* (1970a). The effects of parental consanguinity and inbreeding in Hirado, Japan. II. Physical development, tapping rate, blood pressure, intelligence quotient, and school performance. *American Journal of Human Genetics*, **22**, 263–86.

Nevo, Y., Kutai, M., Jossiphov, J. *et al.* (2004). Childhood macrophagic myofasciitis – consanguinity and clinicopathological features. *Neuromuscular Disorders*, **14**, 246–52.

NCSL (2011). *State Laws Regarding Marriages Between First Cousins.* Washington, DC: National Conference of State Legislatures.

NDSS (2011). *Down Syndrome: Incidences and Maternal Age.* Washington, DC: National Down Syndrome Society.

Nelson, J., Smith, M. & Bittles, A.H. (1997). Consanguineous marriage and its clinical consequences in migrants to Australia. *Clinical Genetics*, **52**, 142–6.

Nemeskéri, J. & Thoma, A. (1961). Ivád: an isolate in Hungary. *Acta Genetica et Statistica Medica, Basel*, **11**, 230–50.

Neto, R.M., Castilla, E.E., & Paz, J.E. (1981). Hypospadias: an epidemiological study in Latin America. *American Journal of Medical Genetics*, **10**, 5–19.

Nettleship, E. (1914). Consanguineous marriages. *Annals of Eugenics*, **6**, 131–9.

New Advent Catholic Encyclopedia (2011). *Consanguinity in Canon Law.* Available at http:/www.newadvent.org/cathen/04264a.htm [21 August 2011].

Newman, R. (1869). On the result of consanguineous marriage. *Transactions of the Medical Society of the State of New York*, pp. 129–30.

NFHS (1995). *National Family Health Survey, India 1992–93.* Bombay: International Institute for Population Sciences.

NHS (2010). *Data Report: 2008/09, Informing Policy & Improving Quality. NHS Screening Programmes for Sickle Cell and Thalassaemia.* London: King's College London.

Nielsen, B.B., Liljestrand, J., Hedegaard, M., Thilsted, S.H. & Joseph, A. (1997). Reproductive pattern, perinatal mortality, and sex preferences in rural Tamil Nadu, south India: community based cross sectional study. *British Medical Journal*, **314**, 1521–4.

NIH (2011). *Fetal Alcohol Spectrum Disorders.* Washington, DC: National Institutes of Health.

Nimri, R., Lebenthal, Y., Lazar, L. *et al.* (2011). A novel loss-of-function mutation in GPR54/KISS1R leads to hypogonadotropic hypogonadism in a highly consanguineous family. *Journal of Clinical Endocrinology and Metabolism*, **96**, E536–45.

Noble, M. & Mason, J.K. (1978). Incest. *Journal of Medical Ethics*, **4**, 64–70.

Nolan, D.K., Chen, P., Das, S., Ober, C. & Waggoner, D. (2008). Fine mapping of a locus for nonsydromic mental retardation on chromosome 19p13. *American Journal of Medical Genetics A*, **146A**, 1414–22.

Nonaka, K., Miura, T. & Peter, K. (1994). Recent fertility decline in Dariusleut Hutterites: an extension of Eaton and Mayer's Hutterite fertility study. *Human Biology*, **66**, 411–20.

Noor, A., Windpassinger, C., Patel, M. *et al.* (2008). CC2D2A, encoding a coiled-coil and C2 domain protein, causes autosomal-recessive mental retardation with retinitis pigmentosa. *American Journal of Human Genetics*, **82**, 1011–8.

Noor, A., Windpassinger, C., Vitcu, I. *et al.* (2009). Oligodontia is caused by mutation in *LTBP3*, the gene encoding latent TGF-β binding protein 3. *American Journal of Human Genetics*, **84**, 519–23.

Norenzayan, A. & Shariff, A.F. (2008). The origin and evolution of religious prosociality. *Science*, **322**, 58–62.

Norio, R. (2003a). Finnish disease heritage. I: characteristics, causes, background. *Human Genetics*, **112**, 441–56.

Norio, R. (2003b). Finnish disease heritage. III: the individual diseases. *Human Genetics*, **112**, 470–526.

Oates, K. & Wilson, M. (2001). Nominal kinship cues facilitate altruism. *Proceedings of the Royal Society Series B*, **269**, 105–9.

Obeidat, B.R., Khader, Y.S., Amarin, Z.O., Kassawneh, M. & Al Omari, M. (2010). Consanguinity and adverse pregnancy outcomes: the North of Jordan experience. *Maternal and Child Health Journal*, **14**, 283–9.

Ober, C. (1998). HLA and pregnancy: the paradox of the fetal allograft. *American Journal of Human Genetics*, **62**, 1–5.

Ober, C., Elias, S., O'Brien, E. *et al.* (1988). HLA sharing and fertility in Hutterite couples: evidence for prenatal selection against compatible fetuses. *American Journal of Reproductive Immunology and Microbiology*, **18**, 111–5.

Ober, C., Hyslop, T. & Hauck, W.W. (1999). Inbreeding effects on fertility in humans: evidence for reproductive compensation. *American Journal of Human Genetics*, **64**, 225–31.

Ober, C., Simpson, J.L., Ward, M. *et al.* (1987). Prenatal effects of maternal-fetal HLA compatibility. *American Journal of Reproductive Immunology and Microbiology*, **15**, 141–9.

Ober, C., Weitkamp, L.R., Cox, N. *et al.* (1997). HLA and mate choice in humans. *American Journal of Human Genetics*, **61**, 497–504.

Ober, C.L., Martin, A.O., Simpson, J.L. *et al.* (1983). Shared HLA antigens and reproductive performance among Hutterites. *American Journal of Human Genetics*, **25**, 994–1004.

O'Callaghan, C., Chetcuti, P. & Moya, E. (2010). High prevalence of primary ciliary dyskinesia in a British Asian population. *Archives of Diseases in Childhood*, **95**, 51–2.

O'Connor, K.A., Holman, D.J. & Wood, J.W. (1998). Declining fecundity and ovarian ageing in natural fertility populations. *Maturitas*, **30**, 127–36.

Oksenberg, J.R., Persitz, E., Amar, A. & Brautbar, C. (1984). Maternal-paternal histocompatibility: lack of association with habitual abortion. *Fertility and Sterility*, **42**, 389–95.

Olusanya, B.O. & Okolo, A.A. (2006). Adverse perinatal conditions in hearing-impaired children in a developing country. *Paediatic and Perinatal Epidemiology*, **20**, 366–71.

OMIM (2011). *Online Mendelian Inheritance in Man*. Available at www.ncbi.nlm.nih.gov/omim [10 May 2011].

ORGCC (2006). *Population Projections for India and States 2001–2006. Report of the Technical Group on Population Projections Constituted by the National Commission on Population*. New Delhi: Office of the Registrar General & Census Commissioners.

Ottenheimer, M. (1990). Lewis Henry Morgan and the prohibition of cousin marriage in the United States. *Journal of Family History*, **15**, 325–33.

Ottenheimer, M. (1996). *Forbidden Relatives – the American Myth of Cousin Marriage*. Chicago: University of Illinois Press.

Ounsted, M. (1974). Familial trends in low birth weight. *British Medical Journal*, **4**, 163.

Overall, A.D.J. (2009). The influence of the Wahlund effect on the consanguinity hypothesis: consequences for recessive disease incidence in a socially structured Pakistani population. *Human Heredity*, **67**, 140–4.

Overall, A.D.J., Ahmad, M. & Nichols, R.A. (2002). The effect of reproductive compensation on recessive disorders within consanguineous human populations. *Heredity*, **88**, 474–9.

Overall, A.D.J., Ahmad, M., Thomas, M.G. & Nichols, R.A. (2003). An analysis of consanguinity and social structure within the UK Asian population using microsatellite data. *Annals of Human Genetics*, **67**, 525–37.

Overall, A.D.J. & Nichols, R.A. (2001). A method for distinguishing consanguinity and population substructure using multilocus genotype data. *Molecular Biology and Evolution*, **18**, 2048–56.

Øyen, N., Poulsen, G., Boyd, H.A. *et al.* (2009). Recurrence of congenital heart defects in families. *Circulation*, **120**, 295–301.

Ozand, P.T., Al Odaib, A., Sakati, N. & Al-Hellani, A.M. (2005). Recently available techniques applicable to genetic problems in the Middle East. *Community Genetics*, **8**, 44–7.

Pakrasi, K. & Sasmal, B. (1971). Infanticide and variation of sex-ratio in a caste population of India. *Acta Medica Auxologica*, **3**, 217–28.

Pandey, B.N., Jha, A.K. & Das, P.K.L. (1994). Effects of consanguinity on blood groups and intelligence quotient among Muslim children of Purnia, Bihar. *Journal of Human Ecology*, **5**, 221–3.

Panicker, S.G., Reddy, A.B.M., Mandal, A.K. *et al.* (2002). Identification of novel mutations causing familial primary congenital glaucoma in Indian pedigrees. *Investigative Ophthalmology & Visual Science*, **43**, 1358–66.

Pannain, S., Weiss, R.E., Jackson, C.E. *et al.* (1999). Two different mutations in the thyroid peroxidase gene of a large inbred Amish kindred: power and limits of homozygosity mapping. *Journal of Clinical Endocrinology and Metabolism*, **84**, 1061–71.

Panter-Brick, C. (1991). Parental responses to consanguinity and genetic disease in Saudi Arabia. *Social Science and Medicine*, **33**, 1295–302.

Paranaiba, L.M.R., de Miranda, R.T., Martelli, D.R.B. *et al.* (2010). Cleft lip and palate: series of unusual clinical cases. *Brazilian Journal of Otorhinolaryngology*, **76**. 649–53.

Pariacote, F., Van Vleck, L.D. & MacNeil, M.D. (1998). Effects of inbreeding and heterozygosity on preweaning traits in a closed population of Herefords under selection. *Journal of Animal Science*, **76**, 1303–10.

Park, C.B. & Horiuchi, B.Y. (1993). Ethnicity, birth weight and maternal age in infant mortality: Hawaiian experience. *American Journal of Human Biology*, **5**, 101–9.

Parker, C.C. & Palmer, A.A. (2011). Dark matter: Are mice the solution to missing heritability? *Frontiers in Genetics*, **2**, 32.

Paul, D.B. & Spencer, H.G. (2008). "It's OK, we're not cousins by blood": the cousin marriage controversy in historical perspective. *PLoS Biology*, **6**, 2627–30.

Paula, L.M., Melo, N.S., Silva Guerra, E.N., Mestrinho, D.H. & Acevedo, A.C. (2005). Case report of a rare syndrome associating amelogenesis imperfecta and nephrocalcinosis in a consanguineous family. *Archives of Oral Biology*, **50**, 237–42.

Paulshock, B.Z. (1980). Tutankhamun and his brothers: familial gynecomastia in the Eighteenth Dynasty. *Journal of the American Medical Association*, **244**, 160–4.

Poulton, J., Chiaratti, M.R., Meirelles, F.V. *et al.* (2010). Transmission of mitochondrial DNA diseases and ways to prevent them. *PLoS Genetics*, **6**, e1001066.

Pearl, R. (1920). Book review: *Inbreeding and Outbreeding, their Genetic and Sociological Significance. Science*, **51**, 415–7.

Pearson, K. (1908). Cousin marriages. *British Medical Journal*, **1**, 1395.

Pearson, K. (1909). Heredity and disease. *Proceedings of the Royal Society of Medicine*, **2**, 54–60.

Pearson, K. (1924). *The Life, Letters and Labours of Francis Galton*, vol. 2, Cambridge: Cambridge University Press, p. 188.

Pedersen, J. (2000). Determinants of infant and child mortality in the West Bank and Gaza strip. *Journal of Biosocial Science*, **32**, 527–46.

Pedersen, J. (2002). The influence of consanguineous marriage on infant and child mortality among Palestinians in the West Bank and Gaza, Jordan, Lebanon and Syria. *Community Genetics*, **5**, 179–81.

Pemberton, T.J., Wang, C., Li, J.Z. & Rosenberg, N.A. (2010). Inference of unexpected genetic relatedness among individuals in HapMap Phase III. *American Journal of Human Genetics*, **87**, 457–64.

Penn, D. & Potts, W. (1998). MHC-disassortative mating preferences reversed by cross-fostering. *Proceedings of the Royal Society of London Series B*, **265**, 1299–306.

Penn, D.J., Damjanovich, K. & Potts, W.K. (2002). MHC heterozygosity confers a selective advantage against multiple-strain infections. *Proceedings of the National Academy of Sciences of the United States of America*, **99**, 11260–4.

Pennings, G. (2001). Incest, gamete donation by siblings and the importance of the genetic link. *Reproductive BioMedicine Online*, **4**, 13–5.

Perveen, F. & Tyyab, S. (2007). Frequency and pattern of distribution of congenital anomalies in the newborn and associated maternal risk factors. *Journal of the College of Physicians and Surgeons, Pakistan*, **17**, 340–3.

Petukhova, L., Shimomura, Y., Wajid, M. *et al.* (2009). The effect of inbreeding on the distribution of compound heterozygotes: a lesson from Lipase H mutations in autosomal recessive woolly hair/hypothrichosis. *Human Heredity*, **68**, 117–30.

Pettener, D. (1985). Consanguineous marriages in the Upper Bologna Appennine (1565–1980): microgeographic variations, pedigree structure and correlation of inbreeding secular trend with changes in population size. *Human Biology*, **57**, 267–88.

Pew Research Center (2011). *The Future of the Global Muslim Population: Projections for 2010–2030*. Washington, DC: Pew Research Center.

Peyvandi, F., Asselta, R. & Mannucci, P.M. (2001). Autosomal recessive deficiencies of coagulation disorders. *Review of Clinical and Experimental Hematology*, **5**, 369–88.

Peyvandi, F., Duga, S., Akhavan, S. & Mannucci, P.M. (2002). Rare coagulation disorders. *Haemophilia*, **8**, 308–21.

Philippe, O., Rio, M., Carioux, A. *et al.* (2009). Combination of linkage mapping and microarray-expression analysis identifies NF-kappaB signaling defect as a cause of autosomal-recessive mental retardation. *American Journal of Human Genetics*, **85**, 903–8.

Philippe, P. (1974). Amenorrhea, intrauterine mortality and parental consanguinity in an isolated French Canadian population. *Human Biology*, **46**, 405–24.

Phillip, M.J. & Caurdy, J. (1985). Inheritance of hypodontia in consanguineous families of Arabic descent. A case report. *Annals of Dentistry*, **44**, 39–41.

Phillipe, P. & Gomila, J. (1972). Inbreeding effects in a French Canadian isolate. I. Evolution of inbreeding. *Zeitscrift für Morphologie und Anthropologie*, **64**, 54–9.

Phillips, C.M. (1961). Blindness in Africans in Northern Rhodesia. *Central African Journal of Medicine*, **7**, 153–8.

Piel, F.B., Patil, A.P., Howes, R.E. *et al.* (2010). Global distribution of the sickle cell gene and geographical confirmation of the malaria hypothesis. *Nature Communications*, **1**, 104.

Pierpont, M.E., Basson, C.T., Benson, D.W. *et al.* (2007). Genetic basis for congenital heart defects: current knowledge. *Circulation*, **115**, 3015–38.

Ping, P. Zhu, W-B., Zhang, X-Z. *et al.* (2011). Sperm donation and its application in China: a 7-year multicenter retrospective study. *Asian Journal of Andrology*, **13**, 644–8.

Pingle, U. (1983). Comparative analysis of mating systems and marriage distance patterns in five tribal groups of Andhra Pradesh. *Social Biology*, **30**, 67–74.

Pociot, F., Akolkar, B., Coccannon, P. *et al.* (2010). Genetics of type 1 diabetes: What's next? *Diabetes*, **59**, 1561–71.

Polašek, O., Leutenegger, A.L., Gornik, O. *et al.* (2011). Does inbreeding affect N-glycosylation of human plasma proteins? *Molecular Genetics and Genomics*, **285**, 427–32.

Pollack, M.S., Wysocki, C.J., Beauchamp, G.K. *et al.* (1982). Absence of HLA association or linkage for variations in sensitivity to the odor of androstenone. *Immunogenetics*, **15**, 579–89.

Port, K.E. & Bittles, A.H. (2001). A population-based estimate of the prevalence of consanguineous marriage in Western Australia. *Community Genetics*, **4**, 97–101.

Port, K.E., Mountain, H., Nelson, J. & Bittles, A.H. (2005). Changing profile of couples seeking genetic counseling for consanguinity in Australia. *American Journal of Medical Genetics*, **132A**, 159–63.

Potts, W.K. & Wakeland, E.K. (1993). Evolution of MHC genetic diversity: a tale of incest, pestilence and sexual preference. *Trends in Genetics*, **9**, 408–12.

PRB (2011). *World Population Data Sheet*. Washington, DC: Population Reference Bureau.

Price, A.L., Patterson, N.J., Plenge, R.M. *et al.* (2006). Principal components analysis corrects for stratification in genome-wide association studies. *Nature Genetics*, **38**, 904–9.

Price, A.L., Zaitlen, N.A., Reich, D. & Patterson, N.J. (2010). New approaches to population stratification in genome-wide association studies. *Nature Reviews. Genetics*, **11**, 459–63.

Pritchard, J.K., Stephens, M., Rosenberg, N.A. & Donnelly, P. (2000). Association mapping in structured populations. *American Journal of Human Genetics*, **67**, 170–81.

Prokopenko, I., Montomoli, C., Ferrai, R. *et al.* (2003). Risk for relatives of patients with multiple sclerosis in Central Sardinia, Italy. *Neuroepidemiology*, **22**, 290–6.

Proulx, E.A. (1993). *The Shipping News*. London: Fourth Estate.

Pugliatti, M., Solinas, G., Sotgiu, S., Castiglia, P. & Rosati, G. (2002). Multiple sclerosis distribution in northern Sardinia: spatial cluster analysis of prevalence. *Neurology*, **58**, 277–82.

Purcell, S.M., Wray, N.R., Stone, J.L. *et al.* and the International Schizophrenia Consortium (2009). Common polygenic variation contributes to risk of schizophrenia and bipolar disease. *Nature*, **460**, 748–52.

Puri, N., Durham-Pierre, D., Aquaron, R. *et al.* (1997). Type 2 oculocutaneous albinism (OCA2) in Zimbabwe and Cameroon: distribution of the 2.7-kb deletion allele of the P gene. *Human Genetics*, **100**, 651–6.

Puri, R.K., Verma, I.C. & Bhargava, I. (1978). Effects of consanguinity in a community in Pondicherry. In *Medical Genetics in India*, vol. 2, I.C. Verma, ed. Pondicherry: Auroma, pp. 129–39.

Pyia, B. (1889). The enumeration of the deaf. *Science*, **14**, 44–7.

Qidwai, W., Syed, I.A. & Khan, F.M. (2003). Prevalence and perceptions about consanguineous marriage among patients presenting to family physicians, in 2001 at a Teaching Hospital in Karachi, Pakistan. *Asia Pacific Family Medicine*, **2**, 27–31.

Queisser-Luft, A., Stolz, G., Wiesel, A., Schlaefer, K. & Spranger, J. (2002). Malformations in newborn: results based on 30,940 infants and fetuses from the Mainz congenital birth defect monitoring system (1990–1998). *Archives of Gynecology and Obstetrics*, **266**, 163–7.

Qureshi, N., Bethea, J., Modell, B. *et al.* (2005). Collecting genetic information in primary care: evaluating a new family history tool. *Family Practice*, **22**, 663–9.

Qureshi, N., Gilbert, P. & Raeburn, J.A. (2003). Consanguinity and genetic morbidity in a British primary care setting: a pilot study with trained linkworkers. *Annals of Human Biology*, **30**, 140–7.

Radavanovic, Z., Shah, N. & Behbehahi, J. (1999). Prevalence and social correlates of consanguinity in Kuwait. *Annals of Saudi Medicine*, **19**, 206–10.

Radha Rama Devi, A., Appaji Rao, N. & Bittles, A.H. (1982). Inbreeding in the State of Karnataka, South India. *Human Heredity*, **32**, 8–10.

Radha Rama Devi, A., Appaji Rao, N. & Bittles, A.H. (1987). Consanguinity and the incidence of childhood genetic disease in Karnataka, South India. *Journal of Medical Genetics*, **24**, 362–5.

Radvany, R.M., Vaisrub, N., Ober, C., Patel, K.M. & Hecht, F. (1987). The human sex ratio: increase in first-born males to parents with shared HLA-DR antigens. *Tissue Antigens*, **29**, 34–42.

Rafiq, M.A., Ansar, M., Marshall, C.R. *et al.* (2010). Mapping of three novel loci for non-syndromic autosomal recessive mental retardation (NS-ARMR) in consanguineous families from Pakistan. *Clinical Genetics*, **78**, 478–83.

Rafiq, M.A., Kuss, A.W., Puetmann, L. *et al.* (2011). Mutations in the alpha 1,2-mannosidase gene, MAN1B1, cause autosomal-recessive intellectual disability. *American Journal of Human Genetics*, **89**, 176–82.

Rahi, J.S., Sripathi, S., Gilbert, C.E. & Foster, A. (1995). Childhood blindness in India: causes in 1318 blind school students in nine states. *Eye*, **9**, 545–50.

Rahman, M. (1998). The effect of child mortality on fertility regulation in rural Bangladesh. *Studies in Family Planning*, **29**, 268–82.

Rajab, A., Vaishnav, A., Freeman, N.V. & Patton, M.A. (1998). Neural tube defects and congenital hydrocephalus in the Sultanate of Oman. *Journal of Tropical Pediatrics*, **44**, 300–3.

Rajabian, M.H. & Sherkat, M. (2000). An epidemiologic study of oral clefts in Iran: analysis of 1669 cases. *Cleft Palate – Craniofacial Journal*, **37**, 191–6.

Ralls, K. & Ballou, J. (1982a). Effect of inbreeding on infant mortality in captive primates. *International Journal of Primatology*, **3**, 491–505.

Ralls, K. & Ballou, J. (1982b). Effect of inbreeding on juvenile mortality in some small animal species. *Laboratory Animals*, **16**, 159–66.

Ralls, K., Brugger, K. & Ballou, J. (1979). Inbreeding and juvenile mortality in small populations of ungulates. *Science*, **206**, 1101–3.

Ramadevi, R., Savithri, H.S., Radha Rama Devi, A., Bittles, A.H. & Appaji Rao, N. (1994). An unusual distribution of glucose 6-phosphate dehydrogenase deficiency of South Indian newborn population. *Indian Journal of Biochemistry and Biophysics*, **31**, 358–60.

Ramankutty, P., Tikretti, R.A.S., Rasaam, K.W. *et al.* (1983). A study on birth weight of Iraqi children. *Journal of Tropical Pediatrics*, **29**, 5–10.

Ramasundrum, V. & Tan, C.T. (2004). Consanguinity and risk of epilepsy. *Neurology Asia*, **9** Suppl. 1, 10–1.

Ramegowda, S. & Ramachandra, N.B. (2006). Parental consanguinity increases congenital heart diseases in South India. *Annals of Human Biology*, **33**, 519–28.

RamShankar, M., Girirajan, S., Dagan, O. *et al.* (2003). Contribution of connexin26 (*GJB2*) mutations and founder effect to non-syndromic hearing loss in India. *Journal of Medical Genetics*, **40**, e68.

Rantanen, E., Hietala, M., Kristoffersson, U. *et al.* (2008). What is ideal genetic counselling? A survey of current international guidelines. *European Journal of Human Genetics*, **16**, 445–52.

Rao, P.S.S. (1983). Religion and intensity of inbreeding in Tamil Nadu, South India. *Social Biology*, **30**, 413–22.

Rao, P.S.S. (1984). Inbreeding in India: concepts and consequences. In *The People of India*, J.R. Lukacs, ed., New York: Plenum, pp. 239–68.

Rao, P.S.S. & Inbaraj, S.G. (1977a). Inbreeding in Tamil Nadu, South India. *Social Biology*, **24**, 281–8.

Rao, P.S.S. & Inbaraj, S.G. (1977b). Inbreeding effects on human reproduction in Tamil Nadu of South India. *Annals of Human Genetics*, **41**, 87–98.

Rao, P.S.S. & Inbaraj, S.G. (1979a). Inbreeding effects on fertility and sterility in southern India. *Journal of Medical Genetics*, **16**, 24–31.

Rao, P.S. & Inbaraj, S.G. (1979b). Trends in human reproductive wastage in relation to long-term practice of inbreeding. *Annals of Human Genetics*, **42**, 401–13.

Rao, P.S.S. & Inbaraj, S.G. (1980). Inbreeding effects on fetal growth and development. *Journal of Medical Genetics*, **17**, 27–33.

Rao, P.S.S., Inbaraj, S.G. & Kaliaperumal, V.G. (1971). An epidemiological study of consanguinity in a large South Indian town. *Indian Journal of Medical Research*, **59**, 294–301.

Rao, V. (1993). Dowry 'inflation' in rural India: a statistical investigation. *Population Studies*, **47**, 283–93.

Rapp, G.E., Pineda-Trujillo, N., McQuillin, A. & Tanetti, M. (2011). Genetic power of a brazilian three-generation family with generalized aggressive periodontitis. II. *Brazilian Dental Journal*, **22**, 68–73.

Rasmussen, M., Guo, X., Wang, Y. *et al.* (2011). An Aboriginal Australian genome reveals separate human dispersals into Asia. *Science*, **334**, 94–8.

Rasmussen, S.A., Whitehead, N., Collier, S.A. & Frías, J.L. (2008). Setting a public health research agenda for Down syndrome. *American Journal of Medical Genetics Part A*, **146A**, 2998–3010.

Rastogi, A. (2010). When cousins marry (Dispatches): the producer's perspective. *BioNews*, No. 574. London: Progress Educational Trust.

Ravikumar, M., Dheenadhayalan, V., Rajaram, K. *et al.* (1999). Association of HLA DRB1, DQB1 and DPB1 alleles with pulmonary tuberculosis in south India. *Tubercle and Lung Disease*, **74**, 309–17.

Raz, A.E. & Atar, M. (2003). Nondirectiveness and its lay interpretations: the effect of counseling style, ethnicity and culture on attitudes towards genetic counseling among Jewish and Bedouin respondents in Israel. *Journal of Genetic Counseling*, **12**, 313–32.

Raz, A.E. & Atar, M. (2004). Cousin marriage and premarital carrier matching in a Bedouin community in Israel; attitudes, service development and educational intervention. *Journal of Family Planning & Reproductive Health Care*, **30**, 49–51.

Raz, A.E., Atar, M., Rodnay, M., Shoham-Vardi, I. & Carmi, R. (2003). Between acculturation and ambivalence: knowledge of genetics and attitudes towards genetic testing in a consanguineous Bedouin community. *Community Genetics*, **6**, 88–95.

Reda, S.M., Afifi, H.M. & Amine, M.M. (2009). Primary immunodeficiency diseases in Egyptian children: a single-center study. *Journal of Clinical Immunology*, **29**, 343–51.

Reddy, B.M. (1992). Inbreeding effects on reproductive outcome: a study based on a large sample from endogamous Vadde of Kolleru Lake, Andhra Pradesh, India. *Human Biology*, **64**, 659–82.

Reddy, M.A., Purbrick, R. & Petrou, P. (2009). The prevalence of patients with ocular genetic disorders attending a general paediatric ophthalmology clinic in the East End of London. *Eye (London)*, **23**, 1111–4.

Reddy, S.G., Reddy, R.R., Bronkhorst, E.M. *et al.* (2010). Incidence of cleft lip and palate in the state of Andhra Pradesh, South India. *Indian Journal of Plastic Surgery*, **43**, 184–9.

Reich, D., Thangaraj, K., Patterson, N., Price, A.L. & Singh, L. (2009). Reconstructing Indian population history. *Nature*, **461**, 489–94.

Reid, R.M. (1976). Effects of consanguineous marriage and inbreeding on couple fertility and offspring mortality in rural Sri Lanka. *Human Biology*, **48**, 139–46.

Reid, R.M. (1988). Church membership, consanguineous marriage, and migration in a Scotch-Irish frontier population. *Journal of Family History*, **13**, 397–414.

Reniers, G. (1998). *Postmigration Survival of Traditional Marriage Patterns: Consanguineous Marriage Among Turkish and Moroccan Immigrants in Belgium.* Inter-university Papers in Demography, PPD-1 Working Paper 1998–1. Gent: Department of Population Studies, University of Gent.

Republic of China. (2002). *Laws and Regulations Database of the Republic of China.* Available at http://www.law.moj.gov.tw/eng/ [21 March 2011].

Resta, R., Biesecker, B.B., Bennett, R.L. *et al.* (2006). A new definition of genetic counseling: National Society of Genetic Counselors' Task Force report. *Journal of Genetic Counseling*, **15**, 77–83.

Retherford, R.D. & Roy, T.K. (2003). *Factors Affecting Sex-Selective Abortion in India.* Mumbai & Honolulu: International Institute for Population Sciences & East-West Center.

Rezaei, N., Pourpak, Z., Aghamohammadi, A. *et al.* (2006). Consanguinity in primary immunodeficiency disorders; the report from Iranian Primary Immunodeficiency Registry. *American Journal of Reproductive Immunology*, **56**, 145–51.

Reznikoff-Etievant, M.F., Edelman, P., Muller, J.Y., Pinon, F. & Sureau, C. (1984). HLA-DR locus and maternal-foetal relation. *Tissue Antigens*, **24**, 30–4.

Rice, T., Nirmala, A., Reddy, P.C. *et al.* (1992). Familial resemblance of blood pressure with residual household environmental effects in consanguineous and nonconsanguineous families from Andhra Pradesh, India. *Human Biology*, **64**, 869–89.

Richter, K. & Adlakha, A. (1989). The effect of infant and child mortality on subsequent fertility. *Journal of Population and Social Studies*, **2**, 43–62.

Rinat, C., Wanders, R.J.A., Drukker, A., Halle, D. & Frishberg, Y. (1999). Primary hyperoxaluria type 1: a model for multiple mutations in a monogenic disease within a distinct ethnic group. *Journal of the American Society of Nephrology*, **10**, 2352–8.

Rittler, M., Liascovich, R., López-Camelo, J. & Castilla, E.E. (2001). Parental consanguinity in specific types of congenital anomalies. *American Journal of Medical Genetics*, **102**, 36–43.

Rittler, M., Lopez-Camelo, J.S., Castilla, E.E. *et al.* (2008). Preferential associations between oral clefts and other major congenital anomalies. *Cleft Palate-Craniofacial Journal*, **45**, 525–32.

Roberts, D.F. (1968). Genetic effects of population size reduction. *Nature*, **220**, 1084–8.

Roberts, D.F. & Bonné-Tamir, B. (1973). Reproduction and inbreeding among the Samaritans. *Social Biology*, **20**, 64–70.

Roberts, D.F., Roberts, M.J. & Cowie, J.A. (1979). Inbreeding levels in Orkney Islanders. *Journal of Biosocial Science*, **11**, 391–5.

Roberts, D.F., Roberts, M.J. & Johnston, A.W. (1991). Genetic epidemiology of Down's syndrome in Shetland. *Human Genetics*, **87**, 57–60.

Roberts, E., Hampshire, D.J., Pattison, L. *et al.* (2002). Autosomal recessive primary microcephaly: an analysis of locus heterogeneity and phenotypic variation. *Journal of Medical Genetics*, **39**, 718–21.

Roberts, S.C., Gosling, L.M., Carter, V. & Petrie, M. (2008). MHC-correlated odour preferences in humans and the use of oral contraceptives. *Proceedings of the Royal Society of London Series B*, **275**, 2715–22.

Roberts, S.C., Little, A.C., Gosling, L.M. *et al.* (2005). MHC-assortative facial preferences in humans. *Biology Letters*, **1**, 400–3.

Robinson, A.P. (1983). Inbreeding as measured by dispensations and isonymy on a small Hebridean island, Eriskay. *Human Biology*, **55**, 289–95.

Rödelsperger, C., Krawitz, P., Bauer, S. *et al.* (2011). Identity-by-descent filtering of exome sequence data for disease-gene identification in autosomal recessive disorders. *Bioinformatics*, **27**, 829–36.

Roeleveld, N., Zielhuis, G.A. & Gabreels, F. (1997). The prevalence of mental retardation: a critical review of recent literature. *Developmental Medicine and Child Neurology*, **39**, 125–32.

Rogers, N.K., Gilbert, C.E., Foster, A., Zahkidov, B.O. & McCollum, C.J. (1999). Childhood blindness in Uzbekhistan. *Eye*, **13**, 65–70.

Roguin, N., Du, Z.D., Barak, M. *et al.* (1995). High prevalence of muscular septal defect in neonates. *Journal of the American College of Cardiology*, **26**, 1545–8.

Roldan, E.R., Cassinello, J., Abaigar, T. & Gomendio, M. (1998). Inbreeding, fluctuating asymmetry, and ejaculate quality in an endangered ungulate. *Proceedings of the Royal Society of London Series B*, **265**, 243–8.

Romeo, G., Menozzi, P., Ferlini, A. *et al.* (1983a). Incidence of Friedreich ataxia in Italy estimated from consanguineous marriages. *American Journal of Human Genetics*, **35**, 523–39.

Romeo, G., Menozzi, P., Ferlini, A. *et al.* (1983b). Incidence of classic PKU in Italy estimated from consanguineous marriages and from neonatal screening. *Clinical Genetics*, **24**, 339–45.

Romeo, G., Menozzi, P., Mastella, G. *et al.* (1981). Studio genetica ed epidemiologica della fibrosi cistica in Italia. Risultati di un'indagine policentrica del gruppo di lavoro per la fibrosi cistica della Società Italiana di Pediatria. *Revista Italiana di Pediatria*, **7**, 201–9. [In Italian.]

Roodpeyma, S., Kamali, Z., Afshar, F. & Naraghi, S. (2002). Risk factors in congenital heart disease. *Clinical Pediatrics*, **41**, 653–8.

Rose, P., Humm, E., Hey, K., Jones, L. & Huson, S.M. (1999). Family history taking and genetic counselling in primary care. *Family Practice*, **16**, 78–83.

Rosenberg, L.T., Cooperman, D. & Payne, R. (1983). HLA and mate selection. *Immunogenetics*, **17**, 89–93.

Rosenberg, N., Yatuv, R., Orion, Y. *et al.* (1997). Glanzmann thrombasthenia caused by an 11.2-kb deletion in the glycoprotein IIIa (β3) is a second mutation in Iraqi Jews that stemmed from a distinct founder. *Blood*, **89**, 3654–62.

Rosenfeld, A.A. (1979). Incidence of a history of incest among 18 female psychiatric patients. *American Journal of Psychiatry*, **136**, 791–5.

Rosner, F. (1969). Hemophilia in the Talmud and Rabbinic writings. *Annals of Internal Medicine*, **70**, 833–7.

Rötig, A. & Munnich, A. (2003). Genetic features of mitochondrial respiratory chain disorders. *Journal of the American Society of Nephrology*, **14**, 2995–3007.

Rötig, A. & Poulton, J. (2009). Genetic causes of mitochondrial DNA depletion in humans. *Biochimica et Biophysica Acta*, **1792**, 1103–8.

Roudi-Rahimi, F. (2004). *Islam and Family Planning*. Washington, DC: Population Reference Bureau.

Roychoudhury, A.K. (1980). Consanguineous marriages in Tamil Nadu. *Journal of the Indian Anthropological Society*, **15**, 167–74.

Rubio-Cabezas, O., Diaz Gonsález, F., Aragonés, A., Argente, J. & Campos-Barros, A. (2008). Permanent neonatal diabetes caused by a homozygous nonsense mutation in the glucokinase gene. *Pediatric Diabetes*, **9**, 245–9.

Rudan, I. (1999). Inbreeding and cancer incidence in human isolates. *Human Biology*, **71**, 173–87.

Rudan, I., Biloglav, Z., Vorko-Jović, A. *et al.* (2006). Effects of inbreeding, endogamy, genetic admixture, and outbreeding on human health: a '1001 Dalmatians' study. *Croatian Medical Journal*, **47**, 601–5.

Rudan, I., Carothers, A.D., Polasek, O. *et al.* (2008). Quantifying the increase in average human heterozygosity due to urbanisation. *European Journal of Human Genetics*, **16**, 1097–102.

Rudan, I., Padovan, M., Rudan, D. *et al.* (2002a). Inbreeding and nephrolithiasis in Croatian Island isolates. *Collegium Antropologicum*, **26**, 11–21.

Rudan, I., Rudan, D., Campbell, H. *et al.* (2002b). Inbreeding and learning disability in Crotian Island isolates. *Collegium Antropologicum*, **26**, 421–8.

Rudan, I., Rudan, D., Campbell, H. *et al.* (2003b). Inbreeding and risk of late onset complex disease. *Journal of Medical Genetics*, **40**, 925–32.

Rudan, I., Škarić-Jurić, T., Smolej-Narancic, N. *et al.* (2004). Inbreeding and susceptibility to osteoporosis in Croatian Island isolates. *Collegium Antropologicum*, **28**, 585–601.

Rudan, I., Smolej-Narancić, N., Campbell, H. *et al.* (2003a). Inbreeding and the genetic complexity of human hypertension. *Genetics*, **163**, 1011–21.

Ruffer, M.A. (1921). On the physical effects of consanguineous marriages on the royal families of Ancient Egypt. In *Studies in the Paleopathology of Egypt*, L.R. Moodie, ed., Chicago: University of Chicago Press, pp. 322–66.

Rukanuddin, A.R. (1982). Infant-child mortality and son preference as factors influencing fertility in Pakistan. *Population and Development Review*, **21**, 297–328.

Rushton, A.R. & Genel, M. (1981). Hereditary ectodermal dysplasia, olivopontocerebellar degeneration, short stature, and hypogonadism. *Journal of Medical Genetics*, **18**, 335–9.

Ryan, K., Bain, B.J., Worthington, D. *et al.* (2010). Significant haemoglobinopathies: guidelines for screening and diagnosis. *British Journal of Haematology*, **149**, 35–49.

Saad, F.A. & Jauniaux, E. (2002). Recurrent early pregnancy loss and consanguinity. *Reproductive Biomedicine Online*, **5**, 167–70.

Saadallah, A.A. & Rashed, M.S. (2007). Newborn screening: experiences in the Middle East and North Africa. *Journal of Inherited Metabolic Disease*, **30**, 482–9.

Saadat, M., Ansari-Lari, M. & Farhud, D.D. (2004). Consanguineous marriage in Iran. *Annals of Human Biology*, **31**, 263–9.

Saadat, M., Khalili, M., Omidvari, S. & Ansari-Lari, M. (2011). Parental consanguineous marriages and clinical response to chemotherapy in locally advanced breast cancer patients. *Cancer Letters*, **302**, 109–12.

Sachar, R.K., Verma, J., Prakash, V. *et al.* (1990). The unwelcome sex – female feticide in India. *World Health Forum*, **11**, 309–20.

Sadovnick, A.D., Yee, I.M.L., Ebers, G.C., and the Canadian Collaborative Group (2001). Recurrence risks to sibs of MS index cases: impact of consanguineous matings. *Neurology*, **56**, 784–5.

Saedi-Wong, S. & al-Frayh, A.R. (1989). Effects of consanguineous matings on anthropometric measurements of Saudi newborn infants. *Family Practice*, **6**, 217–20.

Saedi-Wong, S., Al-Frayh, A.R. & Wong, H.Y.H. (1989). Socio-economic epidemiology of consanguineous matings in the Saudi Arabian population. *Journal of Asian and African Studies*, **24**, 247–51.

Saftlas, A.F., Beydoun, H. & Triche, E. (2005). Immunogenetic determinants of preeclampsia and related pregnancy disorders: a systematic review. *Obstetrics & Gynecology*, **106**, 162–72.

Saha, N. & El Sheikh, F.S. (1988). Inbreeding levels in Khartoum. *Journal of Biosocial Science*, **20**, 333–6.

Saha, N., Hamad, R.E., Mohamed, S. (1990). Inbreeding effects on reproductive outcome in a Sudanese population. *Human Heredity*, **40**, 208–12.

Saha, N.M. & Ali, S.M. (1992). Knowledge and use of family planning. In *Pakistan Demographic and Health Survey, 1990/91*, Islamabad and Columbia, MD: National Institute of Population Studies and Institute for Resource Development/Macro International, pp. 53–72.

Sahni, M., Verma, N., Narula, D. *et al.* (2008). Missing girls in India: infanticide, feticide and made-to-order pregnancies? Insights from hospital-based sex-ratio-at birth over the last century. *PLoS One*, **3**, e2224.

Sajjad, M., Khattak, A.A., Bunn, J.E.G. & McKenzie, I. (2008). Causes of childhood deafness in Pukhtoonkhwa Province of Pakistan and the role of consanguinity. *Journal of Laryngology & Otology*, **122**, 1057–63.

Saleem, R., Gofin, R., Ben-Neriah, Z. & Boneh A. (1998). Variables influencing parental perception of inherited metabolic diseases before and after genetic counselling. *Journal of Inherited Metabolic Disease*, **21**, 769–80.

Saleh, E.A., Mahfouz, A.A.R., Tayel, K.Y., Naguib, M.K. & Bin-Al-Shaikh, N.M.S. (2000). Hypertension and its determinants among primary school-children in Kuwait. *Eastern Mediterranean Health Journal*, **6**, 333–7.

Sales, A.M., Ponce de Leon, A., Düppre, N.C. *et al.* (2011). Leprosy among patient contacts: a multilevel study of risk factors. *PLoS Neglected Tropical Diseases*, **15**, e1013.

Salmon, J.E., Heuser, C., Triebwasser, M. *et al.* (2011). Mutations in the complement regulatory proteins predispose to preeclampsia: a genetic analysis of the PROMISSE cohort. *PLoS Medicine*, **8**, e1001013.

Salzano, F.M., Marcallo, F.A., Freire-Maia, N. & Krieger, H. (1962). Genetic load in Brazilian Indians. *Acta Genetica et Statistica Medica, Basel*, **12**, 212–8.

Samavat, A. & Modell, B. (2004). Iranian national thalassaemia screening programme. *British Medical Journal*, **329**, 1134–7.

Sanghvi, L.D. (1963). Genetic load: three views. 2. The concept of genetic load: a critique. *American Journal of Human Genetics*, **115**, 298–309.

Sanghvi, L.D. (1966). Inbreeding in India. *Eugenics Quarterly*, **13**, 291–301.

Sanghvi, L.D., Varde, D.S. & Master, H.R. (1956). Frequency of consanguineous marriages in twelve endogamous groups in Bombay. *Acta Genetica et Statistica Medica, Basel*, **6**, 41–9.

Sarachana, T., Zhou, R., Chen, G., Manji, H.K. & Hu, V.W. (2010). Investigation of post-transcriptional gene regulatory networks associated with autism spectrum disorders by microRNA expression profiling of lymphoblastoid cell lines. *Genome Medicine*, **2**, 23.

Sariola, H. & Uutela, A. (1996). The prevalence and context of incest abuse in Finland. *Child Abuse and Neglect*, **20**, 843–50.

Sastri, K.A.N. (1976). *A History of South India: from Prehistoric Vijayanagar*, 4th edn. Madras: Oxford University Press, p. 66.

Sathyanarayana Rao, T.S., Prabhakar, A.K., Jagannatha Rao, K.S. *et al.* (2009). Relationship between consanguinity and depression in a south Indian population. *Indian Journal of Psychiatry*, **51**, 50–2.

Sato, T., Nonaka, K., Miura, T. & Peter, K. (1994). Trends in cohort fertility of the Dariusleut Hutterite population. *Human Biology*, **66**, 421–31.

Saugstad, L. & Ødegård, Ø. (1986). Inbreeding and schizophrenia. *Clinical Genetics*, **30**, 261–75.

Sayee, R. & Thomas, I.M. (1998). Consanguinity, non-disjunction, parental age and Down's syndrome. *Journal of the Indian Medical Association*, **96**, 335–57.

Schacter, B., Weitkamp, L.R. & Johnson, W.E. (1984). Parental HLA compatibility, fetal wastage and neural tube defects: evidence for a T/t-like locus in humans. *American Journal of Human Genetics*, **36**, 1082–91.

Scheidel, W. (1996). Brother–sister and parent–child marriage outside royal families in ancient Egypt and Iran: a challenge to the sociobiological view of incest avoidance? *Ethology and Sociobiology*, **17**, 319–40.

Scheidel, W. (1997). Brother–sister marriage in Roman Egypt. *Journal of Biosocial Science*, **29**, 361–71.

Scheidel, W. (2002). Brother–sister and parent–child marriage in premodern societies. In *Human Mate Choice and Prehistoric Marital Networks*, K. Aoki &

T. Akazawa, eds., Kyoto: International Research Center for Japanese Studies, pp. 33–47.

Scheidel, W. (2004). Ancient Egyptian sibling marriage and the Westermarck effect. In *Inbreeding, Incest, and the Incest Taboo*, A. Wolf & W. Durham, eds., Stanford, CA: Stanford University Press, pp. 93–108.

Schipper, R.F., D'Amaro, J. & Oudshoorn, M. (1996). The probability of finding a suitable related donor for bone marrow transplantation in extended families. *Blood*, **87**, 800–4.

Schneider, M.A. & Hendrix, L. (2000). Olfactory sexual inhibition and the Westermarck effect. *Human Nature*, **11**, 65–91.

Schocket, A.L. & Weiner, H.L. (1978). Lymphocytotoxic antibodies in family members of patients with multiple sclerosis. *Lancet*, **1**, 571–3.

Schork, M.A. (1964). The effects of inbreeding on growth. *American Journal of Human Genetics*, **16**, 292–300.

Schork, N.J., Murray, S.S., Frazer, K.A. & Topol, E.J. (2009). Common vs. rare allele hypotheses for complex diseases. *Current Opinion in Genetics and Development*, **19**, 212–9.

Schreider, E. (1967). Body-height and inbreeding in France. *American Journal of Physical Anthropology*, **26**, 1–4.

Schull, W. J. (1958). Empirical risks in consanguineous marriages: sex ratio, malformation, and viability. *American Journal of Human Genetics*, **10**, 294–343.

Schull, W.J., Furusho, T., Yamamoto, M., Nagano, H. & Komatsu, I. (1970b). The effect of parental consanguinity and inbreeding in Hirado, Japan. IV. Fertility and reproductive compensation. *Humangenetik*, **9**, 294–315.

Schull, W.J., Nagano, H., Yamamoto, M. & Komatsu, I. (1970a). The effects of parental consanguinity and inbreeding in Hirado, Japan. I. Stillbirths and prereproductive mortality. *American Journal of Human Genetics*, **22**, 239–62.

Schull, W.J. & Neel, J.V. (1965). *The Effects of Inbreeding on Japanese Children*. New York: Harper and Row.

Schull, W.J. & Neel, J.V. (1966). Some further observations on the effect of inbreeding on mortality in Kure, Japan. *American Journal of Human Genetics*, **18**, 144–52.

Schull, W.J. & Neel, J.V. (1972). The effects of parental consanguinity and inbreeding in Hirado, Japan. V. Summary and interpretation. *American Journal of Human Genetics*, **24**, 425–53.

Scott-Emuakpor, A.B. (1974). The mutation load in an African population. I. An analysis of consanguineous marriages in Nigeria. *American Journal of Human Genetics*, **26**, 674–82.

Scrimshaw, S.C.M. (1978). Infant mortality and behavior in the regulation of family size. *Population and Development Review*, **4**, 383–404.

Seemanová, E. (1971). A study of incestuous matings. *Human Heredity*, **21**, 108–28.

Seidel, H. & Steinlein, O.K. (2008). Compound heterozygosity for three common MEFV mutations in a highly consanguineous family with familial Mediterranean fever. *European Journal of Pediatrics*, **167**, 827–8.

Segalen, M. & Richard, P. (1986). Marrying kinsmen in Pays Bigouten Sud, Brittany. *Journal of Family Studies*, **11**, 109–30.

Segura, J.J. & Jiménez-Rubio, A. (1999). Talon cusp affecting permanent maxillary lateral incisors in 2 family members. *Oral Surgery, Oral Medicine, Oral Pathology, Oral Radiology, and Endodontics*, **88**, 90–2.

Sellick, G.S., Garrett, C. & Houlston, R.S. (2003). A novel gene for neonatal diabetes maps to chromosome 10p12.1-p13. *Diabetes*, **52**, 2636–8.

Seoud, M., Khalil, A., Frangieh, A. *et al.* (1995). Recurrent molar pregnancies in a family with extensive intermarriage: report of a family and review of the literature. *Obstetrics & Gynecology*, **86**, 692–5.

Serre, J.L., Jakobi, L. & Babron, M.-C. (1985). A genetic isolate in the French Pyrenees: probabilities of origin of genes and inbreeding. *Journal of Biosocial Science*, **17**, 405–14.

Serre, J.L., Mayer, F.M., Feingold, N. & Benoist, J. (1982). Étude d'un isolat des Antilles. *Annales de Génétique*, **25**, 43–9. [In French.]

Sezik, M., Ozkaya, O., Sezik, H.T., Yapar, E.G. & Kaya, H. (2006). Does marriage between first cousins have any predictive value for maternal and perinatal outcomes in pre-eclampsia? *Journal of Obstetric and Gynecological Research*, **32**, 475–81.

Shahcheraghi, G.H. & Hobbi, M.H. (1999). Patterns and progression in congenital scoliosis. *Journal of Pediatric Orthopaedics*, **19**, 766–75.

Shahi, K.K. & Mohanty, S. (2006). Alleged dowry deaths: a study of homicidal burns. *Medicine Science & Law*, **46**, 105–10.

Shahin, H., Walsh, T., Sobe, T. *et al.* (2002). Genetics of congenital deafness in the Palestinian population: multiple connexin 26 alleles with shared origins in the Middle East. *Human Genetics*, **110**, 284–9.

Shami, S.A. (1983). Consanguineous marriages in Mianchannu and Muridke (Punjab) Pakistan. *Biologia*, **29**, 19–30.

Shami, S.A., Grant, J.C. & Bittles, A.H. (1994). Consanguineous marriage within social/occupational class boundaries in Pakistan. *Journal of Biosocial Science*, **26**, 91–6.

Shami, S.A. & Hussain, S.B. (1984). Consanguinity in the population of Gujrat (Punjab), Pakistan. *Biologia*, **30**, 93–109.

Shami, S.A. & Iqbal, I. (1983). Consanguineous marriages in the population of Sheikhupura (Punjab), Pakistan. *Biologia*, **29**, 231–44.

Shami, S.A. & Minhas, I.B. (1984). Effects of consanguineous marriages on offspring mortality in the City of Jhelum (Punjab) Pakistan. *Biologia*, **29**, 153–65.

Shami, S.A., Qadeer, T., Schmitt, L.H. & Bittles, A.H. (1991). Consanguinity, gestational period and anthropometric measurements at birth in Pakistan. *Annals of Human Biology*, **18**, 523–7.

Shami, S.A., Qaisar, R. & Bittles, A.H. (1991). Consanguinity and adult morbidity in Pakistan. *Lancet*, **338**, 954–5.

Shami, S.A. & Siddiqui, H. (1984). The effects of parental consanguinity in Rawalpindi City (Punjab), Pakistan. *Biologia*, **30**, 189–200.

Shami, S.A. & Zahida. (1982). Study of consanguineous marriages in the population of Lahore (Punjab) Pakistan. *Biologia*, **28**, 1–5.

Sharkia, R., Azem, A., Kaiyal, Q., Zelnik, N. & Mahajnah, M. (2010). Mental retardation and consanguinity in a selected region of the Israeli Arab community. *Central European Journal of Medicine*, **5**, 91–6.

Sharkia, R., Zaid, M., Athamna, A. *et al.* (2007). The changing pattern of consanguinity in a selected region of the Israeli Arab community. *American Journal of Human Biology* **20**, 72–7.

Sharma, B.R., Harish, D., Gupta, M. & Singh, V.P. (2005). Dowry – a deep-seated cause of violence against women in India. *Medicine Science & Law*, **45**, 161–8.

Sharma, O.P. & Haub, C. (2008). *Sex Ratio at Birth Begins to Improve in India.* Washington, DC: Population Reference Bureau.

Shaw, A. (2000). Kinship, cultural preference and immigration: Consanguineous marriage among British Pakistanis. *Journal of the Royal Anthropological Institute*, **7**, 315–34.

Shaw, A. (2009). *Negotiating Risk: British Pakistani Experiences of Genetics.* Oxford: Berghahn Books.

Shaw, A. & Hurst, J.A. (2008). "What is this genetics, anyway?" Understandings of genetics, illness causality and inheritance among British Pakistani users of Genetic Services. *Journal of Genetic Counseling*, **17**, 373–83.

Shaw, A. & Hurst, J.A. (2009). 'I don't see any point in telling them': attitudes to sharing genetic information in the family and carrier testing of relatives among British Pakistani adults referred to a genetics clinic. *Ethnicity & Health*, **14**, 205–24.

Shaw, B.D. (1992). Explaining incest: brother–sister marriage in Graeco-Roman Egypt. *Man (N.S.)*, **27**, 267–99.

Shearer, A.E., DeLuca, A.P., Hildebrand, M.S. *et al.* (2010). Comprehensive genetic testing for hereditary hearing loss using massively parallel sequencing. *Proceedings of the National Academy of Sciences of the United States of America*, **107**, 21104–9.

Sheehan, N.A., Didelez, V., Burton, P.R. & Tobin, M.D. (2008). Mendelian randomisation and causal inference in observational epidemiology. *PLoS Medicine*, **5**, e177.

Sheets, J.W. (1980). Population structure of depreciated communities: 1. The 1977 genetic demographies of Colonsay and Jura islands, the Scottish Inner Hebrides. *Social Biology*, **27**, 114–29.

Sheiner, E., Shoham-Vardi, I., Sheiner, E.K. *et al.* (1999). Maternal factors associated with severity of birth defects. *International Journal of Gynaecology and Obstetrics*, **64**, 227–32.

Shen, J., Gilmore, E.C., Marshall, C.A. *et al.* (2010). Mutations in *PNKP* cause microcephaly, seizures and defects in DNA repair. *Nature Genetics*, **42**, 245–9.

Shenfield, F. (1998). Recruitment and counselling of sperm donors: ethical problems. *Human Reproduction*, **13** Suppl. 2, 70–5.

Shepher, J. (1971). Mate selection among second generation kibbutz adolescents and adults. *Archives of Sexual Behavior*, **1**, 293–307.

Shields, J. & Slater, E. (1956). An investigation into the children of cousins. *Acta Genetica et Statistica Medica*, **6**, 60–79.

Shields, R. (2011). Common disease: are causative alleles common or rare? *PLoS Biology*, **9**, e1001009.

Shiloh, S., Reznik, H., Bat-Miriam-Katznelson, M. & Goldman, B. (1995). Pre-marital genetic counselling to consanguineous couples: attitudes, beliefs and decisions among counselled, noncounselled and unrelated couples in Israel. *Social Science & Medicine*, **41**, 1301–10.

Shlush, L.I., Behar, D.M., Yudkovsky, G. *et al.* (2008). The Druze: a population genetic refugium of the Near East. *PLoS One*, **3**, e2105.

Shuttleworth, G.E. (1886). The relationship of marriages of consanguinity to mental unsoundness. *British Journal of Psychiatry*, **32**, 353–9.

Sibert, J.R., Jadhav, M. & Inbaraj, S.G. (1979). Fetal growth and parental consanguinity. *Archives of Disease in Childhood*, **54**, 317–9.

Silengo, M., Lerone, M., Martinelli, M. *et al.* (1992). Autosomal recessive microcephaly with early onset seizures and spasticity. *Clinical Genetics*, **42**, 152–5.

Simon, H.A. (1990). A mechanism for social selection and successful altruism. *Science*, **250**, 1665–8.

Simpson, J.L., Martin, A.O., Elias, S., Sarto, G.E. & Dunn, J.K. (1981). Cancers of the breast and female genital system: search for recessive genetic factors through analysis of human isolate. *American Journal of Obstetrics and Gynecology*, **141**, 629–36.

Simpson, J.L., New, M., Peterson, R.E. & German, J. (1971). Pseudovaginal perineoscrotal hypospadias (PPSH) in sibs. *Birth Defects Original Article Series*, **6**, 140–4.

Singh, N., Pripp, A.H., Brekke, T. & Stray-Pedersen, B. (2010). Different sex ratios of children born to Indian and Pakistani immigrants in Norway. *BMC Pregnancy and Childbirth*, **10**, 40.

Sinha, S., Black, M.L., Agarwal, S. *et al.* (2009). Profiling β-thalassaemia mutations in India at state and regional levels: implications for genetic education, screening and counselling programmes. *HUGO Journal*, **3**, 51–62.

Sinha, S., Black, M.L., Agarwal, S. *et al.* (2011). ThalInd, a β-thalassemia and hemoglobinopathies for India: defining a model country-specific and disease-centric bioinformatics resource. *Human Mutation*, **32**, 887–93.

Sirdah, M., Bilto, Y.Y., El Jabour, S. & Najjar, K.H. (1998). Screening secondary school students in the Gaza strip for β-thalassaemia trait. *Clinical and Laboratory Haematology*, **20**, 279–83.

Sivakumaran, T.A. & Karthikeyan, S. (1997). Effects of inbreeding on reproductive losses in Kota tribe. *Acta Genetica et Medica Gemellologia (Roma)*, **46**, 123–8.

Sivaram, M., Richard, J. & Rao, P.S.S. (1995). Early marriage among rural and urban females of South India. *Journal of Biosocial Science*, **27**, 325–31.

Sivertsen, A., Wilcox, A.J., Skjaerven, R. *et al.* (2008). Familial risks of oral clefts by morphological type and severity: population based cohort study of first degree relatives. *British Medical Journal*, **336**, 432–4.

Skeie, A., Froun, J.F., Vege, A. & Stray-Pedersen, B. (2003). Cause and risk of stillbirth in twin pregnancies: a retrospective audit. *Acta Obstetrica et Gynecologica Scandinavica*, **82**, 1010–6.

Slatis, H.M. & Hoene, R.E. (1961). The effect of consanguinity on the variation of continuously variable characteristics. *American Journal of Human Genetics*, **13**, 28–31.

Smedts, H.P., de Vries, J.H., Rakhshandehroo, M. *et al.* (2009). High maternal vitamin E intake by diet or supplements is associated with congenital heart defects in the offspring. *British Journal of Obstetrics & Gynaecology*, **116**, 416–23.

Smith, M.T. (2001). Estimates of cousin marriage and mean inbreeding in the United Kingdom from 'birth briefs'. *Journal of Biosocial Science*, **33**, 55–66.

Smith, M.T., Abade, A. & Cunha, E.M. (1992). Genetic structure of the Azores: marriage and inbreeding in Flores. *Annals of Human Biology*, **19**, 595–602.

Smith, M.T., Asquith-Charlton, R.M., Blodwell, L.M., Clements, C.M. & Ellam, C.J. (1993). Estimating inbreeding from the Faculty Office Registers, 1534–40. *Annals of Human Biology*, **20**, 357–68.

Smits, P., Saada, A., Wortmann, S.B. *et al.* (2010). Mutation in mitochondrial ribosomal protein MRPS22 leads to Cornelia de Lange-like phenotype, brain abnormalities and hypertrophic cardiomyopathy. *European Journal of Human Genetics*, **19**, 394–9.

Smout, T.C. (1980). Scottish marriage, regular and irregular 1500–1940. In *Marriage and Society: Studies in the Social History of Marriage*, R.B. Outhwaite, ed., London: Europe, pp. 204–36.

Soliman, A.S., Bondy, M.L., Levin, B. *et al.* (1998). Familial aggregation of colorectal cancer in Egypt. *International Journal of Cancer*, **77**, 811–6.

Soliman, A.T., El Zalabany, M.M., Bappal, B., Al Salmi, I., de Silva, V. & Asfour, M. (1999). Permanent neonatal diabetes mellitus: epidemiology, mode of presentation, pathogenesis and growth. *Indian Journal of Pediatrics*, **66**, 363–73.

Sommerfelt, K., Kyllerman, M. & Sanner, G. (1991). Hereditary spastic paraplegia with epileptic myoclonus. *Acta Neurologica Scandinavica*, **84**, 157–60.

Spiro, M.E. (1969). *Children of the Kibbutz*, New York: Schoken, pp. 326–35, 347–50.

Spuhler, J.N. & Kluckhorn, C. (1953). Inbreeding coefficients of the Ramah Navaho population. *Human Biology*, **25**, 295–317.

Sridhar, K. (2009). A community-based survey of visible congenital anomalies in rural Tamil Nadu. *Indian Journal of Plastic Surgery*, **42** Suppl. 1, S184–91.

Stefansson, H., Rujescu, D., Cichon, S. *et al.* (2008). Large recurrent microdeletions associated with schizophrenia. *Nature*, **455**, 232–6.

Stein, G.H. (1939). *Marriage in Early Islam*. London: The Royal Asiatic Society, pp. 59–60, 158–68.

Stern, C. & Charles, D.R. (1945). The Rhesus gene and the effect of consanguinity. *Science*, **101**, 305–7.

Sternicki, T., Szablewski, P. & Szwaczkowski, T. (2003). Inbreeding effects on lifetime in Davids' deer (*Elaphurus davidianus*), Milne Edwards (1866) population. *Journal of Applied Genetics*, **44**, 175–83.

Stevenson, A.C. Davison, B.C.C., Say, B. *et al.* (1971). Contribution of feto-maternal incompatibility to aetiology of pre-eclamptic toxaemia. *Lancet*, **II**, 1286–9.

Stevenson, A.C., Johnston, H.A., Stewart, M.I.P. & Golding, D.R. (1966). Congenital malformations: a report of a study of series of consecutive births in 24 centres. *Bulletin of the World Health Organization*, **34** (supplement), 1–125.

Stevenson, A.C., Sya, B., Ustaoglu, S. & Durmas, Z. (1976). Aspects of pre-eclamptic toxaemia of pregnancy, consanguinity, and twinning in Ankara. *Journal of Medical Genetics*, **13**, 1–8.

Stoll, C., Alembik, Y., Dott, B. & Roth, M.P. (1998). Study of Down syndrome in 238,942 consecutive births. *Annales de Génétique*, **41**, 44–51.

Stoll, C., Alembik, Y., Roth, M.P. & Dott, B. (1997). Risk factors in congenital anal atresias. *Annales de Génétique*, **40**, 197–204.

Stoll, C. Alembik, Y., Roth, M.P. & Dott, B. (1999). Parental consanguinity as a cause for increased incidence of birth defects in a study of 238,942 consecutive births. *Annales de Génétique*, **42**, 133–9.

Stoltenberg, C., Magnus, P., Lie, R.T., Daltveit, A.K. & Irgens, L.M. (1997). Birth defects and parental consanguinity in Norway. *American Journal of Epidemiology*, **145**, 439–48.

Stoltenberg, C., Magnus, P., Lie, R.T., Daltveit, A.K. & Irgens, L.M. (1998). Influence of consanguinity and maternal education on risk of stillbirth and infant death in Norway, 1967–1993. *American Journal of Epidemiology*, **148**, 452–9.

Stone, J.L., O'Donovan, M.C., Gurling, H. *et al.* (2008). Rare chromosomal deletions and duplications increase risk of schizophrenia. *Nature*, **455**, 237–41.

Straussberg, R., Kornreich, L., Harel, L. & Varsano, I. (1998). Autosomal recessive microcephaly with neonatal myoclonic seizures: clinical and MRI findings. *American Journal of Medical Genetics*, **80**, 136–9.

Strømme, P., Suren, P., Kanavin, O.J. *et al.* (2010). Parental consanguinity is associated with a seven-fold increased risk of progressive encephalopathy: a cohort study from Oslo, Norway. *European Journal of Paediatric Neurology*, **14**, 138–45.

Su, B. & Macer, D.R.J. (2003). Chinese people's attitudes towards genetic diseases and children with handicap. *Law and Human Genome Review*, **18**, 191–210.

Subbarayan, A., Colarusso, G., Hughes, S.M. *et al.* (2011). Clinical features that identify children with primary immunodeficiency diseases. *Pediatrics*, **127**, 810–6.

Subramanian, S.V. & Selvaraj, S. (2009). Social analysis of sex imbalance in India: before and after the implementation of the Pre-Natal Diagnostic Techniques (PNDT) Act. *Journal of Epidemiology and Community Health*, **63**, 245–52.

Subramanyan, R., Joy, J., Venugopalan, P., Sapru, A. & al Khusaiby, S.M. (2000). Incidence and spectrum of congenital heart disease in Oman. *Annals of Tropical Paediatrics*, **20**, 337–40.

Sudmant, P.H., Kitzman, J.O., Antonacci, F. *et al.* (2010). Diversity of human copy number variation and multicopy genes. *Science*, **330**, 641–6.

Sueyoshi, S. & Ohtsuka, R. (2003). Effects of polygyny and consanguinity on high fertility in the rural Arab population in South Jordan. *Journal of Biosocial Science*, **35**, 513–26.

Sugiura, Y. (1972). Genetic studies in some Japanese populations. VII. The effects of parental consanguinity on skeletal age in Shizuoka children. *Japanese Journal of Human Genetics*, **17**, 199–208.

Sundström, P., Nyström, L. & Forsgren, L. (2001). Prevalence of multiple sclerosis in Västerbotten County in northern Sweden. *Acta Neurologica Scandinavica*, **103**, 214–8.

Sureender, S., Prabakaran, B. & Khan, A.G. (1998). Mate selection and its impact on female marriage age, pregnancy wastages, and first child survival in Tamil Nadu, India. *Social Biology*, **45**, 289–301.

Sutter, J. & Tabah, L. (1953). Structure de la mortalité dans les familles consanguines. *Population*, **8**, 511–26. [In French.]

Swift, G. (1984). *Waterland*. London: Picador.

Swiss Criminal Code (2010). *Felonies and Misdemeanours against the Family*, Article 213. Available at http://www.admin.ch.ch/e/rs/311_0/a213.html [28 June 2011].

Tabbara, K.F. & Badr, I.A. (1985). Changing pattern of childhood blindness in Saudi Arabia. *British Journal of Ophthalmology*, **69**, 312–5.

Tadmouri, G.O., Al Ali, M.T., Al-Haj Ali, S. & Al Khaja. N. (2006). CTGA: the database for genetic disorders in Arab populations *Nucleic Acids Research*, **34** (Database issue), D602–6.

Takata, H., Suzuki, M., Ishii, T., Sekiguchi, S. & Iri, H. (1987). Influence of major histocompatibility complex region genes on human longevity among Okinawan-Japanese centenarians and nonagenarians. *Lancet*, **2**, 824–6.

Talmon, Y. (1965). The family in a revolutionary movement – the case of the kibbutz in Israel. In *Comparative Family Systems*, M.F. Nimkoff, ed., Boston: Houghton Mifflin, pp. 259–86.

Tamim, H., Khogali, M., Beydoun, H., Melki, I. & Yunis, K. (2003). Consanguinity and apnea of prematurity. *American Journal of Epidemiology*, **158**, 942–6.

Tanaka, K. (1973). Genetic studies on inbreeding in some Japanese populations. XI. Effects of inbreeding on mortality in Shizuoka. *Japanese Journal of Human Genetics*, **17**, 319–31.

Tarazi, I., Al Najjar, E., Lulu, N. & Sirdah, M. (2007). Obligatory premarital tests for beta-thalassaemia in the Gaza Strip: evaluation and recommendations. *International Journal of Laboratory Hematology*, **29**, 111–8.

Taylor, G.M., Dearden, S.P., Will, A.M. *et al.* (1995). Infantile osteopetrosis; bone marrow transplantation from a cousin donor. *Archives of Disease in Childhood*, **73**, 453–5.

Tchen, P., Bois, E., Feingold, J., Feingold, N. & Kaplan, J. (1977). Inbreeding in recessive diseases. *Human Genetics*, **38**, 163–7.

Teebi, A.S. & Marafie, M.J. (1988). Uncle–niece/aunt–nephew marriages are not existing in Muslim Arabs. *American Journal of Medical Genetics*, **30**, 981.

Teebi, A.S., Teebi, S.A., Porter, C.J. & Cuticchia, A.J. (2002). Arab genetic disease database (AGDDB): a population-specific clinical and mutation database. *Human Mutation*, **19**, 615–21.

Teeuw, M.E., Henneman, L., Bochdanovits, Z. *et al.* (2010). Do consanguineous parents of a child affected by an autosomal recessive disease have more DNA identical-by-descent than similarly-related parents with healthy offspring? Design of a case-control study. *BMC Medical Genetics*, **11**, 113.

Temtamy, S.A., Kandil, M.R., Demerdash, A.M. *et al.* (1994). An epidemiological/genetic study of mental subnormality in Assuit Governorate, Egypt. *Clinical Genetics*, **46**, 347–51.

Temtamy, S.A., Meguid, A.A., Mazen, I. *et al.* (1998). A genetic epidemiological study of malformations at birth in Egypt. *Eastern Mediterranean Health Journal*, **4**, 252–9.

Tenesa, A., Navarro, P., Hayes, B.J., Duffy, D.L. & Clarke, G.M. (2007). Recent human effective population size estimated from linkage disequilibrium. *Genome Research*, **17**, 520–6.

ten Kate, L.P., Scheffer, H., Cornel, M.C., van Lookeren Campagne, J.G. (1991). Consanguinity sans reproche. *Human Genetics*, **86**, 295–6.

Thapar, R. (1986). *A History of India*, Vol. 1. London: Penguin.

The Economist (2011). The flight from marriage. London: *The Economist*, Vol. 400, No. 8747, pp. 17–20.

The 1000 Genomes Project Consortium (2010). A map of human genome variation from population-scale sequencing. *Nature*, **467**, 1061–73.

Thomas, J.D., Doucette, M.M., Thomas, D.C. & Stoeckle, J.D. (1987). Disease, lifestyle and consanguinity in 58 American Gypsies. *Lancet*, **2**, 377–9.

Thomas, M.L., Harger, J.H., Wagener, D.K., Rabin, B.S. & Gill, T.J. III. (1985). HLA sharing and spontaneous abortion in humans. *American Journal of Obstetrics & Gynecology*, **151**, 1053–8.

Thorburn, D.R. (2004). Mitochondrial disorders: prevalence, myths and advances. *Journal of Inherited Metabolic Disease*, **27**, 349–62.

Thorburn, D.R. & Dahl, H-H.M. (2001). Mitochondrial disorders: genetics, counselling, prenatal diagnosis and reproductive options. *American Journal of Medical Genetics*, **106**, 102–14.

Thornhill, N.W. (1991). An evolutionary analysis of rules regulating human inbreeding and marriage. *Behavior and Brain Sciences*, **14**, 247–93.

Thornton, G.K. & Woods, C.G. (2009). Primary microcephaly: Do all roads lead to Rome? *Trends in Genetics*, **25**, 501–10.

Thornton, T. & McPeek, M.S. (2010). Roadtrips: Case-control association testing with partially or completely unknown population and pedigree structure. *American Journal of Human Genetics*, **86**, 172–84.

Tian, C., Plenge, R.M., Ransom, M. *et al.* (2008). Analysis and application of European genetic substructure using 3000K SNP information. *PLoS Genetics*, **4**, e4.

Tienari, P.J., Sumelahti, M.L., Rantamäki, T. & Wikström, J. (2004). Multiple sclerosis in western Finland: evidence for a founder effect. *Clinical Neurology and Neurosurgery*, **106**, 175–9.

Topaloglu, R., Baskin, E., Bahat, E. *et al.* (2011). Hereditary renal tubular disorders in Turkey: demographic, clinical, and laboratory features. *Clinical and Experimental Nephrology*, **15**, 108–13.

Trasi, S., Shetty, S., Ghosh, K. & Mohanty, D. (2005). Prevalence and spectrum of von Willebrand disease from western India. *Indian Journal of Medical Research*, **121**, 653–8.

Trautmann, T.R. (1987). *Lewis Henry Morgan and the Invention of Kinship*. Berkeley: University of California Press.

Tsafrir, J. & Halbrecht, I. (1972). Consanguinity and marriage systems in the Jewish community in Israel. *Annals of Human Genetics*, **35**, 343–7.

Tucker, J.E. (1988). Marriage and family in Nablus, 1720–1856: toward a history of Arab marriage. *Journal of Family History*, **13**, 165–79.

Tunçbílek, E. & Koc, I. (1994). Consanguineous marriage in Turkey and its impact on fertility and mortality. *Annals of Human Genetics*, **58**, 321–9.

Tunçbílek, E. & Ulusoy, M. (1989). Consanguinity in Turkey in 1988. *Turkish Journal of Population Studies*, **11**, 35–46.

Turan, S., Akin, L., Akcay, T. *et al.* (2010). Recessive versus imprinted disorder: consanguinity can impede establishing the diagnosis of autosomal dominant pseudohypothyroidism type 1b. *European Journal of Endocrinology*, **163**, 489–93.

Twain, M. (1899). *The Great Revolution in Pitcairn. The Works of Mark Twain*, Vol. 20. New York: Harper & Brothers, pp. 341–54.

Tyldesley, J. (1998). *Nefertiti*. New York: Viking Penguin.

UN (1948). *The Universal Declaration of Human Rights*. New York: United Nations.

UN (2001). Department of Economic and Social Affairs Population Division: *Abortion Policies: a Global Review*. New York: United Nations.

Undevia, J.V. & Balakrishnan, V. (1978). Temporal changes in consanguinity among the Parsi and Irani communities of Bombay. In *Medical Genetics in India*, **2**, I.C. Verma, ed., Pondicherry: Aurora Enterprises, pp. 145–50.

UNESCO (1997). *Universal Declaration on the Human Genome and Human Rights*. Paris: United Nations Educational, Scientific and Cultural Organization.

USCB (2011). *Historical Estimates of World Population*. Washington, DC: US Census Bureau.

Uyguner, O., Kayserili, H., Li, Y. *et al.* (2007). A new locus for autosomal recessive non-syndromic mental retardation maps to 1p21.1-p13.3. *Clinical Genetics*, **71**, 212–9.

Vadlamudi, L., Andermann, E., Lombroso, C.T. *et al.* (2004). Epilepsy in twins: insights from unique historical data of William Lennox. *Neurology*, **62**, 1127–33.

Valls, A. (1969). Inbreeding frequencies in the Balearic Islands (Spain). *Zeitscrift für Morphologie und Anthropologie*, **61**, 343–51.

van Adel, B.A. & Tarnopolsky, M.A. (2009). Metabolic myopathies: update 2009. *Journal of Clinical Neuromuscular Disease*, **10**, 97–121.

VanAmerongen, B.M., Dijkstra, C.D., Lips, P. & Polman, C.H. (2004). Multiple sclerosis and vitamin D: an update. *European Journal of Clinical Nutrition*, **58**, 1095–109.

van den Berghe, P.L. (1980). Incest and exogamy: a sociobiological reconsideration. *Ethology and Sociobiology*, **1**, 151–62.

van den Berghe, P.L. (1983). Human inbreeding avoidance: culture in nature. *Behavior and Brain Sciences*, **6**, 91–102.

van den Berghe, P.L. & Mesher, G.M. (1980). Royal incest and inclusive fitness. *American Ethnologist*, **7**, 300–17.

van Eldik, P., van der Waaji, E.H., Ducro, B. *et al.* (2006). Possible negative effects of inbreeding on semen quality in Shetland pony stallions. *Theriogenology*, **65**, 1159–70.

van Soest, S., van den Born, L.I., Gal, A. *et al.* (1994). Assignment of a gene for autosomal recessive retinitis pigmentosa (RP12) to chromosome 1q31-q32.1 in an inbred and heterogeneous disease population. *Genomics*, **22**, 499–504.

Varela, T.A., Lodiero, R. & Fariña, J. (1997). Evolution of consanguinity in the Archbishopric of Santiago de Compostela (Spain) during 1900–1979. *Human Biology*, **69**, 517–31.

Verlaan, D.J., Fantaneanu, T., Meijer, I.A. *et al.* (2007). A PARK2 mutation in a consanguineous Lebanese family affected with early-onset Parkinson disease. *Middle East Journal of Age and Ageing*, **4**, 3–6.

Vézina, H., Heyer, E., Fortier, I., Ouellette, G., Robitaille, Y. & Gauvreau, D. (1999). A genealogical study of Alzheimer disease in the Saguenay region of Québec. *Genetic Epidemiology*, **16**, 412–25.

Visaria, P.M. (1967). Sex ratio at birth in territories with a relatively complete registration. *Eugenics Quarterly*, **14**, 132–42.

Voisin, A. (1865). Contribution à l'histoire des mariages entre consanguins. *Mémoires de la Société d'Anthropologie de Paris*, **2**, 433–59.

von Kleist-Retzow, J-C., Cormier-Daire, V., de Lonlay, P. *et al.* (1998). A high rate (20%–30%) of parental consanguinity in cytochrome-oxidase deficiency. *American Journal of Human Genetics*, **63**, 428–35.

Vujkovic, M., Steegers, E.A., Looman, C.W. *et al.* (2009). The maternal Mediterranean dietary pattern is associated with a reduced risk of spina bifida in the offspring. *British Journal of Obstetrics and Gynaecology*, **116**, 408–15.

Wahab, A. & Ahmad, M. (1996). Biosocial perspective of consanguineous marriages in rural and urban Swat, Pakistan. *Journal of Biosocial Science*, **28**, 305–13.

Wahab, A. & Ahmad, M. (2005). Consanguineous marriages in the Sikh community of Swat, NWFP, Pakistan. *Journal of Social Science*, **10**, 153–7.

Wahab, A., Ahmad, M. & Shah, S.A. (2006). Migration as a determinant of marriage pattern: preliminary report on consanguinity among Afghans. *Journal of Biosocial Science*, **38**, 315–25.

Wahid Saeed, A.A., Al Shammary, F.J., Khoja, T.A. *et al.* (1996). Prevalence of hypertension and sociodemographic characteristics of adult hypertensives in Riyadh City, Saudi Arabia. *Journal of Human Hypertension*, **10**, 583–7.

Walshe, J.M. (1973). Tutankhamun: Klinefelter's or Wilson's? *Lancet*, **1**, 109–10.

Wang, C., Tasi, M-Y., Lee, M-H. *et al.* (2007). Maximum number of live births per donor in artificial insemination. *Human Reproduction*, **22**, 1363–72.

Wang, W., Qian, C. & Bittles, A.H. (2002). Consanguineous marriage in PR China: a study in rural Man (Manchu) communities. *Annals of Human Biology*, **29**, 685–90.

Wang, W., Wise, C., Baric, T., Black, M.L. & Bittles, A.H. (2003). The origins and genetic structure of three co-resident Chinese Muslim populations: the Salar, Bo'an and Dongxiang. *Human Genetics*, **113**, 244–52.

Warady, B.A. & Chadha, V. (2007). Chronic kidney disease in children: the global perspective. *Pediatric Nephrology*, **22**, 1999–2009.

Warburton, D. & Fraser, F.C. (1964). Spontaneous abortion risks in man: data from reproductive histories collected in a medical genetics unit. *Human Genetics*, **16**, 1–25.

Warnock, M. (Chair). (1984). *Report of the Committee of Inquiry into Human Fertilisation and Embryology*. London: Her Majesty's Stationery Office.

Weatherall, D.J. (1997). Thalassaemia and malaria, revisited. *Annals of Tropical Medicine and Parasitology*, **91**, 885–90.

Weatherall, D.J. & Clegg, J.B. (2001). Inherited haemoglobin disorders: an increasing global health problem. *Bulletin of the World Health Organization*, **79**, 704–12.

Wedekind, C. & Füri, S. (1997). Body odour preferences in men and women: Do they aim for specific MHC combinations or simply hetereozygosity? *Proceedings of the Royal Society of London Series B*, **264**, 1471–9.

Wedekind, C., Seebeck, T., Bettens, F. & Paepke, A.J. (1995). MHC-dependent mate preference in humans. *Proceedings of the Royal Society of London Series B*, **260**, 245–9.

Weinreb, A. (2008). Characteristics of women in consanguineous marriages in Egypt, 1988–2000. *European Journal of Population*, **24**, 185–210.

Weisfeld, G.E., Czilli, T., Phillips, K.A., Gall, J.A. & Lichtman, C.M. (2003). Possible-olfaction-based mechanisms in human kin recognition and inbreeding avoidance. *Journal of Experimental Child Psychology*, **85**, 279–95.

Weitzel, W.D., Powell, B.J. & Penick, E.C. (1978). Clinical management of father–daughter incest. A critical reexamination. *American Journal of Diseases of Childhood*, **132**, 127–30.

Westermarck, E. (1937). *The Future of Marriage in Western Civilization*. New York: Macmillan.

Wheeler, D.A., Srinivasan, M., Egholm, M. *et al.* (2008). The complete genome of an individual by massively parallel DNA sequencing. *Nature*, **452**, 872–6.

Whincup, P.H., Nightingale, C.M., Owen, C.G. *et al.* (2010). Early emergence of ethnic differences in type 2 diabetes precursors in the UK: The Child Heart and Health Study in England (CHASE Study). *PLoS Medicine*, **7**, e1000263.

White, K.J.C. (2002). Declining fertility among North American Hutterites: the use of birth control within a Dariusleut Colony. *Social Biology*, **49**, 58–73.

White, L.A. (1957). How Morgan came to write *Systems of Consanguinity and Affinity*. *Papers of the Michigan Academy of Science, Arts, and Letters*, **42**, 257–68.

Whittington, B.R. & Durward, C.S. (1996). Survey of anomalies in primary teeth and their correlation with the permanent dentition. *New Zealand Dental Journal*, **92**, 4–8.

WHO (1997). *Proposed International Guidelines on Ethical Issues in Medical Genetics and Genetic Services*. Report of a WHO Meeting on Ethical Issues in Medical Genetics, Geneva, 15–16 December 1997. Geneva: World Health Organization.

WHO (2002). *Advisory Committee on Health Research, Genomics, and World Health*. Geneva: World Health Organization.

WHO (2006). *Medical Genetic Services in Developing Countries: The Ethical, Legal and Social Implications of Genetic Testing and Screening*. Geneva: World Health Organization.

Wilcox, A.J. & Horney, L.F. (1984). Accuracy of spontaneous abortion recall. *American Journal of Epidemiology*, **120**, 727–33.

Wilcox, A.J., Weinberg, C.R., O'Connor, J.F. *et al.* (1988). Incidence of early loss of pregnancy. *New England Journal of Medicine*, **319**, 189–94.

Wilde, W.R. (1854). *On the Physical, Moral, and Social Condition of the Deaf and Dumb.* London: John Churchill.

Wildt, D.E., Bush, M., Goodrowe, K.L. *et al.* (1987b). Reproductive and genetic consequences of founding isolated lion populations. *Nature*, **329**, 328–31.

Wildt, D.E., Bush, M., Howard, J.G. *et al.* (1983). Unique seminal quality in the South African cheetah and a comparative evaluation in the domestic cat. *Biology of Reproduction*, **29**, 1019–25.

Wildt, D.E., O'Brien, S.J., Howard, J.G. *et al.* (1987a). Similarity in ejaculate-endocrine characteristics in captive versus free-ranging cheetahs of two subspecies. *Biology of Reproduction*, **36**, 351–60.

Williams, E.M. & Harper, P.S. (1977). Genetic study of Welsh gypsies. *Journal of Medical Genetics*, **14**, 172–6.

Williams, T.N., Mwangi, T.W., Roberts, D.J. *et al.* (2005). An immune basis for malaria protection by the sickle cell trait. *PLoS Medicine*, **2**, e128.

Wolański, N., Jarosz, E. & Pyzuk, M. (1970). Heterosis in man: growth in offspring and distance between parents' birthplaces. *Social Biology*, **17**, 1–16.

Wolf, A.P. (2002). Reformulating (yet again) the incest avoidance problem. In *Human Mate Choice and Prehistoric Marital Networks*, K. Aoki & T. Akazawa, eds., Kyoto: International Research Center for Japanese Studies, pp. 49–60.

Wolf, A.P. (2004). Explaining the Westermarck effect, or, what did natural selection select for? In *Inbreeding, Incest, and the Incest Taboo*, A. Wolf & W. Durham, eds., Stanford, CA: Stanford University Press, pp. 76–92.

Wolf, A.P. & Huang. C. (1980). *Marriage and Adoption in China, 1845–1945.* Stanford, CA: Stanford University Press.

Wong, S.S. & Anokute, C.C. (1990). The effect of consanguinity on pregnancy outcome in Saudi Arabia. *Journal of the Royal Society of Health*, **4**, 146–7.

Wood, J.W. (1989). Fecundity and natural fertility in humans. *Oxford Review of Reproductive Biology*, **11**, 61–109.

Woodley, M.A. (2009). Inbreeding depression and IQ in a study of 72 countries. *Intelligence*, **37**, 268–76.

Woods, C.G., Cox, J., Springell, K. *et al.* (2006). Quantification of homozygosity in consanguineous individuals with autosomal recessive disease. *American Journal of Human Genetics*, **78**, 889–96.

Woolf, C.M. & Dukepoo, F.C. (1969). Hopi Indians, inbreeding and albinism. *Science*, **164**, 30–7.

Woolf, C.M., Stephens, F.E., Mulaik, D.D. & Gilbert, R.E. (1956). An investigation of the frequency of consanguineous marriage among the Mormons and their relatives in the United States. *American Journal of Human Genetics*, **8**, 236–52.

Worby, S. (2010). *Law and Kinship in Thirteenth-Century England.* Woodbridge: Boydell & Brower.

World Bank. (2011). *The World Bank Data.* Available at http://web.data.worldbank.org/ [10 July 2011].

Wright, C., Kerzin-Storrar, L., Williamson, P.R. *et al.* (2002). Comparison of genetic services with and without genetic registers: knowledge, adjustment, and attitudes about genetic counselling among probands referred to three genetic clinics. *Journal of Medical Genetics*, **39**, e84.

Wright, S. (1922). Coefficients of inbreeding and relationship. *American Naturalist*, **56**, 330–8.

Wright, S. (1951). The genetical structure of populations. *Annals of Eugenics*, **15**, 321–54.

Wu, C., DeWan, A., Hoh, J. & Wang, Z. (2011). A comparison of association methods correcting for population stratification in case-control studies. *Annals of Human Genetics*, **75**, 418–27.

Wu, L. (1987). Investigation of consanguineous marriages among 30 Chinese ethnic groups. *Heredity and Disease*, **4**, 163–6. [In Chinese.]

Xi, R., Kim, T-M. & Park, P.J. (2010). Detecting structural variations in the human genome using next generation sequencing. *Briefings in Functional Genomics*, **9**, 405–15.

Yamaguchi, M., Yanase, T., Miyake, M., Nagano, H. & Nakamoto, N. (1975). Effects of paternal and maternal inbreeding on mortality and sterility in the Fukuoka population. *Japanese Journal of Human Genetics*, **20**, 123–30.

Yamaguchi, M., Yanase, T., Nagano, H. & Nakamoto, N. (1970). Effects of inbreeding on mortality in Fukuoka population. *American Journal of Human Genetics*, **22**, 145–59.

Yamaguchi-Kabata, Y., Nakazono, K., Takahashi, A. *et al.* (2008). Japanese population structure, based on SNP genotypes from 7003 individuals compared to other ethnic groups: effects on population-based association studies. *American Journal of Human Genetics*, **83**, 445–56.

Yanase, Y., Fujiki, N., Handa, Y. *et al.* (1973). Genetic studies on inbreeding in some Japanese populations. XII. Studies of isolated populations. *Japanese Journal of Human Genetics*, **17**, 332–6.

Yaqoob, M., Cnattingius, S., Jalil, F. *et al.* (1998). Risk factors for mortality in young children living under various socio-economic conditions in Lahore, Pakistan: with particular reference to inbreeding. *Clinical Genetics*, **54**, 426–34.

Yaqoob, M., Gustavson, K.-H., Jalil, F., Karlberg, J. & Iselius, L. (1993). Early child health in Lahore, Pakistan: II. Inbreeding. *Acta Paediatrica,* **390**, Suppl., 17–26.

Yari, K., Kazemi, E., Yarani, R. & Tajehmari, A. (2011). Islamic bioethics for fetus abortion in Iran. *American Journal of Scientific Research*, Issue 18, 18–121.

Yasmin, Naidu, J.M. & Mascie-Taylor, C.G. (1997). Consanguinity and its relationship to differential fertility and mortality in the Kotia: a tribal population of Andhra Pradesh, India. *Journal of Biosocial Science*, **29**, 171–80.

Yearsley, M. (1911). Eugenics and congenital deaf-mutism. *Eugenics Review*, **2**, 299–312.

Yonemitsu, N., Mori, K., Mitsuoka, M. *et al.* (1988). Testicular tumors in non-twin brothers from a consanguineous marriage. *Acta Pathologica Japonica*, **38**, 1077–86.

Yunis, K., Mumtaz, G., Fadi, B. *et al.* (2006). Consanguineous marriage and congenital heart defects: a case-control study in the neonatal period. *American Journal of Medical Genetics Part A*, **140A**, 1524–30.

Yusuf, F. & Rukanuddin, A.R. (1989). Correlates of fertility behaviour in Pakistan. *Biology and Society*, **6**, 61–88.

Zakzouk, S. (2002). Consanguinity and hearing impairment in developing countries: a custom to be discouraged. *Journal of Laryngology & Otology*, **116**, 811–6.

Zakzouk, S., El-Sayed, Y. & Bafaqeeh, S.A. (1993). Consanguinity and hereditary hearing impairment among Saudi population. *Annals of Saudi Medicine*, **13**, 447–50.

Zakzouk, S., Fadle, K.A. & Al Anazy, F.H. (1995). Familial hereditary progressive sensorineural hearing loss among Saudi population. *International Journal of Pediatric Otorhinolaryngology*, **32**, 247–55.

Zalloua, P.A., Azar, S.T., Delépine, M. *et al.* (2008). WFS1 mutations are frequent monogenic causes of juvenile-onset diabetes mellitus in Lebanon. *Human Molecular Genetics*, **17**, 4012–21.

Zaoui, S. & Biémont, C. (2002). Fréquence et structure des mariages consanguins dans le région de Tlemcen (Ouest algérien). *Cahiers Santé*, **12**, 289–95. [In French.]

Zelnik, N., Konopnicki, M., Bennett-Back, O., Castel-Deutsch, T. & Tirosh, E. (2010). Risk factors for epilepsy in children with cerebral palsy. *European Journal of Paediatric Neurology*, **14**, 67–72.

Zhan, J., Qin, W., Zhou, Y. *et al.* (1992). Effects of consanguineous marriages on hereditary diseases: a study of the Han ethnic group in different geographic districts of Zejiang province. *National Medical Journal of China*, **172**, 674–6. [In Chinese.]

Zhang, S., Tang, Y.P., Wang, T. *et al.* (2011). Clinical assessment and genomic landscape of a consanguineous family with three Kallmann syndrome descendants. *Asian Journal of Andrology*, **13**, 166–71.

Zhaoxiong, Q. (2001). Rethinking cousin marriage in China. *Ethnology*, **40**, 347–60.

Zhivotovsky, L.A., Rosenberg, N.A. & Feldman, M.W. (2003). Features of evolution and expansion of modern humans, inferred from genomewide microsatellite markers. *American Journal of Human Genetics*, **72**, 1171–86.

Zlotogora, J. (1995). Major gene is responsible for anencephaly among Iranian Jews. *American Journal of Medical Genetics*, **56**, 87–9.

Zlotogora, J. (1997). Genetic disorders among Palestinian Arabs: 1. Effects of consanguinity. *American Journal of Medical Genetics*, **68**, 472–5.

Zlotogora, J. (2002a). What is the birth defect risk associated with consanguineous marriage? *American Journal of Medical Genetics*, **109**, 70–1.

Zlotogora, J. (2002b). Parental decisions to abort or continue a pregnancy with an abnormal finding after an invasive prenatal test. *Prenatal Diagnosis*, **22**, 1102–6.

Zlotogora, J. (2007). Multiple mutations responsible for frequent genetic diseases in isolated populations. *European Journal of Human Genetics*, **15**, 272–8.

Zlotogora, J. & Barges, S. (2003). High incidence of profound deafness in an isolated community. *Genetic Testing*, **7**, 143–5.

Zlotogora, J., Barges, S., Bisharat, B. & Shalev, S.A. (2006). Genetic disorders among Palestinian Arabs. 4: Genetic clinics in the community. *American Journal of Medical Genetics Part A*, **140A**, 1644–46.

Zlotogora, J., Carmi, R., Lev, B. & Shalev, S.A. (2009). A targeted population carrier screening program for severe and frequent genetic diseases in Israel. *European Journal of Human Genetics*, **17**, 591–7.

Zlotogora, J., Habiballa, H., Odatalla, A. & Brages, S. (2002). Changing family structure in a modernizing society: a study of marriage patterns in a single Muslim village in Israel. *American Journal of Human Biology*, **14**, 680–2.

Zlotogora, J., Hujerat, Y., Barges, S., Shalev, S.A. & Chakravarti, A. (2006). The fate of 12 recessive mutations in a single village. *Annals of Human Genetics*, **71**, 202–8.

Zlotogora, J., Hujerat, Y., Zalman, L. *et al.* (2005). Origin and expansion of four different beta globin mutations in a single Arab village. *American Journal of Human Biology*, **17**, 659–61.

Zlotogora, J. & Shalev, S.A. (2010). The consequences of consanguinity on the rates of malformations and major medical conditions at birth and in early childhood in inbred populations. *American Journal of Medical Genetics*, **152A**, 2023–8.

Zlotogora, J., van Baal, S. & Patrinos, G.P. (2007). Documentation of inherited disorders and mutation frequencies in the different religious communities in Israel in the Israeli National Genetic Database. *Human Mutation*, **28**, 944–9.

Zuhar, S. (2005). *Gender, Sexuality and the Criminal Laws in the Middle East and North Africa: a Comparative Study*, Istanbul: WWHR, pp. 50–3.

Zwerling, A., Behr, M.A., Verma, A., Brewer, T.F., Menzies, D., & Pai, M. (2011). The BCG World Atlas: a database of global BCG vaccination policies and practices. *PLoS Medicine*, **8**, e1001012.

Index

Printed in the United States
By Bookmasters